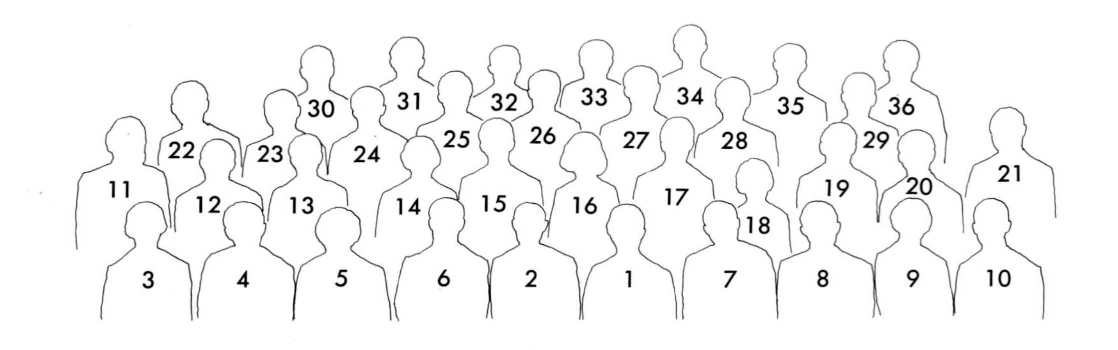

PARTICIPANTS

1. H.I.H. Princess Takamatsu
2. H.I.H. Prince Takamatsu
3. Dr. Martin
4. Dr. Gelboin
5. Mrs. Gelboin
6. Dr. Upton
7. Dr. MacMahon
8. Dr. Nelson
9. Mrs. Miller
10. Dr. Miller
11. Dr. Day
12. Dr. Setlow
13. Dr. Kripke
14. Mrs. Tulinius
15. Dr. Tulinius
16. Mrs. German
17. Dr. German
18. Mrs. Schull
19. Dr. Schull
20. Dr. Li
21. Dr. Vesell
22. Dr. Matsunaga
23. Dr. Hirohata
24. Dr. Kolonel
25. Dr. Harnden
26. Dr. Kersey
27. Dr. Lyon
28. Dr. Sasazuki
29. Dr. Takebe
30. Dr. Takayama
31. Dr. Sugano
32. Dr. Sugimura
33. Dr. Kato
34. Dr. Omura
35. Dr. Ishikawa
36. Dr. Matsushima
37. Dr. Estabrook
38. Dr. Cederlöf

GENETIC AND ENVIRONMENTAL FACTORS IN EXPERIMENTAL AND HUMAN CANCER

Proceedings of the 10th International Symposium of
The Princess Takamatsu Cancer Research Fund, Tokyo, 1979

GENETIC AND ENVIRONMENTAL FACTORS IN EXPERIMENTAL AND HUMAN CANCER

Edited by
HARRY V. GELBOIN, BRIAN MACMAHON,
TAIJIRO MATSUSHIMA, TAKASHI SUGIMURA,
SHOZO TAKAYAMA, and HIRAKU TAKEBE

JAPAN SCIENTIFIC SOCIETIES PRESS, Tokyo

Published by:
JAPAN SCIENTIFIC SOCIETIES PRESS
2-10 Hongo, 6-chome, Bunkyo-ku, Tokyo 113, Japan

Sole distributor for the outside Japan:
BUSINESS CENTER FOR ACADEMIC SOCIETIES JAPAN
20-6 Mukogaoka, 1-chome, Bunkyo-ku, Tokyo 113, Japan

JSSP No. 02402-1104
ISBN 4-7622-02401

Printed in Japan

Organizing Committee of the 10th International Symposium

Harry V. GELBOIN
Laboratory of Molecular Carcinogenesis, National Cancer Institute, Bethesda, Md., U.S.A.

Brian MACMAHON
Department of Epidemiology, School of Public Health, Harvard University, Boston, Mass., U.S.A.

Taijiro MATSUSHIMA
Department of Molecular Oncology, Institute of Medical Science, University of Tokyo, Tokyo, Japan

Takashi SUGIMURA
Biochemistry Division, National Cancer Center Research Institute, Tokyo, Japan

Shozo TAKAYAMA
Department of Experimental Pathology, Cancer Institute, Tokyo, Japan

Hiraku TAKEBE
Radiation Biology Center, Kyoto University, Kyoto, Japan

Contributors

Rune CEDERLÖF	Swedish Twin Registry, Department of Environmental Hygiene, The Karolinska Institute, Stockholm, Sweden
Rufus S. DAY, III	Nucleic Acids Section, Laboratory of Molecular Carcinogenesis, National Cancer Institute, Bethesda, Md., U.S.A.
Ronald W. ESTABROOK	Department of Biochemistry, Southwestern Medical School, Health Science Center at Dallas, The University of Texas, Dallas, Tex., U.S.A.
Harry V. GELBOIN	Laboratory of Molecular Carcinogenesis, National Cancer Institute, Bethesda, Md., U.S.A.
James L. GERMAN, III	Laboratory of Human Genetics, The New York Blood Center, New York, N.Y., U.S.A.
David G. HARNDEN	Department of Cancer Studies, The Medical School, The University of Birmingham, Birmingham, United Kingdom
Tomio HIROHATA	Department of Public Health, School of Medicine, Kurume University, Fukuoka, Japan
Takatoshi ISHIKAWA	Department of Experimental Pathology, Cancer Institute, Tokyo, Japan
W.F.H. JARRETT	Department of Veterinary Pathology, Veterinary School, University of Glasgow, Bearsden, Glasgow, United Kingdom
Ryuichi KATO	Department of Pharmacology, School of Medicine, Keio University, Tokyo, Japan
John H. KERSEY	Laboratory Medicine and Pathology, University Hospitals and Clinics, University of Minnesota, Minneapolis, Minn., U.S.A.
Laurence N. KOLONEL	Epidemiology Program, Cancer Center of Hawaii, Uni-

versity of Hawaii at Manoa, Honolulu, Hawaii, U.S.A.
Margaret L. KRIPKE — Cancer Biology Program, Frederick Cancer Research Center, Frederick, Md., U.S.A.
Min-Hsin LI — Department of Chemical Etiology and Carcinogenesis, Cancer Institute, Chinese Academy of Medical Center, Peking, People's Republic of China
Joseph L. LYON — Division of Epidemiology, Department of Family and Community Medicine, University of Utah Medical Center, Salt Lake City, Utah, U.S.A.
Brian MACMAHON — Department of Epidemiology, School of Public Health, Harvard University, Boston, Mass., U.S.A.
Alice O. MARTIN — Laboratory of Human Genetics, Prentice Women's Hospital and Materinary Center, Northwestern Memorial Hospital, Northwestern University, Chicago, Ill., U.S.A.
Ei MATSUNAGA — Department of Human Genetics, National Institute of Genetics, Shizuoka, Japan
Taijiro MATSUSHIMA — Department of Molecular Oncology, Institute of Medical Science, University of Tokyo, Tokyo, Japan
Robert W. MILLER — Clinical Epidemiology Branch, National Cancer Institute, Bethesda, Md., U.S.A.
Norton NELSON — Institute of Environmental Medicine, New York University Medical Center, New York, N.Y., U.S.A.
Tsuneo OMURA — Department of Biology, Faculty of Science, Kyushu University, Fukuoka, Japan
Tetsuo ONO — Department of Biochemistry, The Tokyo Metropolitan Institute of Medical Science, Tokyo, Japan
Takehiko SASAZUKI — Department of Human Genetics, Medical Research Institute, Tokyo Medical and Dental University, Tokyo, Japan
William J. SCHULL — Epidemiology and Statistics Department, Radiation Effects Research Foundation, Hiroshima, Japan
Richard B. SETLOW — Department of Biology, Brookhaven National Laboratory, Associated Universities, Inc., Upton, N.Y., U.S.A.
Takashi SUGIMURA — Biochemistry Division, National Cancer Center Research Institute, Tokyo, Japan
Shozo TAKAYAMA — Department of Experimental Pathology, Cancer Institute, Tokyo, Japan
Hiraku TAKEBE — Radiation Biology Center, Kyoto University, Kyoto, Japan
Hrafn TULINIUS — Icelandic Cancer Registry, Faculty of Medicine, University of Iceland, Reykjavik, Iceland
Elliot S. VESELL — Department of Pharmacology, The Milton S. Hershey Medical Center, The Pennsylvania State University, Hershey, Pa., U.S.A.

Opening Address

H.I.H. Princess KIKUKO TAKAMATSU

For the past nine years it has been my privilege to witness the development in various aspects of cancer research through the annual International Symposia held under the auspices of the Cancer Research Fund which bears my name. The first Symposium was held in 1970 and it set the academic standard for the other symposia which have followed. It is a matter of great satisfaction to me to welcome you to the Tenth Annual International Symposium.

This morning it is my pleasant duty to attend the opening meeting of the Symposium. The subject chosen this time is "Genetic and Environmental Factors in Experimental and Human Cancer." My scientific advisors are of the opinion that we are now in a position to discuss the relationship between the environmental and genetical factors on the basis of our advancing knowledge of the carcinogenic mechanism. The occurrence of cancer in human beings is intimately influenced by their surroundings and family history. Fundamental knowledge of the function of each of the potential factors may eventually point to practical ways and means of preventing cancer.

I am especially pleased to have this opportunity of extending my cordial welcome to the participants from overseas whose cooperation makes the Symposium international. In this connection I would like to thank Dr. Harry V. Gelboin and Dr. Brian MacMahon, who have assumed special responsibility for the Symposium by joining the Organizing Committee, so making the Organizing Committee itself international.

I now declare open the Tenth International Symposium of our series, with every confidence in its fruitful outcome.

Dr. Robert W. Miller

Your Imperial Highness, coming to Japan at this time of year reminds me of my first arrival here in mid-November 1953. That was 26 years ago this week. Then, as now, the sights and sounds and all perceptions were so different from what I had known. They were and are all-absorbing. The many new experiences made me feel as if I were 7 years old again, or as if a previously unused portion of my brain had been tapped.

Times have changed of course. For example, designer clothing can be seen everywhere in the cities, and signs in English no longer have the same creative use of the language and spelling that they once had. I recall one that dealt with cancer prevention well ahead of its time: it said sternly, No Smorking Please! Fortunately Japanese English still freshens conventional English. A few years ago I saw an example stencilled on the purse of a woman in the subway. It said, "We must everytime challange impossibility and take a chance." That may well be the guiding principle of research on cancer and other things.

The subject of this meeting, Interactions in Carcinogenesis, invites us to challenge impossibility in every area of cancer etiology, from studies of molecules, chromosomes and cells—to studies of whole organisms in the laboratory, or at large as individuals or populations. We must try to relate findings in these separate areas to one another. Interactions in Carcinogenesis could refer to scientists from different disciplines, who, by interacting can lead to new understanding of multiple factors in the origins of cancer.

New ideas can arise when scientists from different cultures come together. We will have in the next few days a scientific Rashomon, in which a given concept will be seen differently because of major cultural variations. That this should happen we know from the experience of the U.S.-Japan Cooperative Cancer Research Program, now in its sixth year. Cancer research and its human benefits have been enhanced by the interaction of scientists on both sides of the Pacific.

The Japanese excel in organizing scientific meetings, Olympic games and other group activities. My first experience with this talent was in 1965 at a binational workshop in teratology and pediatric oncology, sponsored by JSPS and the U.S. National Science Foundation. This single small-group meeting accelerated the internationalization of Japanese activities in both fields. So well did the Japanese organize the workshop and the massive International Pediatrics Congress that followed, that I know a major contribution could be made to science if the Japanese were to hold workshops on How to Hold Workshops.

We can already sense from last night's reception and from your opening remarks that this annual symposium is an example of this talent. I was fortunate enough to be present once before, at the Third International Symposium, and it was a high point in my career. So well known has this conference become that the participants feel honored to be invited.

In some areas of cancer research, honors are not as common as in others, but this series of symposia has covered a wide spectrum. The same is true of this particular meeting. Through your leadership and participation in these conferences, have I feel contributed as much as any scientist has to the growth and quality of cancer research in Japan and internationally. We thank you for honoring us and for making an important contribution to science.

Contents

Lectures

Mixed-Function Oxidases

Pharmacogenetics and Immunogenetics

Genetic Factors

DNA Repair

Epidemiology

LECTURES

GENETIC AND ENVIRONMENTAL FACTORS IN EXPERIMENTAL AND HUMAN CANCER,
H. V. GELBOIN ET AL. (EDS.), JAPAN SCI. SOC. PRESS, TOKYO, PP. 3–16, 1980

KEYNOTE LECTURE

Search for Sources of Cancer: Some Examples from Laboratory and Epidemiological Studies

Norton NELSON

Institute of Environmental Medicine, New York University Medical Center, New York, N.Y. 10016, U.S.A.

Abstract: The sources of cancer are partly external and partly internal and involve a multiplicity of agents, both directly causative and enhancing. The disentanglement of this interplay of internal and host factors from external agents will be a difficult and complex one, which will need to be studied both in the laboratory and epidemiologically. Advances in our understanding of the biology of cancer and in the geographic distribution of cancer have been large in the last several decades. This knowledge can be the basis for new and inventive approaches to the search for sources of cancer.

Internal factors include those in which cancer is determined purely by genetic factors, those in which host susceptibility is altered by genetic factors, and those in which the activation and inactivation of external agents is involved. Internal factors also include hormonal modulation and "spontaneous" cancer occurrence.

We now have reason to believe that external factors play a significant part in cancer causation both as causal agents and as modulating or enhancing agents. These can be considered in terms of a high level exposure pattern as in industry and low level exposure pattern as in dietary, cultural and lifestyle factors. In each case there is an interplay of direct action, indirect action and modulation of these by internal and external circumstances. Multi-factorial interactions in cancer will be discussed and examples cited on the interplay of laboratory with epidemiological research.

The advances of the last three decades in science relevant to the uncovering of the sources of cancer have been immense. I believe, however, we are just now at the beginning of capitalizing on those advances. A major opportunity for fuller exploitation and application of this wave of new knowledge is now here. If we are successful in the efficient use of these new sources of knowledge, the advances in the next three decades should be even more dramatic than in the last three. These advances have been of two kinds. One has to do with our understanding of the biological mechanisms of the conversion of normal to malignant cells. There has been a parallel increase in our understanding of cancer epidemiology. It should be remembered, however, that the dividing line between "clinical alertness" and formal epidemiological studies is not always sharp. The clinical detection of a small cluster of cases has often given us the decisive signal.

In both areas there has been a growing recognition and documentation of the interaction of agents in the production of cancer; that is, that in many instances, perhaps in most, cancer is multi-source in its origin and that several, perhaps many, different factors can contribute to the eventual outcome. Equally important has been the growing information on genetic factors in cancer. Very striking has been the knowledge gained by the study of geographic distribution of cancer by organ site through migrant studies.

It is convenient for some purposes to think of cancer sources as arising from "high level" or "high intensity" exposures to distinguish them from instances in which the exposures, whether single or multiple, are apparently less intense and the frequency of cancer occurrence in the exposed population is generally lower. The first group, that is the "high intensity" group, are epitomized especially by occupational exposures. Both laboratory and epidemiological studies have combined with the high degrees of success to identify source factors in a substantial number (some 18) (*1*) of occupational circumstances in which cancer from work place exposure is involved.

In the group of what I call for the present purposes "low intensity" cancer sources, the frequency of occurrence in the exposed population is generally lower than in occupational groups. Still before us for resolution in this low intensity group are the cancers which have been attributed by a number of writers, preeminently Higginson (*2*) to "lifestyle," cultural, and other external factors.

In the first group, that is the " high intensity " group, although we know that interaction through cocarcinogenic and promoting agents plays a role, they are probably of less importance here than in the second group of low intensity. Indeed, in this latter group it may well be that the enhancing factors may be dominant and perhaps more easily accessible to control than the often ubiquitous carcinogenic agents themselves. In each case, of course, individual host factors play decisive roles in the final outcome. Presumably, however, they could play a more important role in the "low intensity" group of cancers than they would in the "high intensity" group. These host factors are among the important internal determinants as to whether cancer occurs or does not and are, of course, decisive for individual predisposition.

It is now accepted by nearly all workers in the field that cancer is multifactorial in origin. The range of factors and their varying participation is extremely wide and in very few cases have they been identified. In this discussion I would like to review some of these factors and the differing ways in which they may contribute to malignant outcome. Although in restricted instances, with a sufficiently high dose of a carcinogen, *e.g.*, experimentally and in some occupational settings, the occurrence of malignancy is completely dominated by the carcinogen exposure; in almost all other circumstances a variety of factors contribute to the outcome.

In this presentation, I will emphasize those factors which lead to cancer and

TABLE 1. Some Contributing Factors in Cancer Causation

(a) Summary
Internal:
  Heredity decisive
  Heredity important
    Enzymatic capability
    Anatomic features
    Strong hormoral influence
    Repair or misrepair of DNA
    " Spontaneous " (or unknown)
  Disease——prior or concurrent
External:
  Carcinogens (Mutagens)
  Cocarcinogenic and promoting agents
The above neglects inhibitory factors.

(b) Internal: Hereditary factors decisive
  Examples: bilateral retinoblastoma
    bilateral Wilms tumor
  These tumors additionally require spontaneous mutation or exposure to external mutagen/carcinogen (Knudson).

(c) Internal
  Enzymatic capability, balance between activation and inactivation
  Example: arylhydroxylase in activation of polynuclear aromatic hydrocarbons; cigarette smoking and lung cancer; activation of aromatic amines in bladder cancer; inactivation of many kinds

(d) Internal
  Anatomic features important
  Example: airway dimensions vary widely in normal population and are important determinants of extent and locus of deposition of inhaled particles (dosage); could be important in cigarette smoking, chromium chemical inhalation, *etc.*

(e) Internal
  Strong hormonal influence
  Example: breast, possibly thyroid

(f) Internal
  Repair of misrepair of DNA *unusually* important.
  It is assumed that some degree of repair (and mistakes) are normal; in other cases incomplete or misrepair would be of *decisive* importance.
  Example: xeroderma pigmentosum, possible metal carcinogenesis, *e.g.*, arsenic and chromium chemicals

(g) Internal
  " Spontaneous " (or unknown): it is assumed that in the ordinary process of somatic cell division some viable mutants occur potentially contributing to malignancy; possibly endogenous carcinogens.

(h) Internal
  Disease, prior or concurrent
  Example: hepatitis and aflatoxin in liver cancer; possibly non-malignant lung disease and lung cancer

(i) External
  Carcinogen (mutagen)
  Example: chemicals, ionizing and ultraviolet radiation

(j) External
  Cocarcinogenic and promoting agents
  Includes agents active in enhancing tumor occurrence (initiation, promotion); interaction between carcinogens
  Example: cigarette smoking with asbestos and radon daughter exposure; SO_2 and benzo(a)pyrene or As exposure, catechol (cocarcinogenesis) and phorbol ester (promotion)

TABLE 2. Possible Contributory Factors in Cancer Causation[a]

Tumor types and agent	Direct germ cell	Internal — Other host: Enzyme	Anatomic	Hormonal
Retinoblastoma				
—Bilateral	++++	?	–	?
—Unilateral	++	?	–	?
Lung— Cigarette	–	++	++ ?	–
Lung— BCME	–	–	+ ?	–
Lung— Cr. chemicals	–	–	++ ?	–
Lung— As (smelter)	–	–	++ ?	–
Brain— Ionizing radiation	–	–	–	–
Thyroid Ionizing radiation	–	–	–	++
Breast-hair dyes				
"high natural risk"	+	?	– ?	++++
"low natural risk"	+ ?	++	– ?	++

[a] The evaluations are largely speculative.

give little attention to the role of countering or inhibitory factors, which, of course, must be very important in the final outcome. Inhibitory factors have, of course, received extended study. Examples include the role of protease inhibitors in delaying the time of skin tumor appearance (*3*); and the inhibition of a number of tumors with retinoids (*4*). Such agents have much promise in tumor prevention, but would lead us beyond the scope of this paper.

In this presentation, which includes much speculation, I will attempt to identify some of the factors, both internal and external, contributing to the cancer outcome. My main purpose in this is not to venture a definitive identification of such contributory factors but rather to illustrate the varying ways in which such contributions are modulated and interact with each other. A prominent objective in the analysis illustrated in Tables 1 and 2, and to be discussed now, is to suggest that "dissection" of the multiple contributory factors may be a fruitful approach to the development of the knowledge required for the intervention and prevention of cancer stemming from cultural or "lifestyle" patterns.

Although, as noted, interaction and modulating factors may be more important with "low intensity" than "high intensity" exposures, even with the latter, as in some occupations, it is clear that interaction can be very important. Examples of this are the enhancing effect of cigarette smoking with exposure to asbestos (*5*) and as well with exposure to radon daughters (*6*). Another example is the concurrent exposure to arsenic chemicals and to sulfur dioxide (*7*). The first two involve interaction of carcinogens with carcinogens, and possibly non-carcinogen enhancing

factors			External		Usefulness to date	
Repair	"Spontaneous"	Disease	Carcinogen (mutagen)	Cocarcinogens/ promoters	Epidemiology "experience"	Laboratory "experiment"
−	±	?	±	?	++	−
+ ?	±	?	++	+ ?	++	−
+ ?	−	+ ?	+++	+++	++	+
+ ?	−	?	++++	− ?	+	++
++	−	+ ?	++++	+ ?	++	+
++	−	+ ?	+++	+++	++	+ ?
+ ?	−	−	++++	−	++	+
+ ?	−	−	+++	− ?	++	+
+ ?	?	+ ?	+ ?	+ ?	++	+
+ ?	?	+ ?	++	+ ?	++	+

agents; the last involves interaction of a carcinogen (not confirmed in the laboratory) with a non-carcinogen.

In Table 1a I summarize a number of possible contributory factors to cancer causation. In this table and the succeeding one the examples chosen and their analysis are intended only to be illustrative; I hope they will stimulate a more searching examination.

In Table 1a the factors are separated into internal and external. It should be noted that while in a few cases hereditary determinants are wholly decisive, in most other cases heredity will be of contributory significance in determining host susceptibility, that is, individual predisposition. Among the external factors of significance are exposure to carcinogens or mutagens or to enhancing factors such as cocarcinogenic or promoting agents. Viral agents are legitimate additions to the list of external factors; for simplicity, I have omitted them here.

In the succeeding parts of Table 1 the contributory factors are examined in more detail. Table 1b suggests that in some instances hereditary factors appear to be decisive determinants in tumor occurrence. Examples of this would include bilateral retinoblastoma and bilateral Wilms tumor; Knudson (*8*) has suggested that in these cases the outcome is overwhelmingly determined by inherited genes.

Table 1c illustrates enzymatic contributions to tumor outcome. In many cases, perhaps most, involving external chemicals, activation is required to convert the pre-carcinogen into an active carcinogen. However, inactivation can also occur, so one is concerned with the balance between activation and inactivation. Aryl hydrocarbon hydroxlase is an example of an enzyme involved in the activation of aromatic polynuclear hydrocarbons which may alter susceptibility to lung cancer in cigarette smokers (*9*). We should, however, remember that in some cases non-enzymatic processes are also involved; namely, the nitrosation of secondary amines in the stomach and the hydrolytic inactivation of carcinogens such as the alkylating agents.

In some cases anatomic features may be important as noted in Table 1d. Airway dimensions in normal persons vary widely. Evidence in our laboratory has clearly shown that this leads to wide variation among individuals in the locus and extent of particle deposition in the respiratory tract (*10*). Accordingly, dosage to the respiratory tissue is controlled to a significant extent by the architecture of the airways. This must be a modulating factor in the outcome when particulate carcinogenic agents are inhaled.

In some cases it is clear that hormonal status plays a very strong role (*11*). This is presumed to be especially important in breast and possibly thyroid tumors as noted in Table 1e.

Continuing the review of some of the internal contributory factors with Table 1f, it is recognized that DNA repair occurs normally as well as some degree of misrepair. However, repair defects may in some cases be unusually important and perhaps dominant as for example in xeroderma pigmentosum (*8*). There is some evidence that with chromium chemicals, and perhaps arsenic, interference with normal repair mechanisms may be very important steps in the development of malignancy from these agents (*12*, *13*).

Among internal factors we come to so-called "spontaneous" tumors (Table 1g). In this instance we may be simply collecting together tumors for which at present we have no assignable origins and as more knowledge is developed this category would diminish. On the other hand, there is reason to suppose that mutations do occur normally and some of these may lead to a malignant outcome. Aging may itself play a part. There is also some evidence that endogenous tumorigenic/mutagenic agents are normally produced (*14*).

Table 1h suggests that prior or concurrent non-malignant disease may be significant. It has been suggested that non-malignant lung disease may predispose to lung cancer (*15*). It has also been suggested that hepatitis may interact with ingested aflatoxin in the production of liver cancer (*16*, *17*). The latter could be a special case since a viral agent is involved.

Preeminent among the external factors which may be important in tumor outcome is exposure to carcinogens, many of which are mutagens—Table 1i. Carcinogens will include chemicals, direct or indirect action, ionizing and ultra-violet radiation. Most occupational cancer will fall into the category where external agents are of preeminent concern.

We have increasing evidence that amongst external factors of great importance are those which produce enhancement of the outcome without themselves being carcinogenic (Table 1j). There are at least two modes that have been studied in respect to such enhancement. In our group we distinguish those occurring in a sequential two stage process (initiation followed by promotion) from those in which there is concurrent exposure to the carcinogen and the enhancing agent (cocarcinogenic). A possible example of an enhancing agent is sulfur dioxide with arsenic chemicals (*7*). Cigarette smoking with asbestos (*5*) and with radon daughters (*6*) constitute examples of interaction between carcinogenic agents. Experimentally, catechol, present in cigarette smoke (*18*), has been shown to be a co-carcinogenic agent and, of course, phorbol ester (*19*) is a widely studied promoting agent.

In Table 2 I present a simplistic and speculative examination of some possible contributory factors in cancer causation. This is not intended to be definitive in any sense, but is displayed here primarily to suggest the different ways by which these modifying factors can contribute. Some of the examples have been chosen from those studied by our group; they include several from other sources.

As noted earlier, retinoblastoma appears to have a decisive genetic component. It has been suggested by Knudson (*8*) that bilateral retinoblastoma is essentially hereditary, but, at least in the case of monolateral tumors, an additional mutation is required. It appears that such tumors constitute only a modest portion, 1 to 2%, of the total tumor occurrence (*8*). In making this distinction in respect to purely hereditary tumors, one must, of course, recognize that genetic constitution is a very important determinant of many other host characteristics such as enzymatic endowment, repair capability, anatomic structure and so forth.

The important features in the next line, which relates to lung cancer from cigarette smoking, are the contributions of external carcinogens, enzymatic capability, anatomic factors and possible external co-acting factors. Another pattern of lung cancer, that resulting from exposure to chloromethyl ethers (*20–22*) is in strong contrast to lung cancer from cigarette smoking. Here it appears that in the case of high exposure of industrial workers to the chloroethers the direct acting alkylating agents are predominant in producing lung cancer. Limited evidence suggests that the role of cigarette smoking is negligible with these chemicals; Weiss (*23*), as will be noted later, has even suggested that there is an inverse relationship. Although this seems doubtful, there appears to be no enhancement due to cigarette smoking.

Lung cancer in chromium chemical exposure has been thoroughly studied in the laboratory and epidemiologically (*24*, *25*). Clearly, in this case high exposure to the chemicals themselves are decisive; pulmonary anatomic features may be important here as they may be in any inhaled particulate chemical where dosage is altered by such characteristics. There is no evidence to support or deny a role of cocarcinogenic or promoting factors.

The next line relating to arsenic, especially as studied in smelter workers, suggest that arsenic chemicals interact with sulfur dioxide exposure in a very important way; the work of Lee and Fraumeni (*7*) shows graded outcomes with increase in exposure to either component.

In each of the two preceding cases, that is, chromium and arsenic chemicals, it seems plausible that effects on DNA repair are of special importance. Work in this institute on *E. coli* has suggested that in each case repair of mutations produced by other agents plays a special role. In one case excision repair is inhibited and in the other post-replicative repair is altered (*12*, *13*).

The next two lines identify ionizing radiation as a source of cancer. These are chosen from many possible examples of the production of cancer from ionizing radiation only because they have received some study in our group (*26*, *27*). In this first instance, tumors of the brain, there is no evidence one way or the other that co-acting factors are involved, though, of course, they may be. An important consideration with ionizing radiation is that in almost all instances dosage to the susceptible cellular unit is direct and can be estimated with considerable precision. Presumably, enzymatic factors as such normally play no direct role. In this way, ionizing radiation resembles direct acting alkylating agents. In contrasting tumors of the brain with tumors of the thyroid, it seems possible that hormonal status of the thyroid may play a role in the latter. Evidence for this is not at present available, although there does appear to be an increased sensitivity in the young (*28*).

Breast cancer clearly has a very important relationship to hormonal status—time of first pregnancy, tumor before or after menopause, *etc.*—in fact, detailed studies have been undertaken to relate breast cancer occurrence to hormonal status as displayed by estrogen levels (*11*). Later a study will be presented which suggests that the "low natural risk" group of breast cancer cases shows a response to exposure to external carcinogens as found in hair dyes. This response is not detectable in the "high natural risk" group.

This table serves merely to illustrate questions which need to be answered and ways in which various factors may influence the malignant outcome. There are, of course, many question marks in the table; perhaps there should be more.

This analysis is obviously not definitive, in fact, it could not be at this time. The objective is merely to suggest that complexities need to be dissected. If it stimulates inquiry, it will have fulfilled its purpose.

As noted in the final two columns, it appears probable that epidemiological as well as laboratory studies have played useful roles in most cases, but in different degrees. If this multi-factorial approach to the analysis of cancer causation is helpful, it will be the challenge of future years to use both of these tools to study systematically the relative roles of these and other factors.

With these comments on the multi-factorial nature of cancer, I would like to turn to some examples largely drawn from our own studies to illustrate some of the ways in which epidemiology and laboratory studies can contribute to the mutual support of each.

Van Duuren in our group is among those who played a major role in early identification of carcinogenic polynuclear hydrocarbons (some 10) in cigarette smoke (*29*). These studies left him unconvinced that the amounts found were adequate to explain the skin cancer produced (and by analogy, lung cancer) through exposure to tobacco tars. This then led him to look for tumor enhancing factors in cigarette smoke. He found a series of agents which when administered concurrently (*i.e.*, in a cocarcinogenic mode) significantly enhanced tumor production. These included phenols, especially polyphenols such as catechol and some long chain hydrocarbons (*18*). This data taken together with the known presence of the polynuclear aromatic carcinogens brings the laboratory evidence closer to supporting the observed epidemiological occurrence of lung cancer in response to cigarette smoking.

A series of laboratory and epidemiological studies of chloromethyl ethers led first to the identification of these as animal carcinogens and then as human carcinogens. Sequentially, the work started from structure-activity relationships postulated by Van Duuren which in turn led to animal studies of a series of chloroethers. These pointed to bischloromethyl ether as a very potent carcinogen and to chloromethyl ether as a less active member of the family (*30*). Inasmuch as these compounds have a fairly extensive industrial usage, particularly in the production of ion exchange resins, we undertook an inhalation study of the carcinogenicity of these compounds using rats. This study confirmed the eariler work on skin and subcutaneous injection. Bischloromethyl ether produced a substantial incidence of bronchiogenic carcinoma and neuroesthesioepithelial cancer; chloromethyl ether was substantially less potent in the inhalation studies as it was in the other studies (*20*). On the basis of these results we undertook an industry wide epidemiological study in plants where these chemicals were used as intermediates.

Bischloromethyl ether is not used directly, but normally constitutes an impurity of 6 to 8% in the chloromethyl ether used in these processes. This study involved the description of work patterns and the categorization of levels of exposure through the examination of plant processes and study of job records; this permitted the classification of workers according to the intensity, duration and interval since the start of exposure. The worker cohort included some 9,000 individuals with exposures ranging from none to intense thus permitting internal comparison of workers with graded patterns of exposure. This investigation led to the identification of some 19 lung cancers above those in the unexposed (*21*, *22*). These were clustered in the highly exposed groups and especially in those who had had high exposure for a fairly long earlier period of time. In short, the human dose-response fitted the expected pattern of increase in incidence with increase in intensity and duration of exposure. In this instance, laboratory studies provided a direct basis for the epidemiological investigation.

Our data were inadequate to evaluate the role of cigarette smoking. However, Weiss (*23*) was able to secure information on smoking habits from a number of heavily exposed workers with lung cancer. Four out of 14 of these were nonsmokers. In fact, Weiss postulated an inverse relationship. This seems not too persuasive; at best, cigarette smoking appears to be a secondary factor.

Interest in another organic compound, epichlorhydrin, also grew out of studies by Van Duuren (*31*). Awareness of its extensive industrial use led to inhalation studies being undertaken. These showed that at concentrations of 100 ppm epichlorhydrin was a nasal carcinogen in the rat (*32*). This agent is extensively used in the United States (550 million pounds in 1975) as an important and widely used intermediate. Studies undertaken by Enterline for one of the industries using epichlorhydrin identified a significant elevation of lung cancer in exposed workers; this study has now been submitted for publication (*33*).

The above examples represent high level exposure of direct acting alkylating agents in which enzymatic (activating) factors are unimportant, and in one instance, at least, smoking is at best a secondary contributor.

The role of chromium chemicals in lung cancer has been studied extensively epidemiologically (*25*); however, these studies failed to identify the important chemical species. Studies in our group using a technique for implanting the test chemical in a segmental bronchus of the rat identified calcium chromate as the most potent of the likely agents. Subsequent inhalation studies with rats led to the production of bronchogenic cancer with exposure to calcium chromate (*24*). Thus, laboratory studies sharpened the epidemiological findings providing leads for improving control.

In a study of inhaled benzpyrene, the role of co-factors was examined. One of these involved concurrent exposure to sulfur dioxide. In this instance, rats exposed to combinations of particulate benzpyrene and the gas, sulfur dioxide, had a substantially increased incidence of lung cancer compared to those exposed to either alone (*24*). This study provided the basis for speculation in a later epidemiological study of smelter workers by Lee and Fraumeni (*7*) in which they suggested the possibility of the interaction of SO_2 and arsenic in the induction of lung cancer in smelter workers.

It has seemed unlikely that inorganic compounds interact with DNA in the way that reactive organic chemicals do. Accordingly, our group has been exploring the possibility that they interfere with DNA repair. These studies (*12*, *13*) (so far confined to repair deficient *E. coli*) do show interference. Chromium chemicals inhibit excision repair while arsenic alters post-replication repair leading to increased mutagenesis.

In a totally different field having to do with ionizing radiation, our group has conducted an integrated series of studies on children exposed to ionizing radiation of the scalp for control of epidemic ring worm. This study involved the identification of a group of school children who had received radiation for the treatment of tinea capitis and another group of approximately the same number also with tinea capitis who did not receive scalp irradiation. Once the groups were identified and located, a series of studies involving both questionnaires and clinical examinations were undertaken. On the laboratory side, our group active in radiation biology undertook, using phantoms, a careful quantitative study of dosage to various tissues of interest. These ranged from a high of approximately 400R to the skin of the scalp itself through 150R to the brain, down to 6R to the thyroid. Each of these tissues was found to have an excess incidence of tumors in comparison to the controls. Perhaps most interestingly, this study led to the identification of a tumor outcome with one of the lowest carefully substantiated radiation doses so far observed; namely, the occurrence of 8 thyroid adenomas following a dosage of 6R to the thyroids of these children (*26*, *27*). In this case, laboratory studies provided information on the tissue dosages for the epidemiological investigation.

Bladder cancer from aromatic amines was one of the earlier forms of cancer identified in the work place. It was also early recognized that the active agents such as β-naphthylamine were not directly carcinogenic, but required metabolic activation. Laboratory investigations led to the identification of active metabolites of aromatic amines such as β-naphthylamine; ring hydroxylation followed by excretion in the unusual form of a bisaryl phosphate (*34*), and *N*-hydroxylation have both been implicated (*35*). This provides another example of the interplay between laboratory and epidemiological studies.

An example of work in our institute involving the group described above as "low intensity" relates to the role of hair dyes in breast cancer. A number of constituents of hair dyes, all presumably requiring activation, have been identified in the laboratory as mutagenic in bacterial assays and several have now been confirmed as carcinogenic in long-term, whole animal studies (*36*). On the basis of these findings, Shore and others in our group undertook a study of the possible relationship between hair dye use and breast cancer. This was a retrospective case-control study, that is, it involved a comparison of the history and habits of women with breast cancer with those without breast cancer; the groups being matched for age, socio-economic status and other factors. Inasmuch as breast cancer is clearly related to hormonal factors, with the possibility of a clear separation between a "high natural" as compared to the "low natural" risk group, these factors were also taken into account. This study failed to identify any association between hair dye use and breast cancer in the "high risk" group, however, there was a strong and clear suggestion of a relationship between breast cancer and the integral use of hair dye in the "low-risk" group. That is to say, when the intensity and duration of use and the time since the start of hair dye use were taken into account, there was a clear graded relationship between the frequency of breast cancer and the extent and time pattern of hair dye use (*36*). Although the study group was small, and needs to be repeated with larger numbers, the internal consistency and the gradation with intensity and time lend additional weight to these observations. In this case, laboratory studies gave the hint leading to the epidemiological studies.

It is clear in these and other studies that linkage between epidemiological findings and laboratory findings can be very fruitful in the search for, and validation of, cancer causation. Wherever possible, of course, direct linkage in concurrent laboratory and epidemiological studies is desirable and has been too infrequently used in the past. This circumstance derives from the fact that it is rather rare that skills of both types are present in the same research group. There is a trend toward linkage of epidemiological with laboratory studies in cancer research; hopefully, such linkages will be more frequent in the future.

Unquestionably, studies of high risk, high intensity cancer occurrences in occupational groups, will continue to inform, guide and provide the basis for control. Indeed, they should be increasingly exploited for the study of contributory and enhancing factors. In this "high intensity" group such linkages may be easier to identify than in "low intensity" groups and can thus be informative.

On the other hand, in the "low intensity" group, which in terms of the overall impact on human health is of considerably greater importance, the circumstances will almost certainly be far more complicated and much more difficult to deal with both in laboratory and in epidemiological investigations. It is this group that very special attention will need to be directed to the role of modulating factors both internal and external.

As suggested earlier, advances in providing more refined tools have been immense in the last several decades. It seems certain that full exploitation of both laboratory and epidemiological research in association with sharpened clinical alertness will be required for the understanding of the role and extent of contributory factors in this preeminently multi-factorial disease. It is now up to the scientific community to put these new resources at the service of the search for cancer causes and as an ultimate basis for improved control. The search will be difficult, but extremely interesting and rewarding.

ACKNOWLEDGMENT

This work was supported by NIEHS Center Grant No. 00260, NCI Center Grant No. 13343, and additional grants identified in the individual references.

REFERENCES

1. Tomatis, L., Agthe, C., Bartsch, H., Huff, J., Montesano, R., Saracci, R., Walker, E., and Wilbourn, J. Evaluation of the carcinogenicity of chemicals: A review of the Monograph Programme of the International Agency for Research on Cancer (1971–1977). Cancer Res., *38*: 877–885, 1978.
2. Higginson, J. and Muir, C. S. Environmental carcinogenesis: Misconceptions and limitations to cancer control. J. Natl. Cancer Inst., *63*: 1291–1298, 1979.
3. Troll, W., Klassen, A., and Janoff, A. Tumorigenesis in mouse skin: Inhibition by synthetic inhibitors of proteases. Science, *169*: 1211–1213, 1970.
4. Sporn, M. B. Prevention of epithelial cancer by vitamin A and its synthetic analogs (retinoids). *In*; H. H. Hiatt, J. D. Watson, and J. A. Winsten (eds.), Origins of Human Cancer, Book B, Mechanisms of Carcinogenesis (Cold Spring Harbor Conferences on Cell Proliferation, vol. 4), pp. 801–807, Cold Spring Harbor Laboratory, New York, 1977.
5. International Agency for Research on Cancer, Working Group on the Evaluation of the Carcinogenic Risk of Chemicals to Man. Asbestos, IARC Monographs on the Evaluation of the Carcinogenic Risk of Chemicals to Man, vol. 14, International Agency for Research on Cancer, Lyons, 1977.
6. Saccomonno, G. Uranium miners health. *In*; Radiation Standards for Uranium Mining. Hearings before the Sub-Committee on Research, Development, and Radiation of the Joint Committee on Atomic Energy. Congress of the United States, Ninty-first Congress, First Session, March 17 and 18, 1969, pp. 301–315, U. S. Government Printing Office, Washington, D.C., 1969.
7. Lee, A. M. and Frameni, J. F. Arsenic and respiratory cancer in man: An occupational study. J. Natl. Cancer Inst., *42*: 1045–1052, 1969.
8. Knudson, A. G., Jr. Genetic predisposition to cancer. *In*; H. H. Hiatt, J. D. Watson, and J. A. Winsten (eds.), Origins of Human Cancer, Book A, Incidence of Cancer in Humans (Cold Spring Harbor Conferences on Cell Proliferation, vol. 4), pp. 45–52, Cold Spring Harbor Laboratory, New York, 1977.
9. Gelboin, H. V., Okuda, T., Selkirk, J. K., Nemoto, N., Yang, S. K., Rapp, H. J., and Bast, R. C., Jr. Benzo(a)pyrene metabolism: Enzymatic and liquid chromatographic analysis and application to human tissues. *In*; P. N. Magee, S. Takayama, T. Sugimura, and T. Matsushima (eds.), Fundamentals in Cancer Prevention, Pro-

ceedings of the 6th International Symposium of the Princess Takamatsu Cancer Research Fund, Tokyo, 1975, pp. 167–190, Japan Sci. Soc. Press, Tokyo/University Park Press, Baltimore, 1976.

10. Palmes, E. D. and Lippmann, M. Influence of respiratory air space dimensions on aerosol deposition. *In*; W. H. Walton (ed.), Inhaled Particles, vol. IV, pp. 127–136, Pergamon Press, New York, 1977.
11. MacMahon, B., Cole, P., Brown, J. B., Aoki, K., Lin, T. M., and Morgan, R. W. Urine estrogen profiles of Asian and North American women. Int. J. Cancer, *14*: 161–167, 1974.
12. Rossmann, T., Meyn, M. S., and Troll, W. Effects of arsenite on DNA repair in *Escherichia coli*. Environ. Health Perspect., *19*: 229–233, 1977.
13. Rossmann, T. Repair in bacterial systems: The effect of chromate. Personal communication, 1979.
14. Friedell, G. H., Burney, S. E., Bell, J. R., and Soto, E. Pathology as related to tryptophan metabolite excretion, occupational history, and smoking habits in patients with bladder cancer. J. Natl. Cancer Inst., *43*: 303–306, 1969.
15. Finke, W. Chronic pulmonary diseases in patients with lung cancer. N. Y. State J. Med., *58*: 3783–3790, 1958.
16. Linsell, C. A. and Peers, F. G. Aflatoxin and liver cell cancer. Trans. Roy. Soc. Trop. Med. Hyg., *71*: 471–473, 1977.
17. Larouze, B., Blumberg, B. S., London, W. T., Lustbader, E. D., Sankale, M., and Payet, M. Forecasting the development of primary hepatocellular carcinoma by the use of risk factors: Studies in West Africa. J. Natl. Cancer Inst., *58*: 1557–1561, 1977.
18. Van Duuren, B. L. and Goldschmidt, B. M. Co-carcinogenic and tumor promoting agents in tobacco carcinogenesis. J. Natl. Cancer Inst., *56*: 1237–1242, 1976.
19. Van Duuren, B. L., Tseng, S.-S., Segal, A., Smith, A. C., Melchionne, S., and Seidman, I. Effects of structural changes on the tumor promotion activity of phorbol myristate acetate on mouse skin. Cancer Res., *39*: 2644–2646, 1979.
20. Kuschner, M., Laskin, S., Drew, R. T., Cappiello, V., and Nelson, N. The inhalation carcinogenicity of alpha halo-ethers. 3. Lifetime and limited period inhalation studies with bis (chloromethyl) ether at 0.1 ppm. Arch. Environ. Health, *30*: 73–77, 1975.
21. Albert, R. E., Pasternack, B. S., Shore, R. E., Lippmann, M., Nelson, N., and Ferris, B. Mortality patterns among workers exposed to chloromethyl ethers: A preliminary report. Environ. Health Perspect., *11*: 209–214, 1975.
22. Pasternack, B. S., Shore, R. E., and Albert, R. E. Occupational exposure to chloromethyl ethers: A retrospective cohort mortality study (1948–1972). J. Occup. Med., *19*: 741–746, 1977.
23. Weiss, W. Chloromethyl ethers, cigarettes, cough and cancer. J. Occup. Med., *18*: 194–199, 1976.
24. Laskin, S., Kuschner, M., and Drew, R. T. Studies in pulmonary carcinogenesis. *In*; M. G. Hanna, Jr., P. Nettesheim, and J. R. Gilbert (eds.), Inhalation Carcinogenesis, Proceedings of a Biology Division, Oak Ridge National Laboratory, conference held in Gatlinburg, Tennessee, October 8–11, 1969 (AEC Symposium Series, 18), pp. 321–351, U. S. Atomic Energy Commission, Office of Information Services, Washington, D. C., 1970.
25. Machle, W. and Gregorius, F. Cancer of respiratory system in United States chromate producing industry. Publ. Health Rep., *63*: 1114–1127, 1948.

26. Shore R. E., Albert, R., and Pasternack, B. Follow-up study of patients treated by X-ray epilation for tinea capitis. IV. Resurvey of post-treatment illness and mortality experience. Arch. Environ. Health, *31*: 21–28, 1976.
27. Harley, N., Albert, R., Shore, R., and Pasternack, B. Follow-up study of patients treated by X-ray epilation for tinea capitis: Estimation of the dose to the thyroid and pituitary glands and other structures of the head and neck. Phys. Med. Biol., *21*: 631–642, 1976.
28. National Council on Radiation Protection and Measurements. Protection of the Thyroid Gland in the Event of Releases of Radioiodine; Recommendation of the National Council on Radiation and Measruement (NCRP Report No. 55), NCRP, Washington, D.C., 1977.
29. Van Duuren, B. L. Some aspects of the chemistry of tobacco smoke. *In*; G. James and T. Rosenthal (eds.), Tobacco and Health, pp. 33–47, Charles C. Thomas, Springfield, Illinois, 1962.
30. Van Duuren, B., Sivak, A., Goldschmidt, B., Katz, C., and Melchionne, S. Carcinogenicity of halo-ethers. J. Natl. Cancer Inst., *43*: 481–486, 1969.
31. Van Duuren, B. L., Goldschmidt, B. M., Katz, C., Seidman, I., and Paul, J. S. Carcinogenic activity of alkylating agents. J. Natl. Cancer Inst., *53*: 695–700, 1974.
32. Laskin, S., Sellakumar, A. R., Kuschner, M., Nelson, N., La Mendola, S., Rusch, G. M., Katz, G. V., Dulak, N. C., and Albert, R. E. Inhalation carcinogenicity of epichlorohydrin. Submitted for publication, 1979.
33. Enterline, P. E., Joyner, R. E., and Henderson, V. Cancer among workers exposed to epichlorohydrin and other chemicals. Submitted for publication, 1979.
34. Troll, W., Belman, S., and Mukai, F. Studies on the nature of the proximal bladder carcinogens. J. Natl. Cancer Inst., *43*: 283–286, 1969.
35. Miller, E. C. and Miller, J. A. Mechanism of chemical carcinogenesis: Nature of proximate carcinogens and interactions with macromolecules. Pharmacol. Rev., *18*: 805–838, 1966.
36. Shore, R. E., Pasternack, B. S., Thiessen, E. U., Sadow, M., Forbes, R., and Albert, R. E. A case study of hair dye use and breast cancer. J. Natl. Cancer Inst., *62*: 277–283, 1979.

GENETIC AND ENVIRONMENTAL FACTORS IN EXPERIMENTAL AND HUMAN CANCER,
H. V. GELBOIN ET AL. (EDS.), JAPAN SCI. SOC. PRESS, TOKYO, PP. 17–28, 1980

NAKAHARA MEMORIAL

Endogenous and Exogenous Steroids in the Etiology of Cancer of the Breast

Brian MacMahon

Walcott Professor of Epidemiology, Harvard School of Public Health, Harvard University, Boston, Mass. 02115, U.S.A.

Abstract: The subject of the etiology of human breast cancer seems a particular appropriate one for this symposium, since it is likely that both genetic and environmental factors are involved and insights are being gained both from studies in humans and in the experimental laboratory. This presentation deals predominantly with work on humans and with the role of steroid hormones, particularly the estrogens.

There is little doubt that the estrogens play an important role in the causation of breast cancer. This conclusion is supported by experimental work, by observation of the effect of surgical or radiologic castration in humans, and by observation of increased risk of breast cancer following long-term therapy with exogenous estrogens. One paradox which, if it were resolved, would go far towards elucidating the etiology of this disease is that there are very large differences in breast cancer risk between certain populations which do not differ remarkably in ovarian function. One hypothesis explains the difference in breast cancer risk in terms of differences in the patterns of estrogen metabolism. These differences are unlikely to be genetically determined since they change on migration. Of considerable practical as well as theoretical interest is the apparent difference between exogenous estrogens alone and estrogen-progesterone combinations in their carcinogenic potential for the breast.

Another important feature of human breast cancer is its association with parity, or rather lack of parity—an association recently understood as resulting from a protective effect of the first full-time pregnancy if it occurs at an early age. This effect too is consistent with observations of a change in estrogen metabolism after pregnancy, but other explanations also fit the observations. The effect of pregnancy and the geographic differences appear to be independent and therefore require different explanations. Nevertheless, both may be accounted for by the importance of steroid hormones in the etiology of the disease.

The theme of this symposium is the interaction of gene and environment in experimental and human cancer. Cancer of the breast provides an outstanding example of the reality of these inter-relationships, for, while it is certain that there are important environmental determinants of the disease, familial patterns make it likely that the environmental determinants operate against a background of genetic susceptibility. Moreover, the disease in humans has been illuminated by a vast body of work in the experimental laboratory extending over a period of more than 50 years.

The disease is also, I believe, an appropriate subject for this particular lectureship, for it illustrates the benefits to be derived from international collaboration in health research, and specifically collaboration between Japan and the United States. There has been no better illustration of these benefits than in the life and accomplishments of Dr. Waro Nakahara. It is my honor to have been asked to give a lecture dedicated to the memory of that distinguished biologist who did so much to advance the art and science of oncology and to establish, in their early years, the quality and reputation of these symposia.

I propose to consider today only one component of the complex etiology of cancer of the breast—the possible role of steroid hormones—but I will approach the topic somewhat indirectly, as epidemiologists are prone to do.

The goals of the experimentalist and the epidemiologist are identical, but their approaches clearly are different. Faced with a disease of unknown etiology, the experimentalist will go to the heart of the matter and observe the effects of interruption of any of the biologic sequences that may conceivably be involved. This approach was notably successful in leading to the control of diabetes and poliomyelitis, even though the principal features of the human disease—the familial nature of diabetes and the seasonal variation of poliomyelitis—remained unexplained. This approach is denied the epidemiologist who must build an edifice of circumstantial evidence brick by brick until the overall structure leads to an inference inescapable by a rational person, even though often falling short of mathematical or philosophical proof. The principal accomplishments of this approach have been the identification of the methods of transmission of cholera and other infectious diseases and, in this century, the identification of the agent responsible for the century's epidemic of carcinoma of the lung.

Knowledge of the etiology of cancer of the human breast is in such a rudimentary stage that we cannot predict which of these two approaches will yield the critical piece or pieces of information. The importance of the disease is such that both approaches must be taken vigorously and in ways which are both independent and complimentary. If this review emphasizes one approach, at the expense of the other, it is only because it is the one with which the reviewer is most familiar.

Epidemiologic Features of Cancer of the Breast

In contrast to the understanding of etiologic mechanisms, significant features of the descriptive epidemiology of breast cancer are well known. This knowledge provides at least three major building blocks which must in some way be fitted into any rational structure describing the etiology of this disease. They are:

1) The major variation in frequency between areas of the world.
2) The association of risk with age at which a woman has her first full-term pregnancy.
3) Observations suggesting an important role for some component of ovarian activity.

These are by no means all the known peculiarities of the distribution of breast cancer in human populations. I have not, for example, included what is probably the single most important determinant of risk age. This is because association with age, in one form or another, is such a ubiquitous feature of human cancer that, while it needs to be understood in terms of the neoplastic process generally, there at present seems no reason to believe that the shape of the age pattern is likely to assist in understanding the etiology of breast cancer specifically. Age does, however, interact with regional variation and Fig. 1 illustrates that interaction. Rates of incidence of and mortality from breast cancer are 5–6 times as high in North America and Northern Europe as in most Asian countries and the indigenous populations of south and central Africa (*2*, *3*). As is seen in Fig. 1, these differences are greatest in the postmenopausal years. In areas of high risk, rates tend to increase with age throughout life; in areas of low risk they increase with age to about age 50 and then decline; in areas where rates are intermediate—typically, countries around the Mediterranean Sea and in South America—the age pattern is also intermediate between those of high- and low-risk areas, with rates tending to plateau after about age 50.

Another feature of the demography of breast cancer that I have not selected as one of my building blocks today is its familial nature. Again, this is not because it may not be important. However, this feature seems capable of so many different

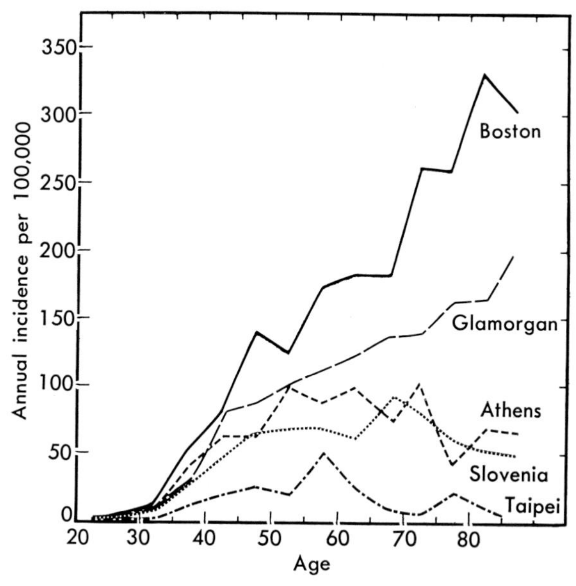

FIG. 1. The pattern of age association of breast cancer in areas of high, low, and intermediate risk (*1*).

TABLE 1. Incidence Rates[a] of Breast Cancer in Populations of Asian or Caucasian Descent in Asia, Hawaii, and the Mainland United

Racial origin	Residence	
	Indigenous	Hawaii[b]
Caucasian[c]	62	63
Chinese[d]	9	44
Japanese[e]	11	23

[a] Age-adjusted annual rates per 100,000 population. [b] Hawaii, 1960–1964 (*5*). [c] Connecticut, 1963–1965 (*5*). [d] Taipei, 1965–1967 (*6*). [e] Miyagi, 1962–1964 (*5*).

interpretations that, at the present time, it does not seem likely to assist in discriminating between prevailing hypotheses; unless, as Pike and others (*4*) have begun to investigate, the familial pattern is associated with some specific hormone profile.

Turning to the three epidemiologic features of breast cancer that do, in my opinion, offer levers for our investigation of the etiology of the disease, the principal point that remains to be made about the international variation is that this clearly depends on some environmental circumstance of life in different parts of the world rather than on genetic characteristics of the inhabitants. The evidence for this is that the ancestors of persons who migrate from low-risk to high-risk areas come to experience the breast cancer risk of the host country. Perhaps the best illustration of this is in the large populations of Japanese and Chinese descent now resident in the United States (Table 1). The rate of change of cancer rates in such populations is much slower for cancer of the breast than for most other sites. Thus, while changes are seen for most sites among the migrants themselves, these are not seen for breast cancer, and it is not until the second and later generations that rates among the offspring of the migrants begin to approach those of the host country. Nevertheless the change is too rapid and too great to be explained in genetic terms, particularly since these migrations have not been accompanied by major miscegenation. The relative slowness of the change in breast cancer rates may be attributable to the influence of some component of early family life which would tend to change less rapidly than would the adult environment.

It has been suspected for over 200 years that some aspect of reproduction plays a role in breast cancer. During this century, evidence developed that the more children a woman had the lower her risk of breast cancer. Recently, it has become apparent that this relationship is principally a function, not of the *number* of children, but of the age at which a woman bears her first full-term child (Fig. 2). The relationship with number of children is in large part an indirect one resulting from the fact that women who bear their first child at an early age tend to have more children subsequently than do women whose first child is delayed. The dramatic effect of early first birth—a birth under age 18, for example, being associated with only one third of the breast cancer risk experienced by women whose first birth occurs after age 30—implies that, in the absence of a pregnancy, events are occurring at these early ages that are favorable to the development of breast cancer. It seems likely that early reproductive life is a period of high risk of tumor initiation

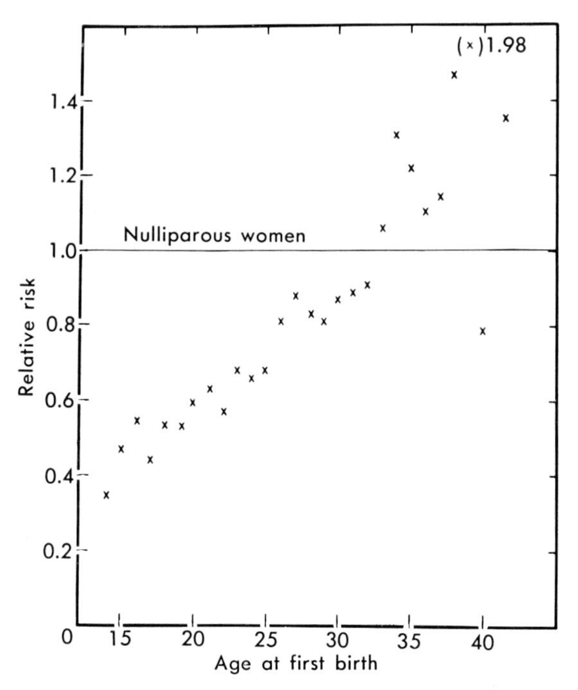

FIG. 2. The relationship between age at first birth and breast cancer risk (*7*).

and that the occurrence of the first pregnancy terminates that period or at least substantially reduces the risk. The inverse association of risk with early first birth is independent of the international variation described earlier, and is of approximately equal magnitude in high- and low-risk areas (*9*, *10*).

The third epidemiologic feature of note is actually a cluster of features all pointing to a pivotal role for an active ovary in the causation of breast cancer. These include association of risk with early menarche and with late natural menopause, but most striking is the reduction in risk associated with the therapeutic induction of menopause either by surgery (*11*) or by radiation (*12*). The younger the age at which the artificial menopause is induced the greater the reduction in breast cancer risk; women whose ovarian function is destroyed prior to age 35 have only one third the breast cancer risk of those who have a natural menopause. However, even the group with menopause induced prior to age 35 have already experienced about their natural life-time of ovarian activity and it is likely that, were castration undertaken at an earlier age, we would see an even greater reduction in risk.

Effect of Exogenous Estrogens

A large body of work on experimental animals makes it a clear inference that the mechanism of reduction of risk associated with interruption of ovarian activity is through interruption of estrogen stimulus. This inference, strengthened by anecdotal reports of breast cancer following exogenous estrogens in humans, has led to

TABLE 2. Published Reports on the Association of Oral Contraceptive Use with Breast Cancer Risk

Year	First author and reference number	Number of Cases	Number of Controls	Minimum duration of use	Use in controls (%)	Relative risk of breast cancer
1971	Arthes (*8*)	119	119	Ever	6	0.9
1972	Vessey (*9*)	90	180	Ever	41	0.8
1973	BCDSP (*10*)	23	842	Ever	20	0.6
1974	RCGP (*11*)	31	46,377[a]	Ever	NA	1.1
1975	Vessey (*12*)	232	232	Ever	63	1.0
				48 months	12	1.0
1975	Fasal (*13*)	452	872	Ever	46	1.1
				2–4 years	22	1.9
1976	Ory (*14*)	137	69,137[a]	Ever	28	0.6
				2 years	?	0.7
1978	Lees (*15*)	301	548	12 months	21	0.6
				60 months	13	1.0
1978	Kelsey (*16*)	99	99	Ever	34	1.4
1979	Brinton (*17*)	286[b]	809	Ever	23	0.9
				10 years	9	0.9
1979	Shore (*18*)	154	227	Ever	8	0.4

[a] Cohort study. Number shown is total cohort size. [b] Exclusive of surgical menopause patients.

considerable concern regarding breast cancer in the very substantial numbers of women now taking estrogens exogenously on a regular basis. There are basically two groups of such women—those taking synthetic estrogen-progestagen combinations for contraceptive purposes and those taking natural conjugated estrogens derived from horse urine for relief of symptoms supposedly attributable to estrogen lack around the time of the menopause.

The users of oral contraceptives have so far received the greater amount of investigative attention, probably because of the very extensive use of these compounds in some parts of the world and because of their introduction at a time when concern was mounting regarding the side effects of therapeutic drugs generally. A summary of the studies published to date on estimates of the risk of breast cancer associated with oral contraceptive use is given in Table 2. In only one study (*13*) is the relatively risk substantially or significantly greater than 1.0, and this is only for the category of users of 2–4 years duration. In this same study, users for periods longer than 4 years were not at excess risk. In individual studies, significantly increased risk has been reported for other sub-groups—for example, women who used oral contraceptive prior to their first child and those who used them in the presence of a history of benign breast disease—but these findings have not been duplicated and may well be attributable to the chance occurrence of statistically significant differences when many comparisons are made.

There is little evidence therefore that oral contraceptive use is associated with increased breast cancer risk. Moreover, it is now quite clear that oral contraceptive use is associated with decrease in the frequency of diagnosis of fibrocystic disease and possibly other benign disorders of the breast, and since such diagnoses are associated with increase in risk of subsequent carcinoma, this piece of evidence

TABLE 3. Published Reports on the Association of Use of Estrogens for Menopausal Symptoms with Breast Cancer Risk

Year	First author and reference number	Number of		Minimum duration of use	Use in controls (%)	Relative risk of cancer
		Cases	Controls			
1971	Arthes (*8*)	119	119	Ever	12	1.3
1974	BCDSP (*19*)	51	774	—[a]	8	1.1
1974	Craig (*20*)	134	260	Ever	10	1.0
1974	Burch (*21*)	19	737	5 years	—[b]	1.3
1975	Mack (*22*)	99	396	Ever	26	1.6
1976	Casagrande (*23*)	90	83	Ever	69	0.8
1976	Hoover (*24*)	49	1,891	6 months	—[b]	1.3
1977	Sartwell (*25*)	284	367	Ever	16	1.0
1979	Brinton (*17*)	160[c]	449	Ever	35	1.0
				5 years	19	1.3
1979	Shore (*18*)	154	227	Ever	31	0.9

[a] Within 3 months prior to diagnosis. [b] Cohort studies. All subjects used estrogens. In these studies, the number shown under "Number of controls" is the total cohort size. [c] Women having had natural menopause.

strengthens the belief that oral contraceptive users do not experience increased risk of breast cancer. We must recognize that oral contraceptives have been in widespread use for less than 20 years, and it is still possible that an increase in risk will appear as the number of women exposed over longer potential induction periods becomes larger. A definitive judgment should probably be delayed for at least another 10 years.

Studies of breast cancer risk associated with the use of estrogens for relief of menopausal symptoms are summarized in Table 3. Again, the relative risks are centered firmly around 1.0. Here, however, the hints in the data tend to point in the opposite direction to those in the oral contraceptive studies. In particular, in one study (*24*) there were twice as many breast cancer cases observed as expected among women followed at least 15 years beyond the time of initial therapy. I know of two other unpublished sets of data containing similar findings. On this issue we must again reserve definitive judgment, but since, in an informal gathering such as this, a certain level of speculation is permissable, I would venture the opinion that it will not be many years before the carcinogenicity of menopausal estrogens will be accepted for the breast as it is now for the endometrium.

Thus, recognizing that this opinion may be changed with the publication of the next study, I believe that the presently available evidence suggests that, while unopposed estrogens in the form given for relief of menopausal symptoms are probably carcinogenic to the human breast, the estrogen-progestagen combinations given for contraception probably are not and may indeed be protective. The discordance of these two effects, if confirmed, has important implications not only to the users of these drugs but to understanding of the endogenous hormonal mechanisms underlying the disease.

Endogenous Steroids

In considering the role of endogenous hormones in the etiology of breast cancer, an almost inescapable starting point is the idea that estrogen stimulus is crucial. As already noted, this idea stems from firm epidemiologic evidence as well as a vast body of experimental work. Epidemiologic evidence—the effect of early first pregnancy and of early oophorectomy—is consonant with the intuitively appealing suggestion that this estrogen effect may be particularly important during early reproductive life when the breast is developing and under greatest endocrine stimulation.

From this starting point, however, we immediately face two significant logical hurdles:

1) Why do populations, such as those of Japan and the United States, which display close comparability in the carrying out of those other activities for which a functioning ovary is required, exhibit such disparate rates of breast cancer?
2) Why is a pregnancy, which produces a massive increase in estrogen stimulus, associated with a decrease, rather than an increase, in breast cancer risk?

The first of these questions was addressed by Zumoff and his colleagues (*26*), who noted that data on urine estrogens collected in Asia, Hawaii, and North America (*27*, *28*) were consistent with the hypothesis that urine levels of active estrogen (estrone and estradiol) were highest in the populations with the highest breast cancer risk. Comparable differences were not found by Bulbrook and his colleagues comparing Japanese and British women with respect both to serum and urine estrogens (*29*). However, cultural differences between the British and Japanese women studied by Bulbrook were probably considerably less than those in the women studied by my colleagues and I, since, in Japan, where diet and other cultural features have changed rapidly in the last 30 years, women were deliberately selected from a remote, rural area where the cultural changes were likely to have been small. The women studied by Bulbrook were from Tokyo and were said to be of relatively high socioeconomic status. Even so, there are other exceptions to the geographic correlation of estrogen levels and breast cancer risk. For example, we have observed that young women from a very low socioeconomic class in Athens, who would be expected to have very low rates of breast cancer, have higher urine levels of estrone and estradiol than women of similar age from a higher socioeconomic class (*30*).

In spite of these exceptions, differences between populations in total active estrogen levels may well explain some of the observed differences in breast cancer rates. The concept must therefore be invoked that there are levels of estrogen production and stimulus that are in excess of those required for normal ovarian function and which, at least in terms of breast carcinogenesis, are undesirable. Such levels presumably obtain in North America and Northern Europe.

Turning to the second question—the paradoxical effect of the first pregnancy—a hypothesis was put forward in 1966 (*31*) and modified in 1969 (*32*) which would explain this effect in terms of the nature of the estrogens to which a woman is exposed during pregnancy. At that time, two of the three principal human estrogens, estrone and estradiol, had been shown to be carcinogenic while the third, estriol, had not. Estriol is present in very much higher concentrations, relative to the other estrogens, during the pregnant than the non-pregnant state. Moreover, this is particular a feature of the latter months of pregnancy which are necessary for the protective effect against breast cancer to be expressed. It was proposed, therefore, that there is competition between estriol and the other estrogens and that women who metabolize a high proportion of their estrogens to estriol are at low breast cancer risk. Although, even at that time, the mechanism of this competition was unclear, this was the only explanation of the pregnancy paradox that was offered.

In the ensuing years, considerable evidence in favor of and against this hypothesis has accumulated. On one hand, striking geographic (*27*, *28*) and socioeconomic (*30*) correlations have been found between the estriol ratios of young women in a population and the population's breast cancer risk, and the first pregnancy has been found to induce a favorable change in the estriol ratio which lasts beyond the termination of the pregnancy (*33*, *34*). On the other hand, the possible mechanism of this effect has become even less clear. In the non-pregnant state, estriol is present in serum in concentrations almost too low to detect, it binds weakly making it an unlikely competitor for receptor sites, and, when given repeatedly, is itself carcinogenic (*26*). In noting the defects of the estriol hypothesis, Zumoff and his colleagues have called attention to the possible anti-estrogenic properties of other estrogen metabolites—notably 2-hydroxyestrone—but little is known of the intra- or inter-population differences in production of these compounds.

A third explanation of the pregnancy effect was put forward by Sherman and Korenman (*35*), who, following the observation by Grattarola (*36*) of a high frequency of anovular cycles in breast cancer patients, suggested deficient corpus luteum function as an explanation of several breast cancer risk factors, including delayed first birth. Sherman and Korenman suggested that the mechanism could be that either women with infrequent ovulation and deficient luteal function were at high breast cancer risk and, coincidently, were delayed in achieving their first pregnancy, or, that the occurrence of the first pregnancy led to establishment of regular ovulation and hormonally-normal cycles. Since there is little evidence that breast cancer patients are in fact biologically infertile, the epidemiologic evidence rather strongly favors the second explanation. An attractive feature of this suggestion is that it offers an explanation of the uniquely protective role of the first, rather than any, pregnancy.

An increase in breast cancer risk associated with anovulatory menstrual cycles —activated perhaps through continued estrogen stimulus unopposed by progesterone—does not rule out an additional role for estriol. Indeed the two effects may be correlated, since estriol ratios are consistently higher during the luteal phase of the menstrual cycle. At the present time, however, it appears that the epidemiologic features of breast cancer can virtually all be explained by the hypothesis that breast cancer risk is causally associated with the estrogen stimulus to which a woman is exposed, particularly when that stimulus is unopposed by estriol, progesterone or some other as yet unidentified concomitant of ovulation. We must now distinguish definitively between the limited number of possibilities. That step alone may lead to practical opportunities for prevention—particularly if a protective role is established for either of the two leading contenders, estriol or progesterone. Beyond that it will remain to identify the genetic, dietary or other determinants of the specific endocrine profile responsible.

REFERENCES

1. MacMahon, B., Cole, P., and Brown, J. Etiology of human breast cancer: A review. J. Natl. Cancer Inst., *50*: 21–42, 1973.
2. Waterhouse, J., Muir, C., Correa, P., and Powell, J. (eds.). Cancer Incidence in Five Continents, vol. III-1976, IARC Scientific Publications No. 15, International Agency for Research on Cancer, Lyons, 1976.
3. Segi, M. and Kurihara, M. Cancer Mortality for Selected Sites in 24 Countries, No. 6 (1966–67), Japan Cancer Society, Nagoya, 1972.
4. Pike, M. C., Casagrande, J. T., Brown, J. B., Gerkins, V., and Henderson, B. E. Comparison of urinary and plasma hormone levels in daughters of breast cancer patients and controls. J. Natl. Cancer Inst., *59*: 1351–1355, 1977.
5. Doll, R., Muir, C., and Waterhouse, J. (eds.). Cancer Incidence in Five Continents, vol. II-1970, International Union against Cancer, Springer-Verlag, Berlin, 1970.
6. Lin, T. M., Chen, K. P., and MacMahon, B. Epidemiologic characteristics of cancer of the breast in Taiwan. Cancer, *27*: 1497–1504, 1971.
7. MacMahon, B., Cole, P., Lin, T. M., Lowe, C. R., Mirra, A. P., Ravnihar, B., Salber, E. J., Valaoras, V. G., and Yuasa S. Age at first birth and breast cancer risk. Bull. WHO, *43*: 209–221, 1970.
8. Arthes, F. G., Sartwell, P. E., and Lewison, E. F. The pill, estrogens and the breast. Cancer, *28*: 1391–1394, 1971.
9. Vessey, M. P., Doll, R., and Sutton, P. M. Oral contraceptives and breast neoplasia: a retrospective study. Br. Med. J., *3*: 719–724, 1972.
10. Boston Collaborative Drug Surveillance Program. Oral contraceptives and venous thromboembolic disease, surgically confirmed gallbladder disease, and breast tumors. Lancet, *i*: 399–1404, 1973.
11. Royal College of General Practitioners. Oral Contraceptives and Health, Pittman, New York, 1974.
12. Vessey, M. P., Doll, R., and Jones, K. Oral contraceptives and breast cancer. Progress report of an epidemiological study. Lancet, *i*: 941–944, 1975.
13. Fasal, E. and Paffenbarger, R. S. Oral contraceptives as related to cancer and benign lesions of the breast. J. Natl. Cancer Inst., *55*: 767–773, 1975.
14. Ory, H., Cole, P., MacMahon, B., and Hoover, R. Oral contraceptives and reduced risk of benign breast diseases. N. Engl. J. Med., *294*: 419–422, 1976.
15. Lees, A. W., Burns, P. E., and Grace, M. Oral contraceptives and breast disease in premenopausal Northern Albertan women. Int. J. Cancer, *22*: 700–707, 1978.
16. Kelsey, J. L., Holford, T. R., White, C., Mayer, E. S., Kilty, S. E., and Acheson, R. M. Oral contraceptives and breast disease: An epidemiological study. Am. J. Epidemiol., *107*: 236–244, 1978.
17. Brinton, L. A., Williams, R. R., Hoover, R. N., Stegens, N. L., Feinleib, M., and

Fraumeni, J., Jr. Breast cancer risk factors among screening program participants. J. Natl. Cancer Inst., *62*: 37–43, 1979.
18. Shore, R. E., Pasternak, B. S., Thiessen, E. U., Sadow, M., Forbes, R., and Albert, R. E. A case-control study of hair dye use and breast cancer. J. Natl. Cancer Inst., *62*: 277–283, 1979.
19. Boston Collaborative Drug Surveillance Program. Surgically confirmed gallbladder disease, venous thromboembolism, and breast tumors in relation to postmenopausal estrogen therapy. N. Engl. J. Med., *290*: 15–19, 1974.
20. Craig, T. J., Comstock, G. W., and Geiser P. B. Epidemiologic comparison of breast cancer patients with early and late onset of malignancy and general population controls. J. Natl. Cancer Inst., *53*: 1577–1581, 1974.
21. Burch, J. C., Byrd, B. F., Jr., and Vaughn, W. K. The effects of long-term estrogen on hysterectomized women. Am. J. Obstet. Gynecol., *118*: 778–782, 1974.
22. Mack, T. M., Henderson, B. E., Gerkins, V. R., Arthur, M., Baptista, J., and Pike, M. C. Reserpine and breast cancer in a retirement community. N. Engl. J. Med., *292*: 1366–1371, 1975.
23. Casagrande, J., Gerkins, V., Henderson, B. E., Mack, T., and Pike, M. C. Exogenous estrogens and breast cancer in women with natural menopause. J. Natl. Cancer Inst., *56*: 839–841, 1976.
24. Hoover, R., Gray, L. A. Sr., Cole, P., and MacMahon, B. Menopausal estrogens and breast cancer. N. Engl. J. Med., *295*: 401–405, 1976.
25. Sartwell, P. E., Arthes, F. G., and Tonascia, J. A. Exogenous hormones, reproductive history, and breast cancer. J. Natl. Cancer Inst., *59*: 1589–1592, 1977.
26. Zumoff, B., Fishman, J., Bradlow, H. L., and Hellman, L. Hormone profiles in hormone-dependent cancers. Cancer Res., *35*: 3365–3373, 1975.
27. MacMahon, B., Cole, P., Brown, J. B., Aoki, K., Lin, T. M., Morgan, R. W., and Woo, N. C. Urine oestrogen profiles of Asian and North American women. Int. J. Cancer, *14*: 161–167, 1974.
28. Dickinson, L. E., MacMahon, B., Cole, P., and Brown, J. B. Estrogen profiles of Oriental and Caucasian women in Hawaii. N. Engl. J. Med., *291*: 1211–1213, 1974.
29. Bulbrook, R. D., Swain, M. C., Wang, D. Y., Hayward, J. L., Kumaoka, S., Takatani, O., Abe, I., and Utsunomiya, J. Breast cancer in Britain and Japan: Plasma oestradiol-17B, oestrone and progesterone, and their urinary metabolites in normal British and Japanese women. Eur. J. Cancer, *12*: 725–735, 1976.
30. Trichopoulos, D., MacMahon, B., and Brown, J. Socioeconomic status and urine estrogens. J. Natl. Cancer Inst., in press, 1980.
31. Lemon, H. M., Wotiz, H. H., Parsons, L., and Mozden, P. J. Reduced estriol excretion in patients with breast cancer prior to endocrine therapy. J. Am. Med. Assoc., *196*: 112–120, 1966.
32. Cole, P. and MacMahon, B. Oestrogen fractions during early reproductive life in the etiology of breast cancer. Lancet, *i*: 604–606, 1969.
33. Cole, P., Brown, J. B., and MacMahon, B. Oestrogen profiles of parous and nulliparous women. Lancet, *ii*: 596–599, 1976.
34. Trichopoulos, D., Cole, P., Goldman, M., MacMahon, B., and Brown, J. B. Estrogen profiles of primiparous and nulliparous women in Athens. J. Natl. Cancer Inst., in press, 1980.
35. Sherman, B. M. and Korenman, S. G. Inadequate corpus luteum function: A pathophysiologic interpretation of human breast cancer epidemiology. Cancer, *33*: 1306–1312, 1974.

36. Grattarola, R. The premenopausal endometrial pattern of women with breast cancer. Cancer, *17*: 1119–1122, 1964.

GENETIC AND ENVIRONMENTAL FACTORS IN EXPERIMENTAL AND HUMAN CANCER,
H. V. GELBOIN ET AL. (EDS.), JAPAN SCI. SOC. PRESS, TOKYO, PP. 29–41, 1980

SPECIAL LECTURE

Thoughts on the Prevention of Cancer

Arthur UPTON*

National Cancer Institute, Bethesda, Md. 20205, U.S.A.

Thank you, Dr. Sugimura, Colleagues, and friends. I am pleased and honored to take part in this symposium. When invited to present a lecture, it seemed to me that an appropriate subject for my remarks, and one related to the broad theme underlying this meeting, would be "Thoughts on the Prevention of Cancer." Probably no other concept has occupied so much of my attention, as Director of the U.S. National Cancer Institute.

The pioneer work of Sir Richard Doll, John Higginson, Cuyler Hammond, Ernst Wynder, and others, suggesting that cancer is a potentially preventable disease, has given us one of the most challenging hypotheses of our time—a view which has revolutionized our attitude toward the cancer problem (*1*).

Since 1900 there has been a dramatic change in the leading causes of death in industrialized countries of the world. The infectious diseases, which were predominant in the 19th century, have been replaced by chronic degenerative diseases, including cancer (Fig. 1). When I was in medical training 35 years ago, every hospital had special isolation wards for patients with infectious diseases, and most sizeable communities had large tuberculosis sanitaria. Now the latter have all but disappeared. In the United States, the death rate for heart disease has also declined dramatically during the last decade, for reasons yet to be disclosed, whereas, in contrast, the age-adjusted death rate for cancer continues to increase (Fig. 2). Among leading causes of death, cancer is the only disease showing an upward trend.

The major reason for the increase in cancer mortality is the steep rise in the frequency of lung cancer (Fig. 3), which can be accounted for, in large measure, by increased consumption of cigarettes. An interval of some 20–30 years elapsed between the dramatic increase in the smoking habits of men and the rise in their lung cancer rates (Fig. 4). More recently, since World War II, the increase in

* Present address: Institute of Environmental Medicine, New York University, New York, N. Y., U.S.A.

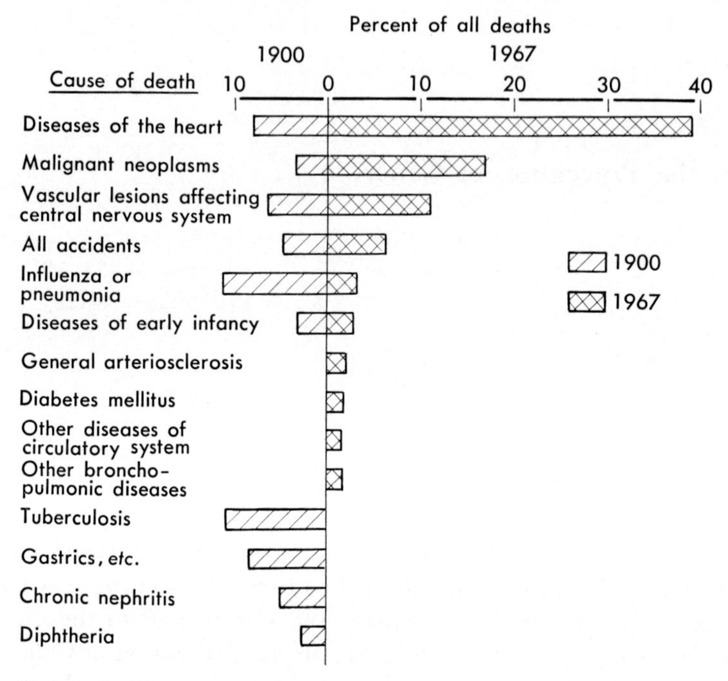

FIG. 1. Leading causes of death in the U.S., 1967, as compared with 1900 (*2*).

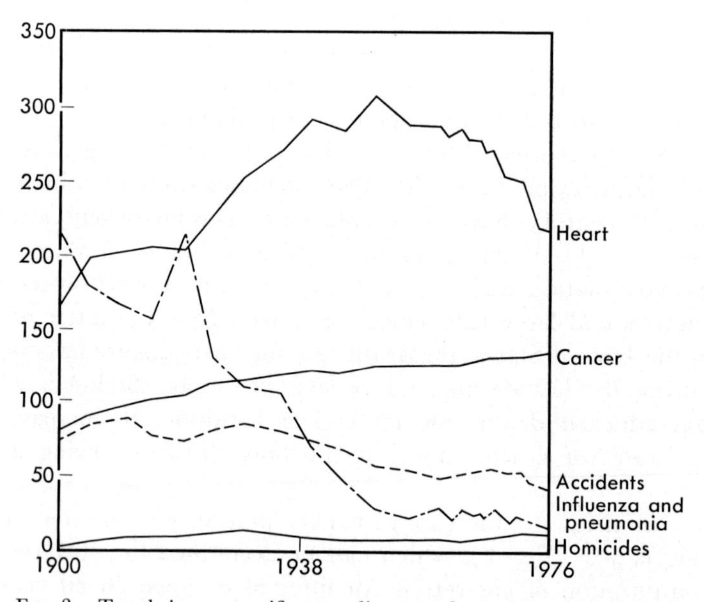

FIG. 2. Trends in age-specific mortality rates for major causes of death in the U.S. (*3*).

smoking among women has, similarly, been followed by an upward trend in deaths from lung cancer in this sex (Fig. 4).

Our experience with lung cancer demonstrates that one of the commonest

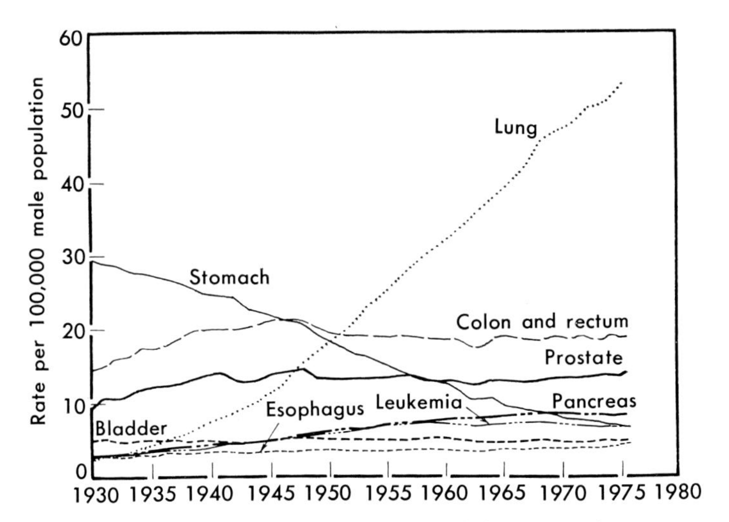

FIG. 3. Trends in age-specific mortality rates* for cancer of commonest types in the U.S., 1930–1976 (*4*). * Rate for the male population standardized for age on the 1940 U.S. population. Sources of data: National Vital Statistics Division and Bureau of the Census, U.S.

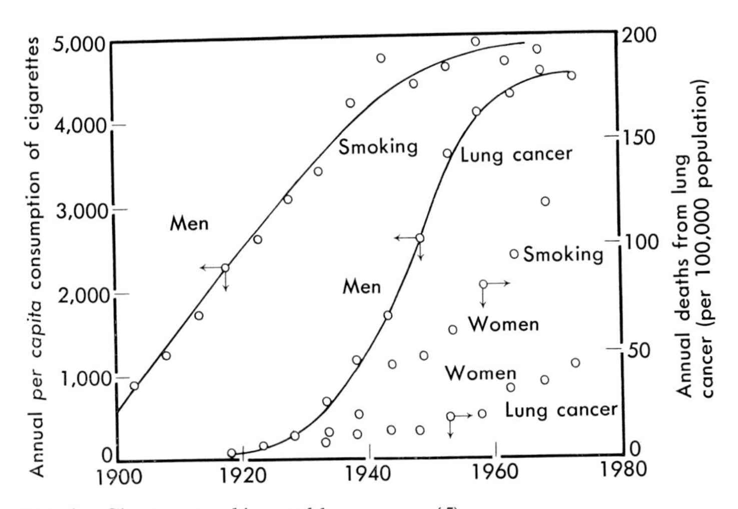

FIG. 4. Cigarette smoking and lung cancer (*5*).

cancers in our population—accounting for perhaps a quarter of all deaths from the disease in American men—is in principle preventable, at least to the extent that cigarette smoking can be eliminated or counteracted. Furthermore, the correlation between the number of cigarettes smoked per day and the associated risk (Fig. 5) is such as to imply that even low-level exposure to cigarette smoke may entail some increase in risk.

Cigarettes are not alone in suggesting a non-threshold type of dose-incidence relation for carcinogenesis. Another example, perhaps equally noteworthy, derives from our experience with ionizing radiation. When Lewis (*7*) first suggested, in 1957, that the incidence of leukemia in radiologists, atomic bomb survivors, and

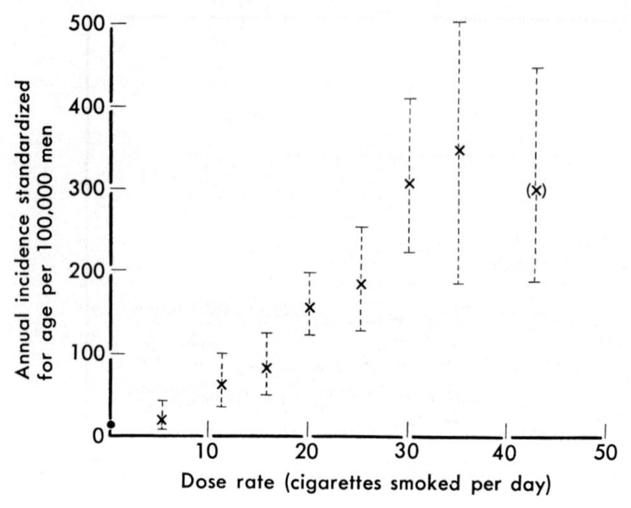

FIG. 5. Incidence of lung cancer in regular cigarette smokers, in relation to number of cigarettes smoked per day (*6*).

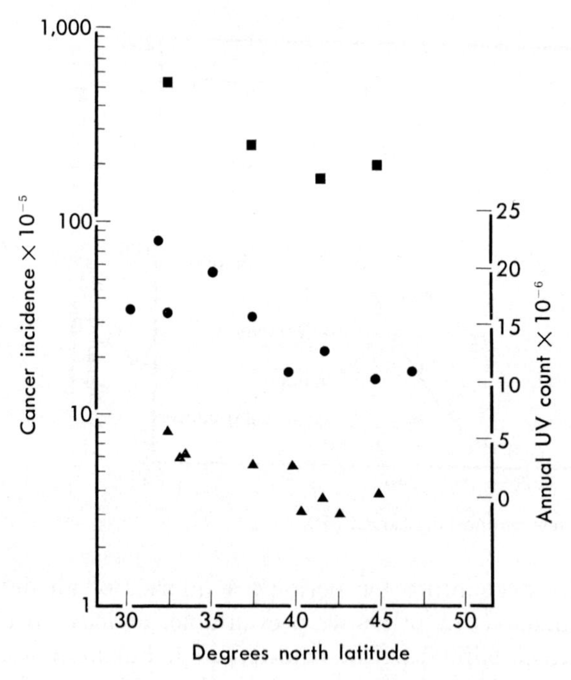

FIG. 6. Variation with latitude in ultraviolet (UV) radiation levels and age-adjusted skin cancer rates for white males in various U.S. cities (*8*). ■ non melanoma; ▲ melanoma; ● UV count.

patients treated with radiotherapy for benign diseases was increased equally for a given dose of radiation, despite widely differing conditions of exposure, his suggestion implied a linear non-threshold dose-incidence relationship. This interpretation led Lewis to infer that leukemia behaved as if it were induced by a somatic mutation, from which it could be deduced that some fraction of the natural incidence of the disease might be attributable to background radiation. His hypothesis was received with skepticism, bordering on hostility at the time. Clearly, the data did not then, nor do they now, suffice to establish conclusively the non-threshold nature of the dose-response, although they point strongly in that direction.

While we have no direct evidence that natural background levels of ionizing radiation contribute significantly to the occurrence of cancer, there is a clear relationship between geomagnetic latitude, or exposure to sunlight, and the incidence of skin cancer (Fig. 6). Furthermore, it was research on injury by ultraviolet light that pointed to DNA as the critical molecular target in carcinogenesis (*9*).

The correlation between carcinogenicity and mutagenicity has become progressively better, as we have learned to test carcinogens under conditions appropriate for demonstrating their mutagenic activity (Figs. 7 and 8). Although far from perfect, the correlation now gives us powerful new approaches for screening compounds in the environment—including crude mixtures—and tissue fluids, as well as for exploring the mechanisms of action of carcinogens at the molecular level (*11*).

We would not expect all carcinogens to be mutagens, since there are unquestionably non-mutational mechanisms contributing to the process of carcinogenesis —for example, promoting effects, to name only one. We clearly need more reliable

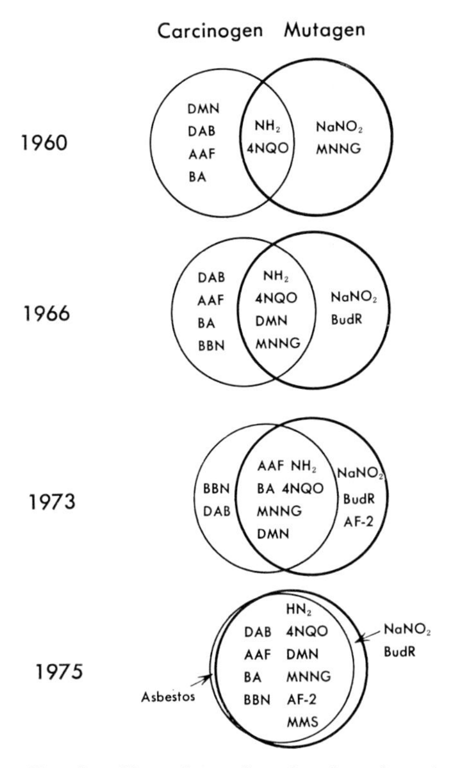

FIG. 7. Chronology of overlapping of carcinogens and mutagens (*10*).

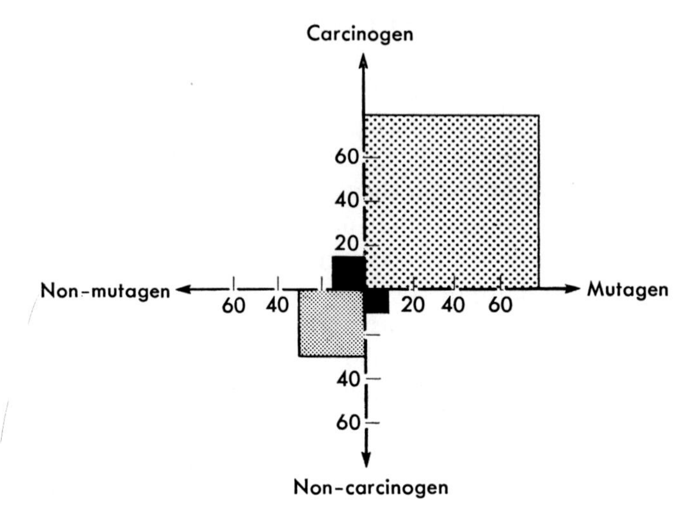

FIG. 8. Correlation between carcinogenic and mutagenic activities among a number of mutagens and non-mutagens (*10*).

methods for carcinogen screening, and it is of no small significance that the *in vitro* transformation of cells is becoming an increasingly reproducible and promising assay system. What we would like to have ultimately, of course, is a battery of short-term tests, coupled with *in vivo* tests in a reliable multi-tier assay system (*1*).

Even with *in vivo* tests there are formilable difficulties in extrapolating the results

TABLE 1. Estimated Human Risks from Saccharin Ingestion of 0.12 g/day[a]

Rat dose adjusted to human dose by surface area rule	Lifetime cases per million exposed	Cases per 50 million per year
Method of high- to low-dose extrapolation		
Single-hit model (Hoel, 1977)	1,200	840
Multi-stage model (with quadratic term) (Hoel, 1977)	0	3.5
Multi-hit model (Scientific Committee of the Food Safety Council, 1978)	0.001	0.0007
Mantel-Bryan probit model (Brown, 1978)	450	315
Rat dose adjusted to human dose by mg/kg/day equivalence		
Single-hit model (Saccharin and its Salts, 1977)	210	147
Multi-hit model (Scientific Committee of the Food Safety Council, 1978)	0.001	0.0007
Mantel-Bryan probit model (Brown, 1978)	21	14.7
Rat dose adjusted to human dose by mg/kg/lifetime equivalence		
Method of high- to low-dose extrapolations		
Single-hit model (Brown, 1977)	5,200	3,640
Multi-hit model (Scientific Committee of the Food Safety Council, 1978)	0.001	0.0007
Mantel-Bryan probit model (Brown, 1978)	4,200	2,940

[a] From Ref. *12*.

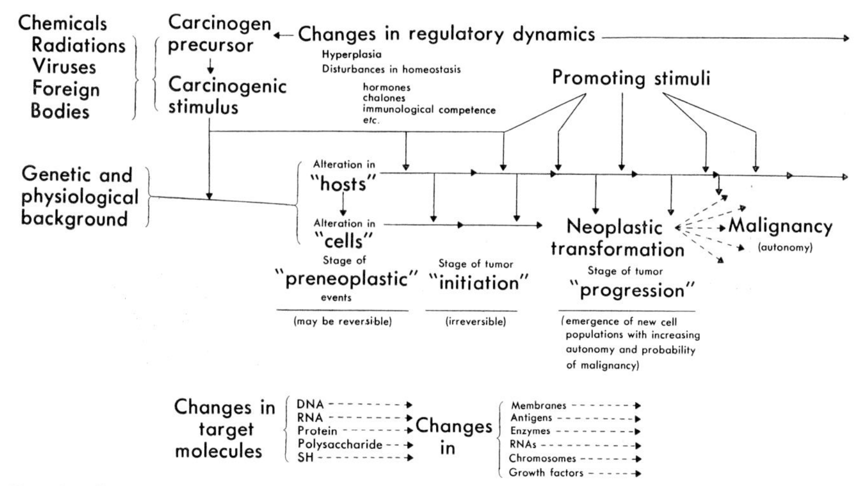

FIG. 9. Sequence of events in the evolution of the cancer process (*1*).

to humans, as brought out by the National Academy of Sciences report on saccharin. Depending upon the model used—single-hit *versus* multi-hit—and on the method of trans-species dose adjustment used, the estimated risk for humans derived from the existing animal data varies by a factor of nearly ten million (Table 1).

Obviously, in order to quantify human risk estimates satisfactorily, we will need to know far more about the mechanisms of carcinogenesis. In approaching this objective, we must envision carcinogenesis as a multicausal, multistage process, in which multiple environmental agents may interact at various stages in the process to favor the development and outgrowth of autonomous cells (Fig. 9). Although we know that promoting factors may be important in this process, we have as yet only limited knowledge of how to analyse the environment for such factors and how to allow for their effects in making risk assessments.

Even in the case of radiation, a well-characterized carcinogen, we are still in doubt as to the appropriate dose-response model for extrapolation into the low dose range. The data for leukemia in atomic bomb survivors of Hiroshima strongly imply a linear dose-incidence regression, whereas the data for Nagasaki survivors imply a sigmoid regression. Since neutrons comprised a large fraction of the dose at Hiroshima but only a negligible fraction of the dose at Nagasaki, the difference in dose-response between the two cities has been interpreted to indicate that the incidence of leukemia varies as a linear function of the neutron dose and as a quadratic function of the γ-ray dose (*13, 14*). This is the type of difference that would be predicted if the leukemias resulted from a chromosome translocation, such as the Philadelphia chromosome. It is noteworthy, therefore, that the frequency of interchange aberrations in circulating lymphocytes in Hiroshima survivors increases as a linear function of the dose, whereas in Nagasaki survivors it increases as a curvilinear function of the dose (Fig. 10).

These differences in response can be rationalized at the molecular level on the basis of differences in the distribution of energy along the tracks of impinging neutrons as compared with γ-rays. In general, a single traversal of the cell nucleus

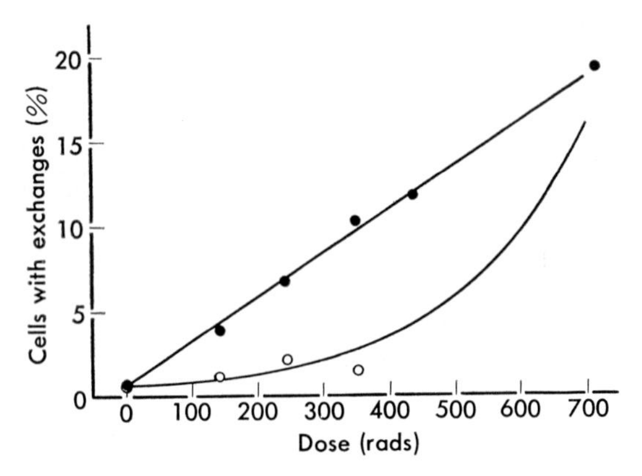

FIG. 10. Frequency of chromosome exchange aberrations in peripheral blood lymphocytes of atomic bomb survivors, as related to kerma dose (*15*). ● Hiroshima; ○ Nagasaki.

by a γ-ray fails to deposit sufficient energy at any one point along its path to cause two chromosome breaks close enough together to give rise to an interchange aberration. In contrast, a single traversal by a densely ionizing particle deposits enough energy at almost any point along its path to cause conditions leading to one or more interchange aberrations. Hence, the yield of interchange aberrations typically varies as a linear function of the dose in the case of densely ionizing radiations, but as linear-quadratic function of the dose in the case of sparsely ionizing radiations (Figs. 11 and 12).

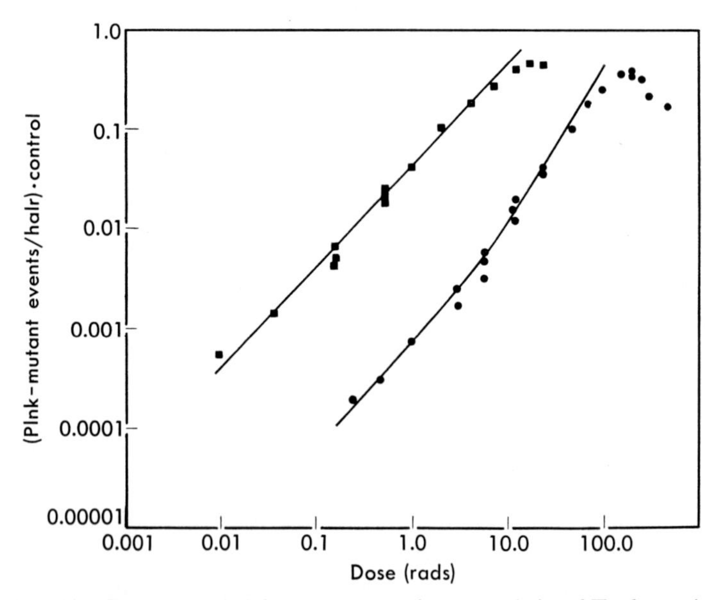

FIG. 11. Frequency of pink-mutant events in stamen hairs of Tradescantia, as influenced by fast neutrons or X-rays (*16*). In the rising portion of the curve, the data for neutrons are consistent with a linear dose-response relation ($I=C+\alpha D$), while those for X-rays are consistent with a linear-quadratic dose-incidence relation ($I+C+\alpha D+\beta D^2$). ■ neutrons (0.43 MeV); ● X-rays (250 kV-peak).

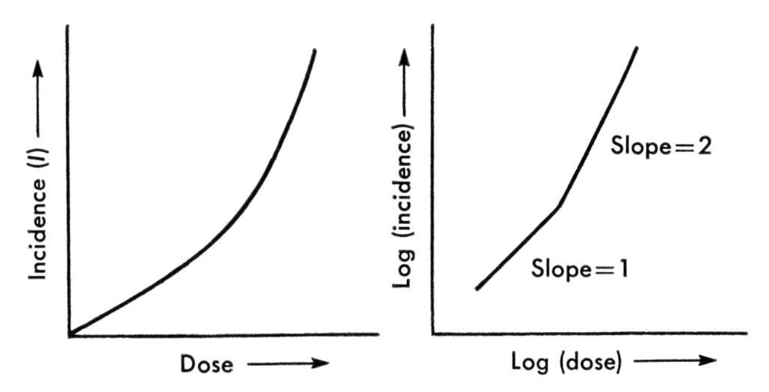

FIG. 12. The linear-quadratic dose response relation (*17*). $I = C + \alpha D + \beta D^2$.

In the case of linear-quadratic dose-response relation, linear extrapolation from observations at high doses tends to overestimate the risks at low doses and low dose rates (*18*). For this reason, the National Academy of Sciences Advisory Committee on the Biological Effects of Ionizing Radiation—the so-called Beir Committee—has questioned the validity of risk estimates for cancer derived on the basis of purely linear extrapolations, at least insofar as may be applicable to sparsely ionizing radiation (*19*).

On the other hand, the similarity in dose-incidence relations for breast cancer among atomic bomb survivors, patients treated with radiation for post-partum mastitis, and patients subjected to multiple fluroscopic examinations of the chest during the treatment of pulmonary tuberculosis, irrespective of wide differences in dose and dose rate, implies that successive small doses are essentially fully additive for this effect (*20*). The apparent additivity in successive doses is a pattern more consistent with a linear response than with a linear-quadratic dose-incidence response (*20*). Whatever the nature of this response, there is no reason to assume

TABLE 2. Estimated Risks of Radiation-induced Cancer[a]

Type of cancer	Population affected	Lifetime cases per million per rem
Leukemia	A-bomb[b] survivors; radiation therapy; thorotrast injections	10–60
Thyroid gland	A-bomb survivors; radiation therapy; Marshall Islanders	20–150
Breast (female)	A-bomb survivors; radiation therapy; multiple fluoroscopies	30–200
Lung	A-bomb survivors; radiation therapy; uranium miners	20–100
Bone	Radium ingestion and injection	5
Brain, salivary glands, stomach, liver, large intestine	A-bomb survivors; radiation therapy; thorotrast injections	10–15
Esophagus, small intestine, pancreas, rectum, bladder, ovary, lymphoid tissues, cranial sinuses	A-bomb survivors; radiation therapy; radium ingestion	Less than 5
Total		100–1,000

[a] From Refs. *21* and *22*. [b] A-bomb: atomic bomb.

TABLE 3. Estimates of Annual Contribution of Radiation Exposure to Lifetime Burden of Fatal Cancer to the U.S. Population[a)]

Source	Lifetime cancer mortality commitment (number of deaths)
Natural background	5,000
Technologically enhanced natural radiation	250
Healing arts	4,250
Nuclear weapons fallout	250–450
Nuclear energy	9
(Three mile Island)	(1.2)
Consumer products	1.5
Total	10,000
=2.7% of cancer mortality[b)]	

[a)] From Ref. *23*. [b)] Total cancer mortality, 1975, 365,000.

a priori that the response should be the same from one type of cancer to another. This is especially true in view of the striking and unexplained variations among tissues in susceptibility to radiation carcinogenesis (Table 2). Why the female breast should be the most sensitive tissue in the adult, insofar as we know, while tissues such as bone are much less sensitive defies explanation on the basis of present knowledge.

Whether we accept the linear hypothesis or assume that it may exaggerate the risks of small doses of ionizing radiation, it is likely that natural background levels contribute only slightly to the natural incidence of cancer (Table 3). Hence, even including all the man-made sources to which the population is exposed, radiation probably accounts for only a small percentage of the total cancer incidence (Table 3).

The dramatic decrease in the incidence of carcinoma of the stomach in the U.S. during recent decades (Fig. 2), which is explicable only in terms of undisclosed changes in environmental factors, attests to the plausibility of the cancer prevention hypothesis. The data for Japan suggest that the high incidence of stomach cancer in that country has in recent years, similarly, begun to decrease. Although it is tempting to ascribe the decrease to some favorable change in the Japanese diet, there have recently been such profound changes in the Japanese diet (Fig. 13) as to complicate explanation of the cancer trend.

To exploit fully the available epidemiologic evidence, we will need far more adequate record-keeping and record-linkage systems, providing data covering the better part of the human lifespan, including the intrauterine period. Such systems will be essential if the role of lifestyle in the development of cancer is to be elucidated in detail.

The suggestion that much of cancer is, in principle, preventable is an important and challenging hypothesis, but one which tends to be exaggerated and oversimplified. Hence, it is important that it be kept in proper perspective. For example, the possibility that working in Grand Central Station for 2 weeks may theoretically carry a one-in-a-million risk of fatal cancer (Table 4) should not dissuade anyone

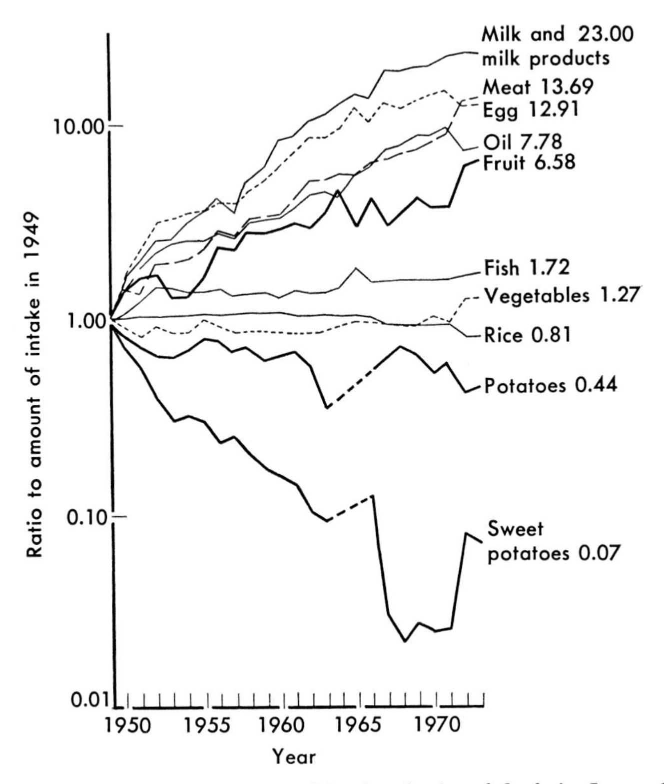

FIG. 13. Change in amount of intake of selected foods in Japan, 1949–1973 (1949=1.00) (*24*).

TABLE 4. Situations Involving an Estimated One-in-a-million Risk of Death[a)]

Travelling 400 miles by air (accident)
Travelling 60 miles by car (accident)
2 weeks of work in Grand Central Station (radiation cancer)
2 weeks of skiing (accident)
1 1/2 weeks of typical factory work (accident)
3 hr of work in a coal mine (accident)
Smoking 1 cigarette (cancer, heart-lung disease)
1 1/2 min of rock climbing (accident)
20 min of being a man aged 60 (all causes)

a) From Refs. *25* and *26*.

from going there to catch a train or even spending a lifetime working there, since one could presumably work 10,000 years there without increasing his natural lifetime cancer risk by as much as 1%.

If we are to succeed in our efforts to prevent cancer, we must develop a strategy which is practical and credible. This strategy will require quantification of risk estimates, which, in turn, will depend on research into all aspects of the cancer process, covering a comprehensive spectrum of basic and applied research activities. This approach must include the social sciences, as well as the biomedical sciences, in order to cope effectively with the societal and behavioral issues to be faced. It must also give high priority to effective education of the health professions and the public, with systematic attention to coordination of efforts by organized labor, industry, the public health establishment, the health professions, and the regulatory agencies.

There has been little attmpt thus far to mount such a comprehensive strategy, but the time is now ripe to do so. This symposium addresses many of the problems that are paramount to cancer prevention, but there are other aspects also to be considered. It is our responsibility to exploit the opportunities at hand, but at the same time it is crucial that we present the prospects for cancer prevention in a realistic timetable and perspective. Otherwise there is the danger that we may raise false hopes and, ultimately, cause disappointment, disillustionment, and rejection.

REFERENCES

1. Upton, A. C. Progress in the prevention of cancer. Pre. Med., *7*: 476–485, 1978.
2. Donabedian, A., Axelrod, S. J., Swearingen, C., and Jameson, J. Medical Care Chart Book, 5th ed., Bureau of Public Health Economics, University of Michigan School of Public Health, Ann Arbor, Michigan, 1972.
3. Schneiderman, M. A. Trends in cancer incidence and mortality in the United States. Testimony before the Subcommittee on Health and Scientific Research, Senate Committee on Human Resources, March 5, 1979.
4. Cancer Facts and Figures, American Cancer Society, Inc., New York, 1979.
5. Cairns, J. The cancer problem. Sci. Am., 1165–1175, Nov. 1975.
6. Doll, R. An epidemiological perspective of the biology of cancer. Cancer Res., *38*: 3573–3583, 1978.
7. Lewis, E. B. Leukemia and ionizing radiation. Science, *125*: 965, 1957.
8. Upton, A. C. Radiation effects. *In*; H. H. Hiatt, J. D. Watson, and J. A. Winsten (eds.), Origins of Human Cancer (Cold Spring Harbor Conference on Cell Proliferation, vol. 4), pp. 477–500, Cold Spring Harbor Laboratory, New York, 1977.
9. Setlow, R. B., Ahmed, F. E., and Grist, E. Xeroderma pigmentosum: Damage to DNA is involved in carcinogenesis. *In*; H. H. Hiatt, J. D. Watson, and J. A. Winsten (eds.), Origins of Human Cancer (Cold Spring Harbor Conference on Cell Proliferation, vol. 4), pp. 889–902, Cold Spring Harbor Laboratory, New York, 1977.
10. Sugimura, T., Sato, S., Nagao, M., Yahagi, T., Matsushima, T., Seino, Y., Takeuchi, M., and Kawachi, T. Overlapping of carcinogens and mutagens. *In*; P. N. Magee *et al.* (eds.), Fundamentals in Cancer Prevention, pp. 191–215, Japan Sci. Soc. Press, Tokyo/University Park Press, Baltimore, 1976.
11. Hooper, N. K., Ames, B. N., Saleh, M. A., and Casida, J. E. Toxaphene, a complex mixture of polychloroterpenes and a major insecticide, is mutagenic. Science, *205*: 591–593, 1979.
12. Saccharin: Technical Assessment of Risks and Benefits. Part 1 of a 2-part study of the Committee for a Study on Saccharin and Food Safety Policy. Panel I: Saccharin and its Impurities. Assembly of Life Sciences/National Research Council and the Institute of Medicine, National Academy of Sciences, Washington, 1978.
13. Rossi, H. H. and Kellerer, A. M. The validity of risk estimate of leukemia incidence based on Japanese data. Radiat. Res., *58*: 131–140, 1974.

14. Ishimaru, T. Otake, M., and Ishimaru, M. Dose-response relationship of neutrons and X-rays to leukemia incidence among atomic bomb survivors in Hiroshima and Nagasaki by type of leukemia, 1950–1971. Radiat. Res., *77*: 377–394, 1979.
15. Awa, A. A. A review of thirty years study of Hiroshima and Nagasaki atomic bomb survivors. II. Biologic effects. G. Chromosome aberrations in somatic cells. J. Radiat. Res., *16* (Suppl.): 122–131, 1975.
16. Sparrow, A. H., Underbrink, A. G., and Rossi, H. H. Mutations induced in tradescantia by small doses of X-rays and neutrons; analysis of dose-response curves. Science, *176*: 916–918, 1972.
17. Brown, J. M. The shape of the dose-response curve for radiation carcinogenesis: Extrapolation to low doses. Radiat. Res., *71*: 34–50, 1977.
18. Upton, A. C. Radiobiological effects of low doses: Implications for radiological protection. Radiat. Res., *71*: 51–74, 1977.
19. Marx, J. L. Research News. Low-level radiation: Just how bad is it? Science, *204*: 160–164, 1979.
20. Upton, A. C., Beebe, G. W., Brown, J. M., Quimby, E., and Shellabarger, C. Report of NCI ad how working group on the risk associated with mammography in mass screening for the detection of breast cancer. J. Natl. Cancer Inst., *59*: 479–493, 1977.
21. The effects on population of exposure to low levels of ionizing radiation. Advisory Committee on the Biological Effects of Ionizing Radiation, National Academy of Sciences, National Research Council, Washington, 1972.
22. Sources of and effects of ionizing radiation. United Nations Scientific Committee on the Effects of Atomic Radiation 1977 Report to the General Assembly, with annexes, United Nations, New York, 1977.
23. Jablow, S. and Bailar, J. The contribution of ionizing radiation to cancer motality in the United States. Prev. Med., in press.
24. Hirayama, T. Changing patterns of cancer in Japan with special reference to the decrease in stomach cancer mortality. *In*; H. H. Hiatt, J. D. Watson, and J. A. Winsten (eds.), Origins of Human Cancer (Cold Spring Harbor Conference on Cell Proliferation, vol. 4), pp. 55–75, Cold Spring Harbor Laboratory, New York, 1977.
25. Pochin, E. E. Why be quantitative about radiation risk estimates? National Council on Radiation Protection and Measurements, Washington, 1978.
26. Wilson, R. Risks caused by low levels of pollution. Yale J. Biol. Med., *51*: 37–51, 1978.

MIXED-FUNCTION OXIDASES

GENETIC AND ENVIRONMENTAL FACTORS IN EXPERIMENTAL AND HUMAN CANCER,
H. V. GELBOIN ET AL. (EDS.), JAPAN SCI. SOC. PRESS, TOKYO, PP. 45–58, 1980

Multiple Products of Polycyclic Hydrocarbon Metabolism

J. Capdevila, R. Renneberg,* R. A. Prough, and R. W. Estabrook
Department of Biochemistry, Southwestern Medical School, The University of Texas Health Science Center, Dallas, Texas 75235, U.S.A.

Abstract: Considerable interest has centered on the oxidative metabolism of the polycyclic hydrocarbon, benzo(a)pyrene (BP), since it is well established that this compounds ubiquitous occurrence in the environment may affect the daily lives of many people and that it must be metabolically activated to its ultimate carcinogenic form. The concept that carcinogens must be metabolically activated prior to manifesting their deleterious effect has been shown to be true for a number of chemicals. The oxidative transformation of BP is effectively catalyzed by an enzyme system, termed microsomal mixed-function oxidase or aryl hydrocarbon hydroxylase, in which a hemeprotein, cytochrome P-450, functions as a terminal oxidase. It is now known that cytochrome P-450 consists of a family of hemeproteins which differ in their physical and enzymatic properties and each is apparently under separate genetic control. The studies on BP metabolism have been greatly facilitated by the introduction of high pressure liquid chromatography which permits accurate measurement of the multiple products formed: dihydrodiols, phenols, and quinones. Studies on BP metabolism *in vitro* clearly show that the metabolite pattern obtained differs depending on the source of enzyme (*i.e.*, the predominant form of cytochrome P-450) utilized in the reaction.

The existence of significant amounts ($<25\%$) of at least three quinone products (1,6-, 3,6-, and 6,12-diones) during NADPH-dependent microsomal metabolism of BP is of particular interest. In addition, quinones appear to be the primary products formed (90–95%) during the peroxidative function of cytochrome P-450 when BP is incubated with the organic peroxide, cumene hydroperoxide. These studies suggest an alternate mechanism of BP oxidation, presumably involving free radical intermediates. This process may function concomitant with the hydroxylase or mixed-function oxidase activity of cytochrome P-450. However, the presence of a highly active quinone reductase in the microsomal fraction of liver and the ability of hydroquinones to be oxidized to form hydrogen peroxide further adds to the

* Present address: Department of Biocatalysis, Central Institute of Molecular Biology of the Academy of Sciences G.D.R., Berlin-Buch, G.D.R.

complexity of interpreting the carcinogenic and toxic characteristics of BP metabolites.

One may envision at least two means of biologically activating polycyclic hydrocarbons so that the resultant products may lead directly to cellular toxicity. For example, activation of oxygen by cytochrome P-450 can generate the reactive diol-epoxides on one hand and the peroxidative function of the hemeprotein may result in the formation of free radicals, superoxide anion radical, and hydrogen peroxide on the other hand. An understanding of the factors influencing the balance between these two processes and other conjugative processes which may occur during the metabolism of polycyclic hydrocarbons may define the extent and pattern of cellular alterations observed.

The manner in which the human body deals with a life-time of exposure to potentially harmful chemicals (natural and man-made) with only a minimum of perceptible perturbation of cellular function serves as a challenge to the limited understanding of xenobiotic metabolism available to present day scientists. The realization that at some critical point in time tumorigenesis may occur as a result of this exposure further complicates the task at hand. One can now construct a series of equations which reduces the sequence of events associated with the metabolic transformation of foreign chemicals to a readily understandable, albeit oversimplified summary of the great advances that have accrued in the last two or three decades. To some scientists, the explanation of chemical carcinogenesis is at hand and only the details require elaboration and to others, the current framework of understanding only now allows significant and penetrating questions to be asked which will require new approaches and methodologies. For the remaining few who are skeptics, the accepted dogma is inconclusive and perhaps other avenues and approaches should be pursued.

This state of affairs brings to mind the plight faced by the alchemist during medieval times. Indeed, one can allegorize the parallelism. The alchemist was concerned either with the transformation of baser metals into gold or the discovery of an elixir for perpetual youth. Many scientists today are concerned with understanding the biological system functional in the transformation of xenobiotic com-

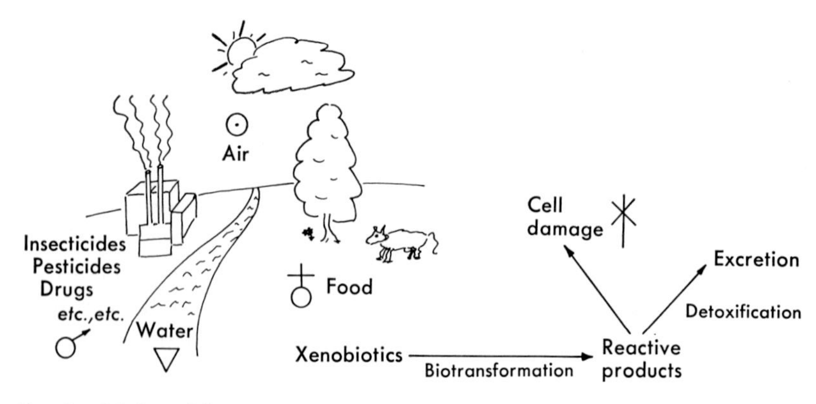

FIG. 1. Modern alchemy.

pounds into derivatives which can be effectively detoxified by the body and thereby removed without causing cellular damage (Fig. 1). Indeed, there are some who feel deeply that concerted attempts to remove all toxic chemicals from our environment, food supply, water, *etc.*, would be the first step in obtaining the equivalent of perpetual youth. (Note in Fig. 1, the allegory has been extended to show the similarity of the alchemists four basic elements of air, water, fire, and earth to our own concerns of potential sources of exposure to such harmful chemicals.)

The Mechanism of Oxidative Metabolism of Xenobiotic Compounds

The existence in many tissues of an enzyme system commonly called either the aryl hydrocarbon hydroxylase, the microsomal mixed-function oxidase, or the cytochrome P-450-containing monooxygenase has now been well established. Although the nature of the "active oxygen" required for the oxidative transformation of substrates such as polycyclic hydrocarbons has eluded chemical definition, significant details of the cyclic function of the hemeprotein cytochrome P-450, required as the catalyst for many monooxygenation reactions, permits an evaluation of potential control sites which could serve as focal points for further investigation.

As shown in Fig. 2, a sequence of reactions occur which are considered by many to be common for the oxidative transformation of many chemicals *via* cytochrome P-450 regardless of the type of substrate studied; *i.e.*, whether it is a drug undergoing *N*- or *O*-dealkylation, a steroid molecule which is transformed by hydroxylation, or a polycyclic hydrocarbon such as benzo(a)pyrene (BP) undergoing epoxidation. The six reaction steps shown in Fig. 2 and currently proposed to be operational in the cyclic function of cytochrome P-450 can be briefly described as follows (*1*):

a) A molecule of the substrate to be metabolized reacts with the ferric form of the hemeprotein, cytochrome P-450;
b) The complex of ferric cytochrome P-450 plus a molecule of substrate undergoes a one electron reduction. The electron is derived from NADPH and is transferred by the flavoprotein, NADPH-cytochrome P-450 reductase;
c) The complex of ferrous cytochrome P-450 and substrate reacts with oxygen to form a ternary complex called oxy-cytochrome P-450;
d) Oxy-cytochrome P-450 can be further reduced by an electron donated by either reduced cytochrome b_5 or the reduced flavoprotein. The resulting two electron reduced state of the complex is termed peroxy-cytochrome P-450. Alternatively, oxy-cytochrome P-450 can decompose giving rise to superoxide anion while regenerating the substrate-bound form of ferric cytochrome P-450 (see below);
e) The two electron reduced form of oxygen in the complex of peroxy-cytochrome P-450 can presumably undergo protonation with the resultant formation of a molecule of water. This would cause the generation of an electrophilic form of atomic oxygen postulated by some to be the oxenoid species (*2*). Evidence for this proposed intermediate is very meager and further experimentation will be required to substantiate its existence;

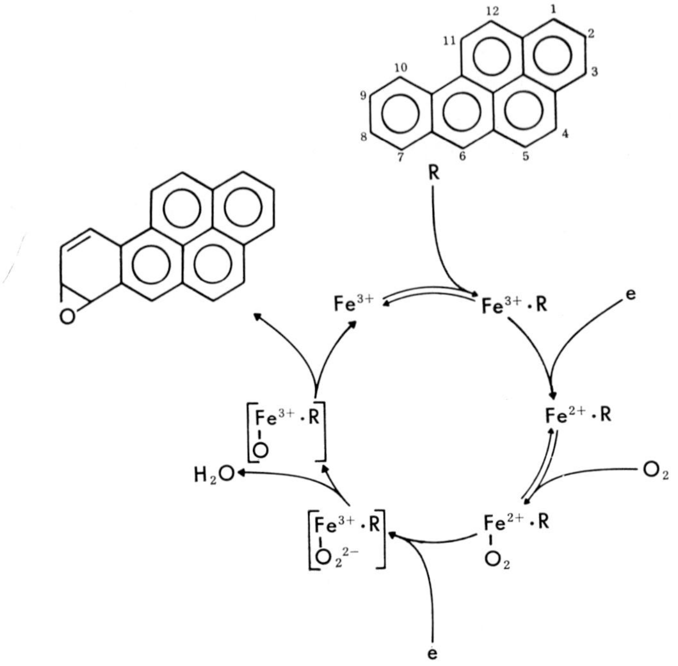

Fig. 2. Proposed catalytic cycle of cytochrome P-450 function during BP epoxidation.

f) Presumably the "activated oxygen" then reacts with the spatially adjacent molecule of substrate to form an epoxide, in the case of polycyclic hydrocarbon metabolism. This product dissociates from the active site of the cytochrome, thereby regenerating the substrate free form of the ferric hemeprotein which can then interact with a new molecule of substrate to be metabolized.

The Possible Role of Superoxide and Hydrogen Peroxide

Studies of liver microsomal electron transport reactions catalyzed by cytochrome P-450 have demonstrated the formation of hydrogen peroxide concomitant with the oxidation of NADPH. The hydrogen peroxide is believed to be formed by dismutation of the superoxide anion generated during the dissociation of oxy-cytochrome P-450. Some substrates of cytochrome P-450 tend to enhance the rate of hydrogen peroxide formation by an undefined "uncoupling" action (*3–5*).

During the metabolism of some substrates, such as polycyclic hydrocarbons, one can demonstrate the additional formation of hydrogen peroxide as a consequence of the oxidation of hydroquinone metabolites (*6*). Nishibayaishi-Yamashita and Sato (*7*) have reported a vitamin K_3-dependent NADPH oxidase activity of liver microsomes which generates appreciable amounts of hydrogen peroxide. Recent studies in our laboratories have shown that liver microsomes contain an NADPH-dependent BP quinone reductase (*8*) which involves the same electron transport system, including cytochrome P-450, as the reaction system functional in mixed function oxidation reactions. Thus, at least two sources of the highly reactive superoxide anion or its dismutation product, hydrogen peroxide, can be envisioned to operate during the time course of NADPH oxidation by liver microsomes.

Multiple Products of BP Metabolism

Although it is recognized that a number of additional hypothetical intermediates may participate in the above described scheme for the cyclic function of cytochrome P-450, it is apparent that the reaction mechanism for the oxidative transformation of a polycyclic hydrocarbon such as BP is even more complex. Incubation of BP with liver microsomes in the presence of NADPH and oxygen results in the concomitant formation of a multitude of products (*9, 10*), including

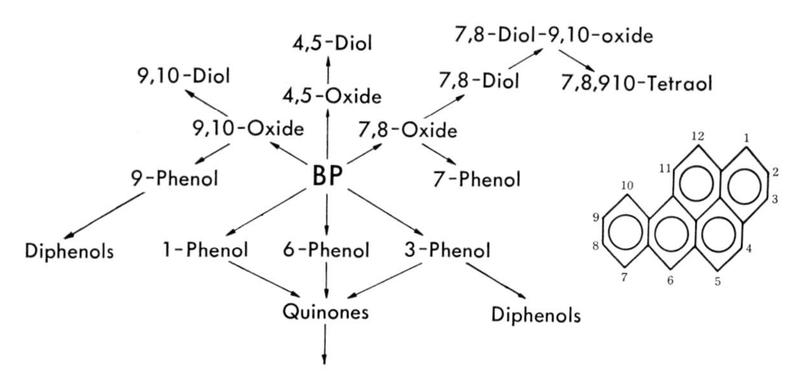

FIG. 3. The known metabolic fates of BP and its oxidative metabolites.

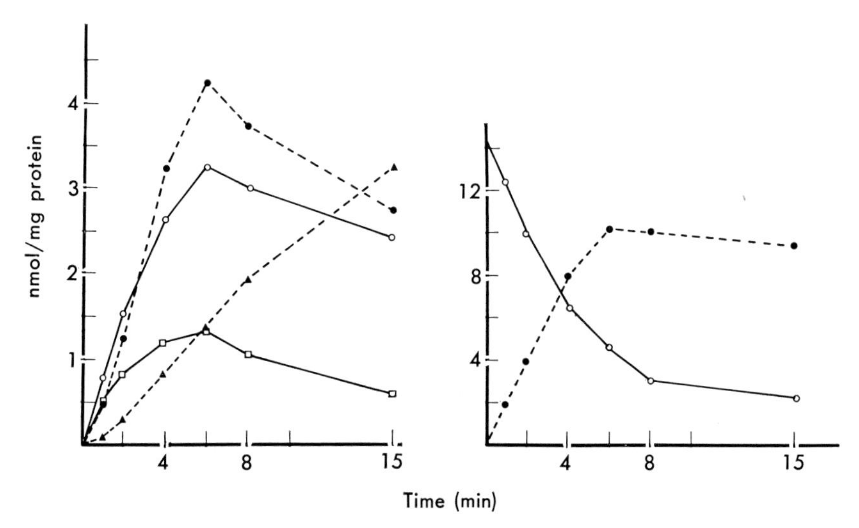

FIG. 4. The time course of NADPH-dependent BP metabolism by liver microsomes from 5,6-benzoflavone-treated rats and limiting concentration of substrate. The reaction mixture contained 0.5 mg/ml microsomal protein and the metabolites were determined using high performance liquid chromatography (HPLC) (*9*). Left: ● phenols; ▲ polar; ○ diols; □ quinones. Right: ● total products; ○ 7 μM BP.

phenols, dihydrodiols, quinones, and their further oxidation products (*11*) (Fig. 3). Indeed, as shown in Fig. 4 where a limiting amount of BP has been added to the incubation media, the initial products formed are further metabolized following nearly complete depletion of the substrate, BP. This experiment is also of interest since it indicates that no significant pool of BP remains sequestered in the lipid mileau of the membrane. This undoubtably reflects the very high affinity of cytochrome P-450 for this substrate molecule and suggests that cytochrome P-450 serves as a scavenger for lipophilic substrates from other cellular pools or reservoirs which could potentially exist.

While a number of investigators have been concerned with the exciting discoveries of diol-epoxides (*12–14*), our laboratories turned their attention to the question of the mechanism of formation of quinones and their possible role in chemical carcinogenesis. In particular, the excellent work of Nagata and his colleagues at the National Cancer Center Research Institute, Tokyo, Japan (*15, 16*) and Ts'o and his group at Johns Hopkins University (*17, 18*) who have studied in detail the formation of the 6-oxo derivitive of BP, served as a stimulus for continuation of these studies. In addition, recent studies with cytochrome P-450 have demonstrated the ability of this hemeprotein to function also as a peroxidase in the metabolism of many substrates, *i.e.*, serve in part as a catalyst for one electron oxidation reactions (*19, 20*). Thus, studies were initiated to evaluate whether BP phenols could serve as substrates for the formation of quinones.

Metabolism of Phenols

Incubation of liver microsomes with 3-hydroxy-BP, oxygen, and NADPH results in the formation of two major metabolites—one of which corresponds to the 3,6-quinone of BP (*21*) and the other to the 3,9-dihydroxy-BP (*22*). Of interest, is the observation that the pattern of metabolites formed during the metabolism of 3-hydroxy-BP was not significantly modified when liver microsomes from rats pretreated with either phenobarbital or 5,6-benzoflavone were used. In contrast, studies on the metabolism of 9-hydroxy-BP showed a marked change in the metabolite patterns (Fig. 5) depending on the animal pretreatment regimen (*23*). Using liver microsomes from phenobarbital-treated rats, one sees three major metabolites formed during the *in vitro* incubation of 9-hydroxy-BP and NADPH. With liver microsomes from 5,6-benzoflavone-treated animals, only two of the three metabolites are seen while with untreated (control) animals a single major product was formed. The fact that peak A of Fig. 5, seen during the metabolism of 9-hydroxy-BP, is identical in physical properties to one of the major metabolites observed during the metabolism of 3-hydroxy-BP identifies this product as 3,9-dihydroxy-BP. The chemical characteristics of the other products derived from 9-hydroxy-BP metabolism suggests that these metabolites are also dihydroxy compounds. The chemical identification of these products remains to be fully established. It is apparent that liver microsomes contain a cytochrome P-450 phenol oxygenase and that dihydroxy compounds, as well as, quinones can be readily formed during the further metabolism of these compounds.

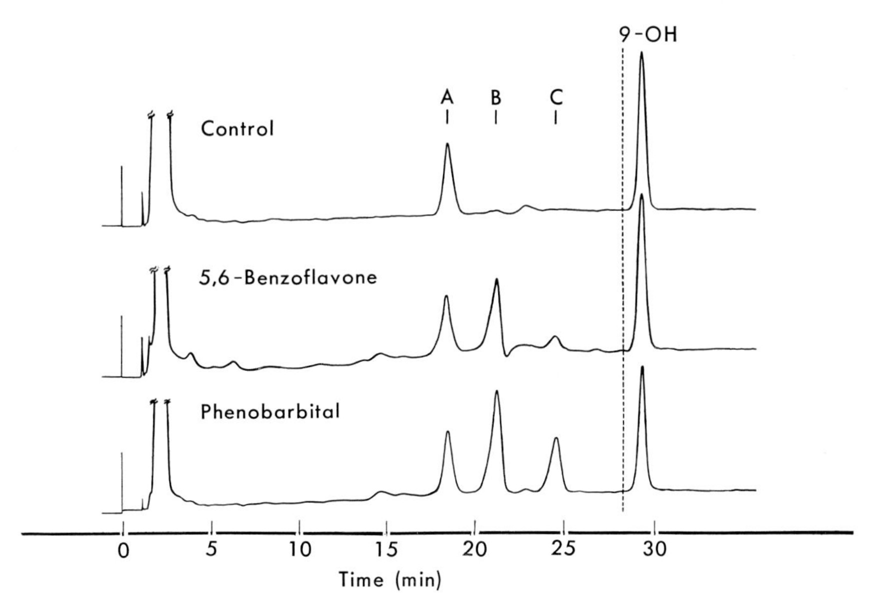

FIG. 5. Effect of animal pretreatment on the chromatographic profile of 9-hydroxy-BP metabolites. The dashed line indicates a change in the ordinate from 0.2 to 2.0 AUFS for the output of the HPLC detector at 254 nm. The HPLC gradient used was similar to that reported by Selkirk *et al.* (*9*).

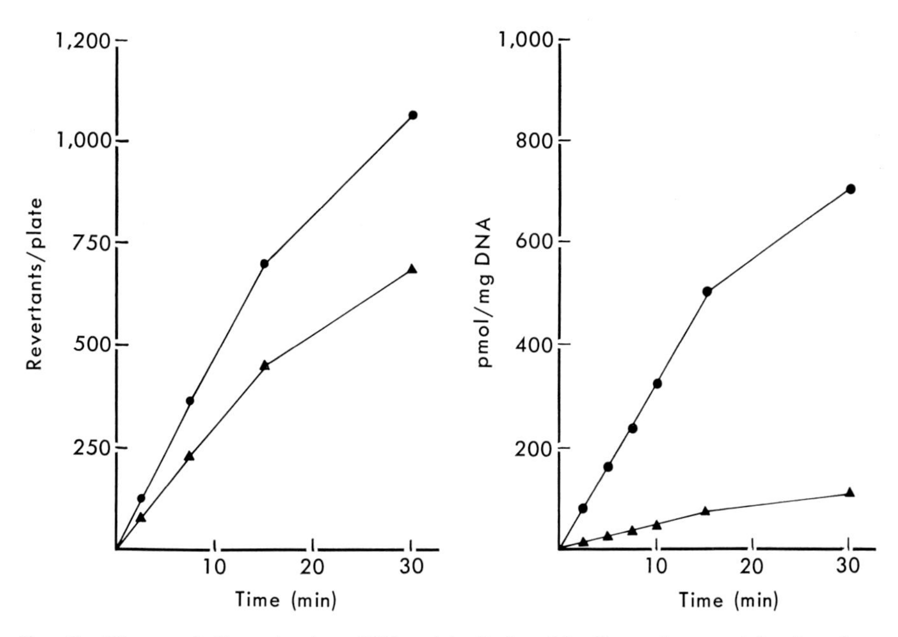

FIG. 6. The metabolic activation of BP and its 9-phenol by liver microsomal fractions from 5,6-benzoflavone-pretreated rats. Left: microsome-mediated mutagenesis of *S. typhimurium*. Right: microsome-mediated alkylation of exogenous DNA. Data taken from Ref. *25*. ● 9-phenol; ▲ BP.

As an added point of interest, an informative series of experiments were carried out by Lubet *et al.* to assess the ability of the 9-phenol to serve as a premutagen in a microsome-mediated *Salmonella* mutagenesis assay (*24, 25*). Upon metabolic activation with rat liver microsomes from 5,6-benzoflavone-treated animals, it was noted that 9-hydroxy-BP was more potent as a premutagen than BP itself. Comparative studies designed to measure the binding of the metabolites of BP and its 9-phenol to exogenous DNA also showed the greater effectiveness of the 9-hydroxy-BP in this assay. This observation demonstrates that mutagenic metabolites other than the 7,8-dihydrodiol-9,10-oxide of BP can be formed *in vitro* (*12*). It is apparent that additional metabolites, in particular those formed during the metabolism of phenols, should be considered as agents capable of modifying cellular genetic information or its expression (*24–28*).

The Peroxide-supported Metabolism of BP

Recent studies have shown that a number of compounds which are oxidatively transformed *via* the microsomal cytochrome P-450 system, in the presence of NADPH and oxygen, can also be metabolized by organic peroxides such as cumene hydroperoxide (*19, 20*). These observations have led to the hypothesis that the pathway of cytochrome P-450 function (*cf.* Fig. 2) share a common series of reactive intermediates when NADPH and oxygen or an organic hydroperoxide serves as the initiator of the reaction. Earlier studies by Rahimtula *et al.* (*29*) had shown that BP could be metabolized to fluorescent phenols in the presence of cumene hydroperoxide. This result was unusual because fluorescent phenol formation was critically dependent on a very narrow range of organic hydroperoxide concentration. Our recent studies indicate that products other than phenols are formed in this reaction (*30*). As shown in Table 1, marked differences exist in the pattern of metabolites formed when comparing the oxygenase function of cytochrome P-450 with its organic peroxide-supported function in the metabolism of the polycyclic hydrocarbon. Some of these differences in product formation are readily apparent when one contrasts the time-dependent absorbance changes, as measured by difference repetitive scan spectrophotometry (*31*) when liver microsomes are incubated with BP in the presence of either NADPH or cumene hydroperoxide (Fig. 7). The dominant feature of the difference spectra is the development of absorbance in the

TABLE 1. NADPH- and Cumene Hydroperoxide-dependent Metabolism of BP

Addition	Animal pretreatment[b]	Total metabolism (nmols/min/mg)	% of metabolites[a]		
			Diols	Quinones	Phenols
NADPH	PB	5.4	22.5	27.6	49.9
NADPH	BF	8.0	25.1	20.6	54.3
CuOOH	PB	5.7	6.4	94.0	0.4
CuOOH	BF	2.0	6.5	73.0	19.0

[a] The reaction mixtures were analyzed as described in Ref. *30*. The conditions for each experiment were identical except either NADPH or cumene hydroperoxide was added to initiate the reaction.
[b] PB, phenobarbital; BF, 5, 6-benzoflavone.

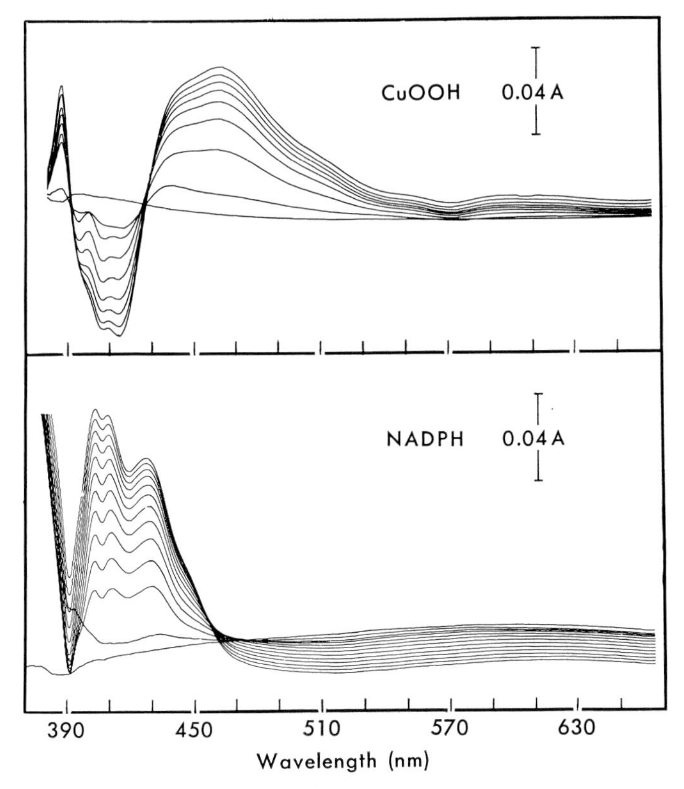

FIG. 7. Difference spectral analysis of the metabolism of BP supported by either cumene hydroperoxide or NADPH. Data taken from Ref. *30*.

optical region around 480 nm when cumene hydroperoxide is added to the reaction system. The presence of this absorbance indicates that increasing concentrations of quinones are the major metabolites formed during the steady-state of the reaction in the presence of cumene hydroperoxide. In contrast, the spectrophotometric studies using NADPH do not show any increase in absorbance near 480 nm with time since liver microsomes have been shown to contain a very active NADPH-dependent BP quinone reductase (*8*, *31*). The small shoulder near 448 nm suggests that the quinones exist as hydroquinones during the steady-state of the reaction in the presence of NADPH.

Analysis of samples by high performance liquid chromatography directly demonstrates the marked difference in the type of products formed (Fig. 8). As shown in the figure, quinones are the major products formed in the presence of the organic peroxide. In addition, Table 1 provides a direct comparison of the distribution of products obtained using liver microsomes from either phenobarbital- or 5,6-benzoflavone-treated animals to support the metabolism of BP in the presence of either NADPH or cumene hydroperoxide. It should be noted that the relative rates of metabolism by either cumene hydroperoxide or NADPH differ depending on the source of microsomes used. These results undoubtably reflect differences in the type and specificity of the cytochromes P-450 present in these microsomal preparations.

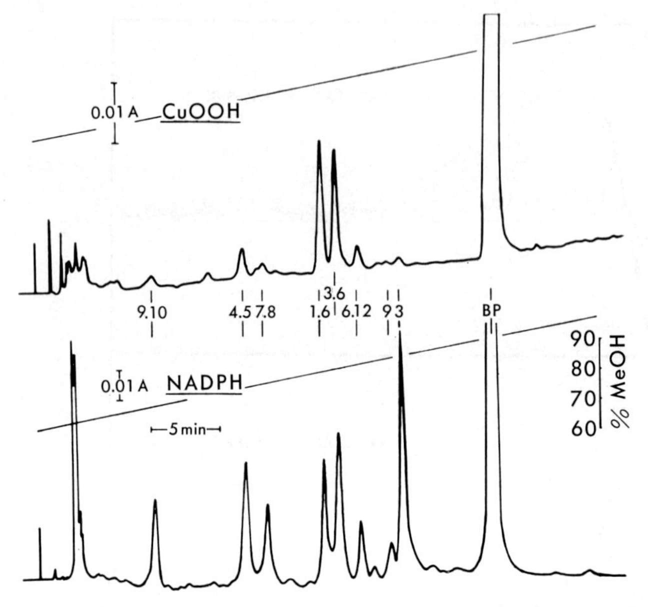

FIG. 8. HPLC separation of BP metabolites produced by either the cumene hydroperoxide- or NADPH-supported oxidative reactions of cytochrome P-450. The HPLC was performed similar to the method of Selkirk *et al.* (*9*). For purposes of comparison, the sample injected from the NADPH-dependent reaction mixture contained a 5 times larger concentration of metabolites based on time and protein concentration. Data taken from Ref. *30*.

The fact that dihydrodiols were detected only in low concentrations when liver microsomes were incubated with cumene hydroperoxide suggests that the epoxide intermediates were formed at lower rates than those observed with NADPH. As shown elsewhere (*30*), this result was not due to the destruction of epoxide hydrase or the dihydrodiol products by the cumene hydroperoxide added to the reaction mixture. Likewise, the presence of only low percentage concentrations of phenols suggested that these products were rapidly further metabolized in the presence of cumene hydroperoxide or the pathway to quinone formation did not involve the intermediary role of such phenols. The latter was proven not to be the case since it could be demonstrated that the 3- and 6-hydroxy-BP derivatives were rapidly metabolized to quinones in the presence of cumene hydroperoxide and liver microsomes. Studies utilizing a wide number of cumene hydroperoxide concentrations indicated that a delicate balance exists between phenol formation from BP and phenol metabolism to quinones. This observation would explain the unusual concentration dependence on organic hydroperoxide seen in the studies reported by Rahimtula *et al.* (*29*).

We were able to isolate and measure by electron paramagnetic resonance spectroscopy a free radical species with a signal identical to that of the 6-oxo radical of BP (*30*). This observation supports the conclusions of Nagata *et al.* (*15*) suggesting that the formation of the 6-oxo radical proceeds by a one electron oxidation process. It would also seem likely that quinones may be formed from phenols by a similar one electron oxidation reaction during the NADPH-supported BP metabolism.

TABLE 2. Hydrogen Peroxide-dependent Metabolism of BP

Total metabolism (nmols/min/mg)	% of metabolites[a]		
	Diols	Quinones	Phenols
0.9	19.0	43.0	38.0

[a] The reaction mixture was identical to that described in Ref. *30* except 5 mM hydrogen peroxide was used in the place of either NADPH or cumene hydroperoxide. Liver microsomes from phenobarbital-pretreated rats were used for this experiment.

In an attempt to investigate whether hydrogen peroxide can also support BP metabolism in a manner similar to cumene hydroperoxide, the pattern of metabolism was determined using high performance liquid chromatography (Table 2). The conditions for optimal hydrogen peroxide-supported metabolism were determined; it was noted that 10 mM hydrogen peroxide was sufficient for maximal reaction. In contrast to the results observed with the cumene hydroperoxide-supported reaction, significant concentrations of dihydrodiols and phenols were formed when BP is oxidized by hydrogen peroxide and liver microsomes. However, the percentage amount of quinone generated is significantly higher than that seen in the NADPH-dependent reaction. These results suggest that in contrast to the cumene hydroperoxide-dependent reactions, the hydrogen peroxide-supported metabolism of BP may be a more equal mixture of an epoxidation reaction mechanism and the one-electron oxidation reaction mechanism.

Theory, Speculation, and Conclusions

Although the role of the "bay region diol-epoxides" of BP and several other polycyclic hydrocarbons seems to be clearly established, other observations indicate that the process of tumor induction *in vivo* could involve the interaction of a multiplicity of active molecules with their respective targets. The predominence of a particular reactive intermediate would be determined by the complex interplay of several factors. These would include an enzymatic factor such as the specificity of the various types of cytochrome P-450 involved in epoxidation or one electron oxidation reactions and a factor reflecting the chemical properties of the polycyclic aromatic hydrocarbon. The presence of regions of high electron density at various

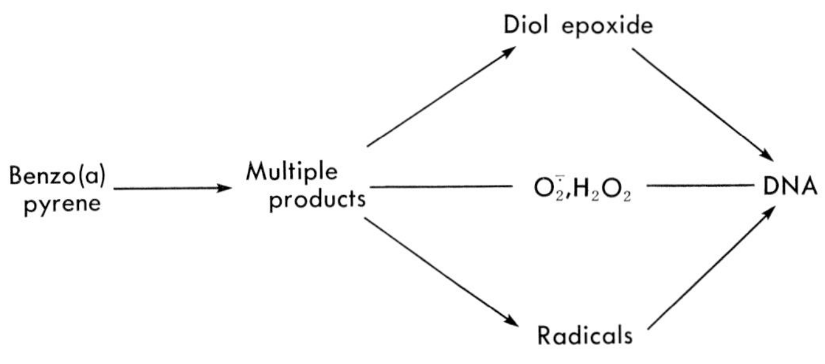

FIG. 9. The possible complexity of BP metabolism and the resultant alteration of genetic material.

carbons of the hydrocarbon (such as the L region proposed by Pullman (*32*)) may direct the metabolism of the molecule away from forming highly mutagenic diol-epoxides or phenol-epoxides. On the other hand, a molecule with a weak L region may be preferentially metabolized through epoxidation reactions. The role of each pathway of activation and the processes which may regulate the enzymes involved in the formation and disposition of the different metabolites postulated in tumor induction can be schematically represented as seen in Fig. 9. A definition of the degree of involvement of each of the pathways shown in Fig. 9 remains the goal of future research effort.

ACKNOWLEDGMENTS

This work was supported in part by research grants from the National Institutes of Health GM 16488 (RWE) and HL 19654 (RAP) and National Cancer Institute Contract No 1 CP 33362. RAP is a USPHS Research Career Development Awardee HLCA 00255.

REFERENCES

1. Griffin, B. W., Peterson, J. A., and Estabrook, R. W. Cytochrome P-450: Biophysical properties and catalytic function. *In*; D. Dolphin (ed.), The Porphyrins, vol. VII, pp. 333–375, Academic Press, New York, 1979.
2. Daly, J. W., Jerina, D. M., and Witkop, B. Arene oxides and the NIH shift: The metabolism, toxicity and carcinogenicity of aromatic compounds. Experientia, *28*: 1129–1264, 1972.
3. Narasimhulu, S. Uncoupling of oxygen activation from hydroxylation in the steroid C-21 hydroxylase of bovine adrenocortical microsomes. Arch. Biochem. Biophys., *147*: 384–390, 1971.
4. Ullrich, V. and Diehl, V. Uncoupling of monooxygenation and electron transport of fluorocarbons in liver microsomes. Eur. J. Biochem., *20*: 509–512, 1971.
5. Hildebrandt, A. G. and Roots, I. Reduced nicotinamide adenine dinucleotide phosphate (NADPH)-dependent formation and breakdown of hydrogen peroxide during mixed function oxidation reactions in liver microsomes. Arch. Biochem. Biophys., *171*: 385–397, 1975.
6. Lorentzen, R. J. and Ts'o, P.O.P. Benzo(a)pyrenedione/benzo(a)pyrenediol oxidation-reduction couples and the generation of reactive reduced oxygen. Biochemistry, *16*: 1467–1473, 1977.
7. Nishibayaishi-Yamashita, H. and Sato, R. Vitamin K_3-dependent NADPH oxidase of liver microsomes. J. Biochem., *67*: 199–210, 1970.
8. Capdevila, J., Estabrook, R. W., and Prough, R. A. The existence of a benzo(a) pyrene-3,6-quinone reductase in rat liver microsomal fractions. Biochem. Biophys. Res. Commun., *83*: 1291–1258, 1978.
9. Sims, P. and Grover, P. L. Epoxides in polycyclic aromatic hydrocarbon metabolism and carcinogenesis. Adv. Cancer Res., *20*: 165–274, 1975.
10. Selkirk, J. K., Croy, R. G., and Gelboin, H. V. Benzo(a)pyrene metabolites: Efficient and rapid separation by high pressure liquid chromatography. Science, *184*: 169–171, 1974.
11. Holder, G. M., Yagi, H., Jerina, D. M., Levin, W., Lu, A.Y.H., and Conney, A. H. Metabolism of benzo(a)pyrene. Effect of substrate concentration and 3-methyl-

cholanthrene pretreatment on hepatic metabolism by microsomes from rats and mice. Arch. Biochem. Biophys., *170*: 557–566, 1975.

12. Sims, P., Grover, P. L., Swaisland, A., Pal, K., and Hewer, A. Metabolic activation proceeds by a diol-epoxide. Nature, *252*: 326–327, 1974.
13. Huberman, E., Sachs, L., Yang, S. K., and Gelboin, H. V. Identification of mutagenic metabolites of benzo(a)pyrene in mammalian cells. Proc. Natl. Acad. Sci. U.S., *73*: 607–611, 1976.
14. Thakker, D. R., Yagi, H., Lu, A.Y.H., Levin, W., Conney, A. H., and Jerina, D. M. Metabolism of benzo(a)pyrene: Conversion of (±)-*trans*-7,8-dihydroxy-7,8-dihydrobenzo(a)pyrene to highly mutagenic 7,8-diol-9,10-epoxides. Proc. Natl. Acad. Sci. U.S., *73*: 3381–3385, 1976.
15. Nagata, C., Tagashira, Y., and Kodama, M. Metabolic activation of benzo(a)-pyrene: Significance of the free radical. *In*; P.O.P. Ts'o and J. DiPaulo (eds.), Chemical Carcinogenesis, part A, vol. 4, pp. 87–111, Dekker, New York, 1974.
16. Nagata, C., Kodama, M., and Ioki, Y. Electron spin resonance study of the binding of the 6-oxybenzo(a)pyrene radical and benzo(a)pyrene-semiquinone radicals with DNA and polynucleotides. *In*; H. V. Gelboin and P.O.P. Ts'o (eds.), Polycyclic Hydrocarbons and Cancer, vol. 1, pp. 247–260, Academic Press, New York, 1979.
17. Ts'o, P.O.P., Caspary, W. J., Cohen, B. I., Leavitt, J. C., Lesko, S. A., Lorentzen, R. J., and Schechtman, L. M. Basic mechanisms in polycyclic hydrocarbon carcinogenesis. *In*; P.O.P. Ts'o and J. DiPaulo (eds.), Chemical Carcinogenesis, part A, vol. 4, pp. 113–147, Dekker, New York, 1974.
18. Lesko, S. A., Lorentzen, R. J., and Ts'o, P.O.P. Benzo(a)pyrene metabolism: One electron pathways and the role of nuclear enzymes. *In*; H. V. Gelboin and P.O.P. Ts'o (eds.), Polycyclic Hydrocarbons and Cancer, vol. 1, pp. 261–269, Academic Press, New York, 1979.
19. Hrycay, E. G. and O'Brien, P. J. Cytochrome P-450 as a microsomal peroxidase utilizing a lipid peroxide substrate. Arch. Biochem. Biophys., *147*: 12–27, 1975.
20. Rahimtula, A. D., O'Brien, P. J., Hrycay, E. G., Peterson, J. A., and Estabrook, R. W., Possible higher valence states of cytochrome P-450 during oxidative reactions. Biochem. Biophys. Res. Commun., *60*: 695–702, 1974.
21. Wiebel, F. J. Metabolism of monohydroxybenzo(a)pyrene by rat liver microsomes and mammalian cells in culture. Arch. Biochem. Biophys., *168*: 609–621, 1975.
22. Capdevila, J., Estabrook, R. W., and Prough, R. A. The microsomal metabolism of benzo(a)pyrene phenols. Biochem. Biophys. Res. Commun., *82*: 518–525, 1978.
23. Capdevila, J., Lubet, R. A., and Prough, R. A. The metabolism of 3- and 9-hydroxybenzo(a)pyrene by rat liver microsomal fractions. *In*; M. J. Coon, A. H. Conney, R. W. Estabrook, H. V. Gelboin, J. R. Gillette, and P. J. O'Brien (eds.), Microsomes, Drug Oxidations, and Chemical Carcinogenesis, pp. 1169–1172, Academic Press, New York, 1980.
24. Prough, R. A., Capdevila, J., and Lubet, R. A., Metabolic activation of benzo(a)-pyrene and 9-hydroxybenzo(a)pyrene by tissue fractions from rat liver and lung. Biochem. Soc. Trans., *7*: 121–124, 1979.
25. Lubet, R. A., Capdevila, J., and Prough, R. A. The metabolic activation of benzo(a)-pyrene and 9-hydroxybenzo(a)pyrene by liver microsomal fractions. Int. J. Cancer, *23*: 353–357, 1979.
26. Capdevila, J., Jernstrom, B., Vadi, H., and Orrenius, S. Cytochrome P-450-linked activation of 3-hydroxybenzo(a)pyrene. Biochem. Biophys. Res. Commun., *65*: 894–900, 1975.

27. King, H.W.S., Thompson, M. H., and Brookes, P. The role of 9-hydroxy-benzo(a)-pyrene in the microsome mediated binding of benzo(a)pyrene to DNA. Int. J. Cancer, *18*: 339–344, 1976.
28. Jernstrom, B., Orrenius, S., Undeman, O., Graslund, A., and Ehrenberg, A. A fluorescence study of DNA-binding metabolites of benzo(a)pyrene formed in hepatocytes isolated from 3-methylcholanthrene treated rats. Cancer Res., *38*: 2600–2607, 1978.
29. Rahimtula, A. D. and O'Brien, P. J. Hydroperoxide catalyzed liver microsomal aromatic hydroxylation reactions involving cytochrome P-540. Biochem. Biophys. Res. Commun., *60*: 440–447, 1974.
30. Capdevila, J., Estabrook, R. W., and Prough, R. A. Differences in the mechanism and NADPH- and cumene hydroperoxide-supported reactions of cytochrome P-450. Arch. Biochem. Biophys., *200*: 186–195, 1980.
31. Prough, R. A., Patrizi, V. W., and Estabrook, R. W. The direct spectrophotometric observation of benzo(a)pyrene phenol formation by liver microsomes. Cancer Res., *36*: 4439–4443, 1976.
32. Pullman, A. and Pullman, B. Electronic structure and carcinogenic activity of aromatic molecules. Adv. Cancer Res., *3*: 117–169, 1955.

GENETIC AND ENVIRONMENTAL FACTORS IN EXPERIMENTAL AND HUMAN CANCER,
H. V. GELBOIN ET AL. (EDS.), JAPAN SCI. SOC. PRESS, TOKYO, PP. 59–66, 1980

Induction of Microsomal Mixed-function Oxidase by Chemical Compounds

Tsuneo OMURA, Nobuhiro HARADA, and Toshiaki ODA
Department of Biology, Faculty of Science, Kyushu University, Fukuoka, Japan

Abstract: The drug-induced increase of the mixed-function oxidase activity of liver microsomes is usually accompanied by pronounced alterations in the substrate specificity of the oxidase system. The stimulated oxidase activity has been correlated with the increase of both cytochrome P-450 and NADPH-cytochrome P-450 reductase or of cytochrome P-450 only, while the altered substrate specificity has been accounted for by a change in the relative amounts of multiple molecular species of cytochrome P-450 in the microsomes. Phenobarbital and 3-methylcholanthrene induce different molecular species of microsomal cytochrome P-450, which can be clearly distinguished by the use of specific antibodies. We studied the effects of these two inducers on the synthesis and turnover of various components of the mixed-function oxidase system of rat liver microsomes, and we confirmed highly selective stimulation of the synthesis of each of these two cytochrome P-450 species by the drugs. Whereas the increase of the total amounts of cytochrome P-450 in the liver microsomes was about 2- to 3-fold even when the animals received maximal doses of the drugs, the content of each specific molecular species of cytochrome P-450 rapidly increased by 10- to 20-fold. The contributions of these two species of cytochrome P-450 to the oxidation of various chemical compounds were also drastically altered by the administration of phenobarbital or 3-methylcholanthrene.

The mixed-function oxidase of the microsomes of animal tissues can be induced by the administration of a number of chemical compounds, including various drugs, to animals. The increase of the oxidase activity is, in many cases, accompanied by significant alterations in the substrate specificity of the oxidase (*1*). This induction phenomenon was first noticed for the mixed-function oxidase of liver microsomes at the end of the 1950's (*2–4*), then the discovery of cytochrome P-450 (*5*, *6*) and its principal role in microsomal mixed-function oxidase reactions (*7*, *8*) enabled us to correlate the induction of the microsomal oxidase activity with selective increases of cytochrome P-450 and its NADPH-linked reductase (*9*). It was also found that both increased synthesis and decreased degradation contributed to the increase of NADPH-cytochrome P-450 reductase in the liver microsomes of phenobarbital-treated animals (*10*, *11*).

The existence of multiple forms of cytochrome P-450 in liver microsomes was first found using 3-methylcholanthrene-treated rats (*12*, *13*). Separation and purification of various forms of microsomal cytochrome P-450 led to the conclusion that those different forms of microsomal cytochrome P-450 were differed molecularly from one another, and that their substrate specificities were also significantly different (*14–18*). The drug-induced alteration in the substrate specificity of the microsomal mixed-function oxidase could be explained by selective induction of one particular type of cytochrome P-450 by each inducer. Therefore, the induction of microsomal cytochrome P-450 by various chemical compounds must be examined for each molecular species of cytochrome P-450, and not for the total as was done previously when the multiple molecular nature of microsomal cytochrome P-450 had not been firmly established.

This presentation is mainly concerned with the induction of two types of cytochrome P-450 in the liver microsomes of drug-treated rats. One type, which will be called PB-P-450 in this paper, is specifically induced by phenobarbital, while the other, MC-P-448, is inducible by 3-methylcholanthrene. We purified these two types of cytochrome P-450 to homogeneity by the method of Imai and Sato (*19*) with some modifications, starting from the liver microsomes of phenobarbital-treated and 3-methylcholanthrene-treated rats, respectively, and prepared rabbit antibodies to each (*20*). Immunological distinction between PB-P-450 and MC-P-448 was clearly shown by Ouchterlony double diffusion tests using two antibody preparations, one specific to PB-P-450 and the other specific to MC-P-448 (Fig. 1). The antibodies also formed a sharp single precipitation line when examined by Ouchterlony double diffusion tests using solubilized microsomes, confirming their immunochemical specificity (*21*).

The immunological difference between PB-P-450 and MC-P-448 was more

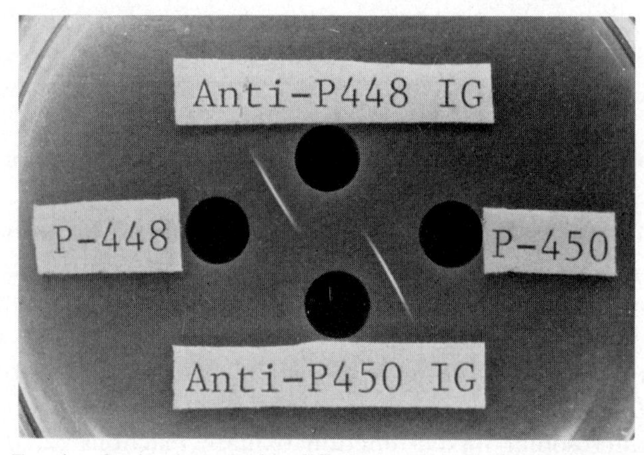

FIG. 1. Ouchterlony double diffusion test of rabbit antibodies against PB-P-450 and MC-P-448 purified from rat liver microsomes. The immunodiffusion test was carried out at 25°C using 1.2% agar containing 50 mM potassium phosphate buffer (pH 7.5), 0.9% sodium chloride, 0.02% sodium azide, and 0.5% sodium cholate.

clearly demonstrated by quantitative immunoprecipitation of purified PB-P-450 or MC-P-448 by its antibody in the presence of the other. When a fixed amount of anti-PB-P-450 antibody was titrated with increasing amounts of purified PB-P-450, in an range between 5 and 50 μg of the cytochrome, and the amounts of immunoprecipitate formation were assayed, the presence of 10 μg of MC-P-448 in the immunoreaction mixture had no effect on the immunoprecipitation of PB-P-450. The same was true for the antibody to MC-P-448. Addition of 15 μg of PB-P 450 did not affect the immunoprecipitation of MC-P-448 by its antibody when the amounts of the cytochrome were varied between 5 μg and 50 μg. These two types of microsomal cytochrome P-450 do not seem to share any common antigenic site, and they could be separately determined by quantitative immunoprecipitation using our antibody preparations.

An immunoprecipitation assay method was therefore used in determining the amounts of PB-P-450 and MC-P-448 in liver microsomes separately. In an experiment shown in Fig. 2, phenobarbital was given to rats at time zero and at 24 hr, and the contents of total cytochrome P-450, PB-P-450, and MC-P-448 were assayed at various times. Total cytochrome P-450 in microsomes was assayed by the conventional spectrophotometric method utilizing the carbon monoxide-difference spectrum of dithionite-reduced microsomes (*22*). As is shown in the figure, the amount of total cytochrome P-450 started to increase soon after the injection of the drug, and increased by about 50% at 24 hr, and about 2-fold at 48 hr. The increase of PB-P-450 was much more marked: about 5-fold at 24 hr and nearly 10-fold at 48 hr, becoming a predominant component of cytochrome P-450 in the liver micro-

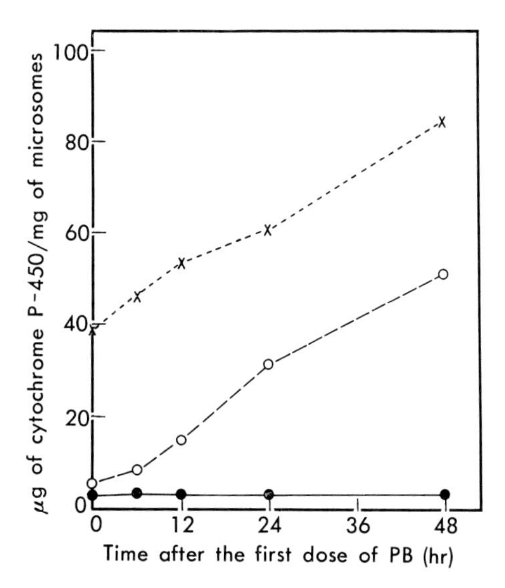

FIG. 2. Induction of liver microsomal PB-P-450 by phenobarbital administration to rats. Phenobarbital was dissolved in saline, and injected intraperitoneally to a group of rats at time 0 and 24 hr at 100 mg per kg body weight. Several rats were killed at each time as shown in the figure, and the contents of total cytochrome P-450 (×), PB-P-450 (○), and MC-P-448 (●) in liver microsomes were assayed as described in the text.

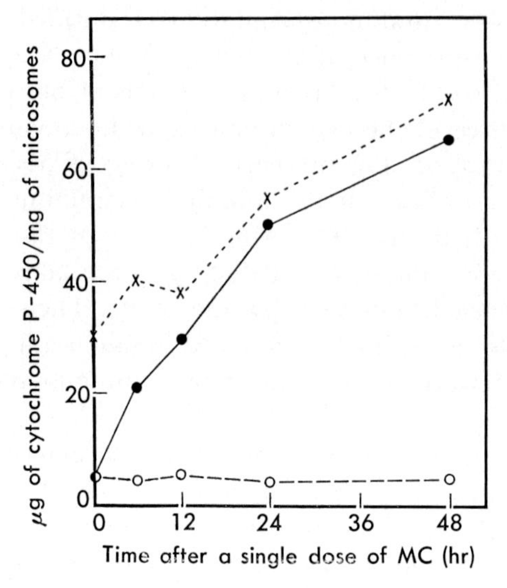

FIG. 3. Induction of liver microsomal MC-P-448 by 3-methylcholanthrene to rats. 3-Methylcholanthrene was dissolved in corn oil, and injected intraperitoneally to a group of rats at time 0 at 20 mg per kg body weight. The contents of total cytochrome P-450 (×), PB-P-450 (○), and MC-P-448 (●) in liver microsomes were determined as described in the legend to Fig. 2.

somes of treated animals. On the other hand, the content of MC-P-448 was not affected by phenobarbital treatment.

The reverse was true in the case of 3-methylcholanthrene treatment of rats. When 3-methylcholanthrene was injected to rats at time zero, MC-P-448 increased very rapidly, and became the predominant molecular species of cytochrome P-450 in the liver microsomes at 48 hr after the injection. The content of PB-P-450 was not obviously affected by 3-methylcholanthrene.

The independent inducing effect of these two chemical compounds on different molecular species of cytochrome P-450 could be clearly demonstrated when 3-methylcholanthrene was given to the rats that had received phenobarbital injections for 5 consecutive days in advance (Fig. 4). Since the animals had already been fully induced by the first inducer, phenobarbital, the initial content of PB-P-450 in liver microsomes was very high. The injection of the second inducer, 3-methylcholanthrene, at time zero (Fig. 4) caused a rapid increase of MC-P-448 as in the case of normal animals, while the content of PB-P-450 started slowly to decline. Therefore, the composition of cytochrome P-450 changed drastically, whereas the total amount of the cytochrome did not show much change.

As expected, the alteration in the composition of cytochrome P-450 resulted in a significant change in the substrate specificity of the microsomal mixed function oxidase. The injection of 3-methylcholanthrene to phenobarbital-treated rats caused an immediate increase of the oxidation of some compounds including benzo(a)-pyrene and 7-ethoxycoumarin concomitant with a gradual decrease of the oxidation of others including benzphetamine and aminopyrine. This alteration of substrate specificity of cytochrome P-450-catalyzed oxidation reaction coincides with

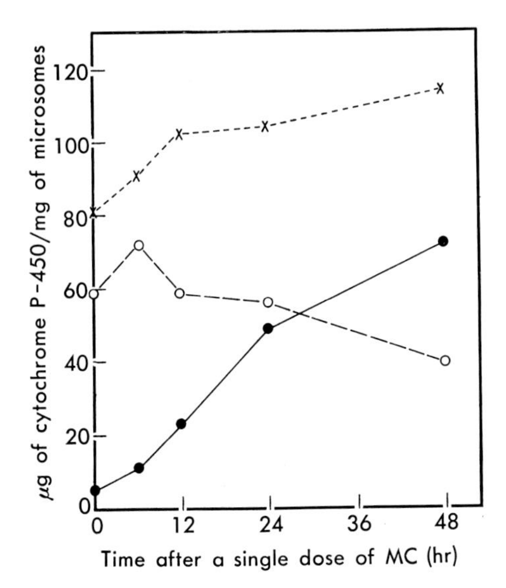

FIG. 4. Induction of liver microsomal MC-P-448 by a single dose of 3-methylcholanthrene given to phenobarbital-treated rats. A group of rats were given a daily phenobarbital injection for 5 consecutive days. On the 6th day, 3-methylcholanthrene was injected to all rats instead of phenobarbital, and the amounts of total cytochrome P-450 (×), PB-P-450 (○), and MC-P-448 (●) in liver microsomes were determined at each time as shown in the figure.

the rapid increase of MC-P-448 accompanied by a gradual decrease of PB-P-450 in the liver microsomes of treated animals, as shown in Fig. 4.

In order to estimate the actual contributions of PB-P-450 and MC-P-448 to the observed drug oxidation activities of microsomes, antibodies against these two types of cytochrome P-450 were utilized as selective inhibitors of the oxidation reactions catalyzed by each type of cytochrome P-450. Figure 5 shows the results of such an inhibition study using the same microsomal samples as used in the experiment shown in Fig. 4. Contributions of PB-P-450 and MC-P-448 to the oxidation of two substrates, benzphetamine and benzo(a)pyrene, are shown in the figure as percentages of their contributions to the total oxidation activities of microsomes. The inhibition experiments were carried out as described in a previous paper (*23*).

Judging from the effect of the antibodies on the oxidation of benzphetamine, this drug is metabolized exclusively by PB-P-450, the contribution of MC-P-448 being insignificant even after its extensive induction by 3-methylcholanthrene. On the other hand, there was a rapid dramatic change in the contribution of these two types of cytochrome P-450 to the oxidation of benzo(a)pyrene when a single dose of 3-methylcholanthrene was given to phenobarbital-treated rats. Before the administration of 3-methylcholanthrene, the oxidation of benzo(a)pyrene was almost exclusively catalyzed by PB-P-450, which was the predominant component of cytochrome P-450 in the microsomes at that time. However, the injection of 3-methylcholanthrene caused a rapid decrease of the contribution of PB-P-450 to the oxidation of benzo(a)pyrene, which was much faster than the observed gradual decrease of this type of cytochrome P-450 in the livers of treated animals, as if the

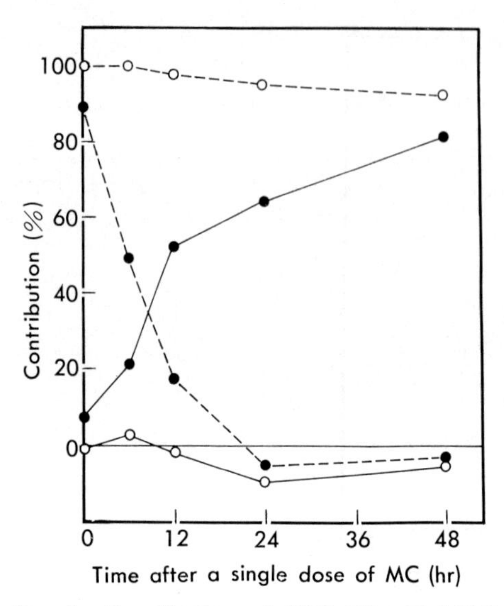

FIG. 5. Contributions of PB-P-450 and MC-P-448 to the microsomal oxidation of benzphetamine and benzo(a)pyrene. NADPH-dependent oxidation of benzphetamine (○) and benzo(a)pyrene (●) was assayed using microsomal preparations whose contents of cytochrome P-450s are shown in Fig. 4. Contributions of PB-P-450 (broken lines) and MC-P-448 (solid lines) at each time point to the oxidation of those two substrates were estimated by inhibiting the activity of each cytochrome P-450 species by the addition of an excess amount of an antibody which was specific to that particular type of cytochrome P-450, and are shown in the figure by the percentages of the total.

capability of PB-P-450 in oxidizing benzo(a)pyrene was almost abolished by the action of 3-methylcholanthrene. We do not have an adequate explanation of why the benzo(a)pyrene-oxidizing activity of PB-P-450 decreased so rapidly in the livers of 3-methylcholanthrene-treated rats, while the same cytochrome P-450 species was fully active in the oxidation of another substrate, benzphetamine.

Our observations on the alterations of the composition of cytochrome P-450 in the liver microsomes of drug-treated rats agree with those reported recently by Thomas *et al.* (*24*). Immunochemical quantitative determination of various forms of cytochrome P-450 in microsomes by the use of specific antibodies will undoubtedly be highly useful in elucidating the mechanism of selective induction of various molecular species of cytochrome P-450 by inducers.

In agreement with the observed rapid increase of one particular type of cytochrome P-450 by the administration of phenobarbital or 3-methylcholanthrene to rats, we could confirm a great increase of the synthesis of PB-P-450 in the livers of phenobarbital-treated animals. As is shown in Table 1, the synthesis of this particular cytochrome P-450 species occupied only about 0.5% of the total microsomal protein synthesis as judged from the *in vivo* incorporation of radioactive leucine to liver proteins. The injection of phenobarbital to the animals caused a marked stimulation of the synthesis of PB-P-450, however, and the synthesis of this particular microsomal protein amounted to nearly 10% of the total at 10 hr after injection of the drug. The increase of PB-P-450 synthesis in the liver is apparently

TABLE 1. Incorporation of ^{3}H-leucine to Microsomal Proteins in the Livers of Phenobarbital (PB)-treated Rats

Time after PB injection (hr)	0	3	10
	(% of total incorporation)		
PB-P-450	0.51	3.7	9.4
P-450 reductase	0.15	0.37	0.68

more marked than that of NADPH-cytochrome P-450 reductase, although the increase of total microsomal cytochrome P-450 is usually in parallel with that of the reductase.

Since alterations in the activity and substrate specificity of the microsomal cytochrome P-450 enzyme system have decisive importance in the metabolism of foreign compounds in animal tissues, the elucidation of the effects of various chemicals on the synthesis and turnover of each molecular species of cytochrome P-450 is an important and interesting subject for our future studies.

REFERENCES

1. Conney, A. H. Pharmacological implications of microsomal enzyme induction. Pharmacol. Rev., *19*: 317–366, 1970.
2. Conney, A. H., Miller, E. C., and Miller, J. A. Substrate-induced synthesis and other properties of benzopyrene hydroxylase in rat liver. J. Biol. Chem., *228*: 753–766, 1957.
3. Conney, A. H., Gillette, J. R., Inscoe, J. K., Trams, E. G., and Posner, H. S. 3,4-Benzopyrene-induced synthesis of liver microsomal enzymes which metabolize foreign compounds. Science, *130*: 1478–1479, 1959.
4. Remmer, H. Die Beschleunigung der Evipan Oxydation und der Methylierung von Methylaminoantipyrin durch Barbiturate. Arch. Exp. Pathol. Pharmakol., *237*: 296–307, 1959.
5. Omura, T. and Sato, R. A new cytochrome in liver microsomes. J. Biol. Chem., *237*: 1375–1376, 1962.
6. Omura, T. and Sato, R. The carbon monoxide-binding pigment of liver microsomes. I. Evidence for its hemoprotein nature. J. Biol. Chem., *239*: 2370–2378, 1964.
7. Estabrook, R. W., Cooper, D. Y., and Rosenthal, O. The light reversible carbon monoxide inhibition of the steroid C21-hydroxylase system of the adrenal cortex. Biochem. Z., *338*: 741–755, 1963.
8. Cooper, D. Y., Levin, S., Narasimhulu, S., Rosenthal, O., and Estabrook, R. W. Photochemical action spectrum of the terminal oxidase of mixed function oxidase system. Science, *147*: 400–402, 1965.
9. Orrenius, S., Ericsson, J.L.E., and Ernster, L. Phenobarbital-induced synthesis of the microsomal drug-metabolizing enzyme system and its relationship to the proliferation of endoplasmic reticulum. A morphometric and biochemical study. J. Cell Biol., *25*: 627–639, 1965.
10. Jick, H. L. and Shuster, L. The turnover of microsomal HADPH-cytochrome c

reductase in the livers of mice treated with phenobarbital. J. Biol. Chem., *241*: 5366–5369, 1966.
11. Kuriyama, Y., Omura, T., Siekevitz, P., and Palade, G. E. Effects of phenobarbital on the synthesis and degradation of the protein components of rat liver microsomal membranes. J. Biol. Chem., *244*: 2017–2026, 1969.
12. Sladek, N. E. and Mannering, G. J. Evidence for a new P-450 hemoprotein in hepatic microsomes from methylcholanthrene-treated rats. Biochem. Biophys. Res. Commun., *24*: 668–674, 1966.
13. Alvares, A. P., Schilling, G., Levin, W., and Kuntzman, R. Studies on the induction of CO-binding pigments in liver microsomes by phenobarbital and 3-methylcholanthrene. Biochem. Biophys. Res. Commun., *29*: 521–526, 1967.
14. Haugen, D. A. and Coon, M. J. Properties of electrophoretically homogeneous phenobarbital-inducible and -naphthoflavon-inducible forms of liver microsomal cytochrome P-450. J. Biol. Chem., *251*: 7929–7939, 1976.
15. Thomas, P. E., Lu, A.Y.H., Ryan, D., West, S. B., Kawalek, J., and Levin, W. Multiple forms of rat liver cytochrome P-450. Immunochemical evidence with antibody against cytochrome P-448. J. Biol. Chem., *251*: 1385–1391, 1976.
16. Hashimoto, C. and Imai, Y. Purification of a substrate complex of cytochrome P-450 from liver microsomes of 3-methylcholanthrene-treated rabbits. Biochem. Biophys. Res. Commun., *68*: 821–827, 1976.
17. Guengerich, F. P. Separation and purification of multiple forms of microsomal cytochrome P-450. Partial characterization of three apparently homogeneous cytochrome P-450 prepared from livers of phenobarbital- and 3-methylcholanthrene-treated rats. J. Biol. Chem., *253*: 7931–7939, 1978.
18. Ryan, D. E., Thomas, P. E., Korzeniowski, D., and Levin, W. Separation and characterization of highly purified forms of liver microsomal cytochrome P-450 from rats treated with polychlorinated biphenyls, phenobarbital, and 3-methylcholanthrene. J. Biol. Chem., *254*: 1365–1374, 1979.
19. Imai, Y. and Sato, R. A gel-electrophoretically homogeneous preparation of cytochrome P-450 from liver microsomes of phenobarbital-treated rabbits. Biochem. Biophys. Res. Commun., *60*: 8–14, 1974.
20. Omura, T., Noshiro, M., and Harada, N. Distribution of electron transfer components on the surface of microsomal membrane. *In*; M. J. Coon, A. H. Conney, R. W. Estabrook, H. V. Gelboin, J. R. Gillette, and P. J. O'Brien (eds.), Microsomes, Drug Oxidations, and Chemical Carcinogenesis, Academic Press, New York, in press, 1980.
21. Omura, T. Biosynthesis and drug-induced increase of microsomal enzymes. *In*; R. W. Estabrook and E. Lindenlaub (eds.), The Induction of Drug Metabolism, pp. 161–175, F. K. Schattauer Verlag, Stuttgart, 1978.
22. Omura, T. and Sato, R. The carbon monoxide-binding pigment of liver microsomes. II. Solubilization, purification, and properties. J. Biol. Chem., *239*: 2379–2385, 1964.
23. Noshiro, M. and Omura, T. Immunochemical study on the electron pathway from NADH to cytochrome P-450 of liver microsomes. J. Biochem., *83*: 61–77, 1978.
24. Thomas, P. E., Korzeniowski, D., Ryan, D., and Levin, W. Preparation of monospecific antibodies against two forms of rat liver cytochrome P-450 and quantitation of these antigens in microsomes. Arch. Biochem. Biophys., *192*: 524–532, 1979.

GENETIC AND ENVIRONMENTAL FACTORS IN EXPERIMENTAL AND HUMAN CANCER,
H. V. GELBOIN ET AL. (EDS.), JAPAN SCI. SOC. PRESS, TOKYO, PP. 67–77, 1980

Biochemical Individuality in Carcinogenesis: Studies in Benzo(a)pyrene Metabolism and Activation

H. V. Gelboin, R. Robinson, H. Miller, and P. Okano

Laboratory of Molecular Carcinogenesis, National Cancer Institute, Bethesda, Md. 20205, U.S.A.

Abstract: The pathways of benzo(a)pyrene (BP) metabolism are diverse and complex and result in the conversion of BP to more than 40 metabolites including phenols, dihydrodiols, quinones, and water-soluble sulfate, glucuronide, and glutathione conjugates. In addition, BP is converted to diol epoxides which are considered the major carcinogenic and mutagenic intermediates of BP metabolism. We have identified and characterized a large number of the above metabolites and have studied the stereospecificity of the above reactions catalyzed by purified enzymes and human blood monocytes and lymphocytes. The balance between detoxification and carcinogen activation routes of metabolism may be elements determining individual susceptibility to the carcinogenic action of BP. The diverse pathways are regulated by the relative amounts of the various forms of the mixed function oxidases, epoxide hydratase, and the conjugating enzymes. Humans differ in their ability to metabolize BP. We will describe results obtained with recently developed high performance liquid chromatography and immunological methods and studies with purified enzymes that may help to define differences in BP metabolism which are relevant to susceptibility to BP carcinogenesis.

It is now generally believed that environmental factors play a significant role in human cancer causation. Although environment obviously includes a large variety of factors including nutrition and lifestyle, a major focus of attention has been brought to bear on environmental chemicals. These foreign compounds or xenobiotics include polycyclic hydrocarbons resulting from energy-producing processes, chemicals from industry, pesticides, food additives, and a variety of drugs. The primary metabolic interface or biological receptor between this large variety of chemicals and biological organisms, including man, is the enzyme system which metabolizes these xenobiotics. This complex is the cytochrome P-450 mixed-function oxidase enzyme system (*1–3*). This enzyme system generally converts foreign chemicals to metabolites which are further metabolized by epoxide hydratase and/or conjugating enzymes to water-soluble metabolites which are safely excreted.

These systems are thus largely detoxification systems which are highly beneficial and necessary for organism survival. These systems, however, have another face. They also catalyze the formation of metabolites which are toxic, mutagenic, teratogenic, and carcinogenic. In other words, during the process of detoxification, hazardous metabolites may be formed.

The mixed-function oxidases system oxygenating the polycyclic aromatic hydrocarbons (PAHs) is commonly called aryl hydrocarbon hydroxylase (AHH). This enzyme system is ubiquitous being present in most tissues of mammals and many lower organisms (*4*, *5*). It is a multi-component enzyme system consisting basically of a cytochrome pigment, cytochrome P-450, NADPH cytochrome P-450 reductase, and a phospholipid. The enzyme system is inducible, *i.e.* its activity is increased to high levels, sometimes several hundred fold, upon exposure of the organism to various drugs and hydrocarbons. This response provides an important mechanism for increasing the rate of metabolism of PAHs and other xenobiotics. A clearer understanding of the factors involved in this mode of enzyme regulation *i.e.* enzyme induction and the factors regulating enzyme activity especially those at the metabolic level may give us insight into the molecular basis for differences in human susceptibility to chemical carcinogens.

The conventional assay used to measure AHH activity is extremely sensitive, easily applicable to small amounts of tissue and depends on the enzymatic oxygenation of benzo(a)pyrene (BP), to phenols (*4*). The phenols are easily detected by their fluorescence in alkali. The effects of a variety of other environmental agents such as specific inducers, inhibitors, and other factors, on BP metabolism can easily be measured in cell culture. One can construct interspecies somatic cell hybrids in order to study the genetics of the enzyme system. In addition, the culturing of easily obtainable cells from humans, such as blood monocytes and lymphocytes, allows us to characterize an individual's ability to metabolize known carcinogens such as BP. Furthermore, the AHH assay or the high performance liquid chromatography (HPLC) assay can easily be applied to different cells, tissues, or purified enzyme preparations in order to investigate the molecular modes of metabolism related to mutation and transformation susceptibility.

Characterization of BP Metabolites

In addition to the AHH assay which is a rough estimate of total activity, we have used HPLC which is an extremely sensitive method for separation and analysis of metabolites of the PAHs. With HPLC, one can analyze and quantitate almost the entire BP metabolite pattern formed by the cells under investigation.

Figure 1 is a summary of our present state of knowledge of BP metabolism. BP is converted by the mixed-function oxidases to four unstable epoxide intermediates and five phenols, the 1, 3, 6, 7, and 9 phenols. The phenols arise largely through epoxide re-arrangement but can also be formed by direct oxygenation. The epoxides can also be enzymatically hydrated to three dihydrodiols by epoxide hydratase. These two reactions are highly stereospecific resulting in production of the optically active (−)-*trans*-7, 8, 9, 10, and 4, 5-dihydrodiols (*7*). The phenols

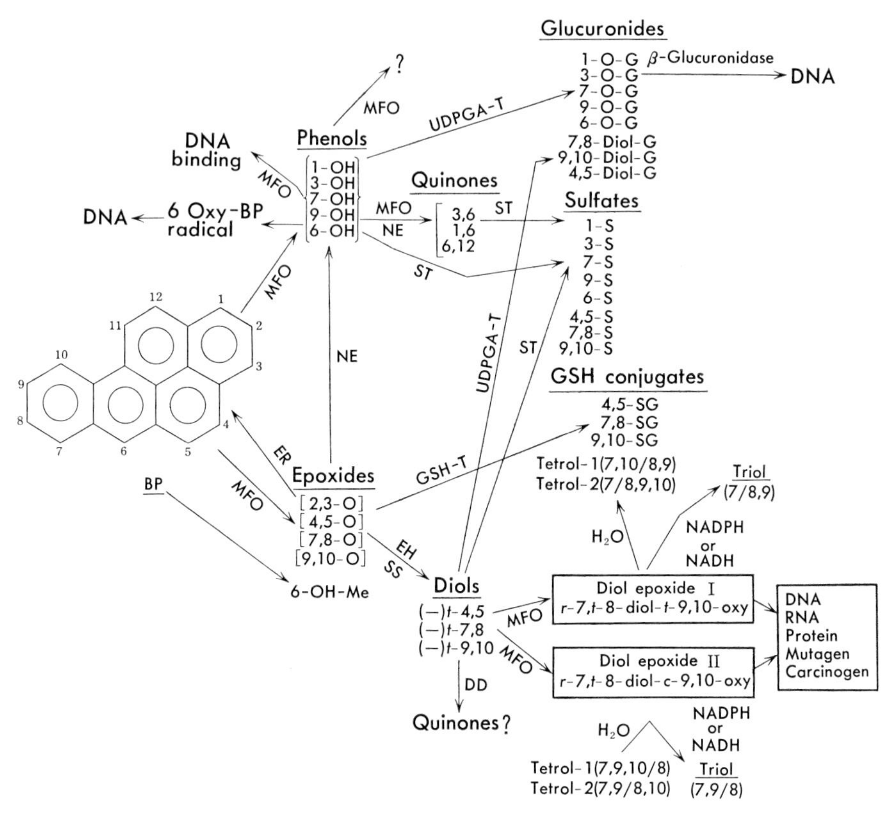

FIG. 1. The metabolism and activation of BP. MFO, mixed-function oxidases; EH, epoxide hydratase; ER, epoxide reductase; DD, dihydrodiol dehydrogenase; GSH-T, glutathione S-epoxide transferase; UDPGA-T, glucuronic acid transferase; ST, sulfotransferase; NE, non-enzymatic; SS, stereospecific; *t*, *trans*.

also give rise to three quinones. All of these oxygenated intermediates can be further metabolized to water-soluble products by conjugation with either glutathione, sulfate, or glucuronic acid. Almost all of these metabolites are detoxification products and exhibit little or no mutagenic, DNA-binding, or carcinogenic activity. Figure 2 shows the major pathway of activation of BP. Thus BP is oxygenated by the mixed-function oxidases to the 7, 8-epoxide which is stereospecifically converted to the (−)-*trans*-7, 8-diol. This product in turn is further metabolized by the mixed-function oxidases at the 9, 10 which we call diol epoxide I (*8*); this diol epoxide has been shown to be the most mutagenic metabolite of BP (*9*). About 10% of the diol is converted to the other diol epoxide. Both of these diol epoxides were fully characterized by isolating and characterizing the respective tetrol hydrolysis products and triol reduction product. The structure of all these are shown in Fig. 2.

All of these tetrols and triols were separated by HPLC and characterized by mass spectrometry and formation of acetonide derivatives. Their structure is shown in Fig. 2. The slash indicates which hydroxyl groups are on the same side of the ring. This figure also indicates that these diol epoxides are the major metabolites

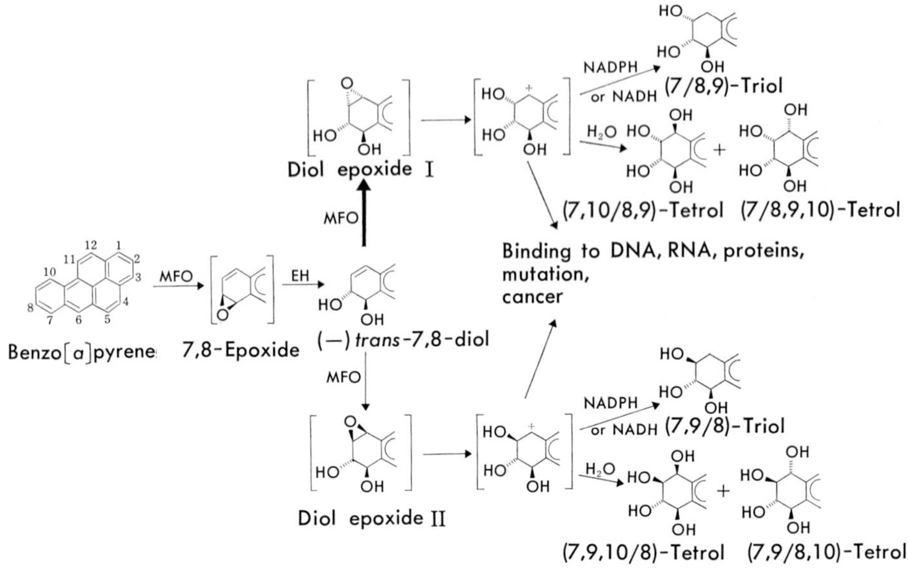

FIG. 2. Metabolic activation of benzo(a)pyrene.

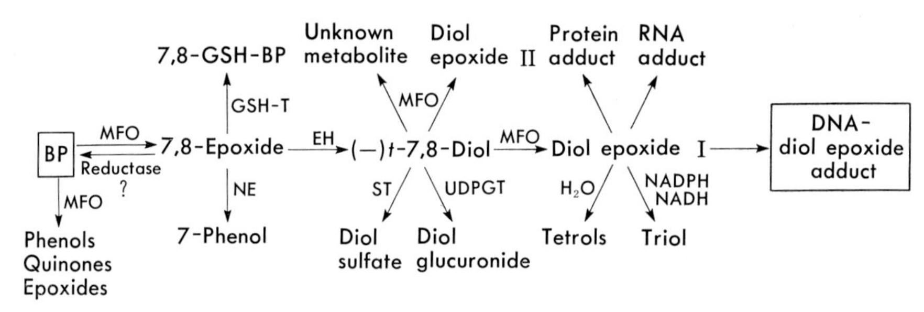

FIG. 3. Alternative pathways to diol epoxide I formation and DNA binding.

of BP that bind to macromolecules including DNA. They also exhibit mutagenic activity which is several orders of magnitude greater than any of the other benzo-(a)pyrene metabolites tested. Figure 3 is a summary of the activation pathway from BP to its interaction as a diol epoxide with DNA. It also shows the possible alternative pathways that can be entered. These are all largely detoxification pathways. Thus there are many alternatives to the diol epoxide activation pathway.

One of the major goals of our research is to understand the regulation of these pathways and eventually to determine whether individuals differ in the relative balance of detoxification and activation pathways; and if they do differ, whether the differences may relate to carcinogen susceptibility.

This brings us to the question of what regulates the amount of diol epoxides formed, relative to the other BP metabolites, which are primarily detoxification products. A primary factor is the activity of the cytochrome P-450 component of the mixed-function oxidase system. The activity of other enzymes, such as epoxide hydratase and the conjugating enzymes, also affects the pathway of BP metabolism.

TABLE 1. Specificity of Various Forms of Cytochrome P-450 from Rabbit Liver Microsomes toward BP and (−)-*trans*-7, 8-diol

$P\text{-}450_{LM}$ preparation	% of substrate metabolized[a]		
	BP	(−)-*trans*-7, 8-diol	Activity ratio[b]
LM_2	37.7	1.8	20.9
$LM_{4(BNF)}$[c]	2.7	8.6	0.3
$LM_{4(PB)}$[c]	4.8	14.5	0.3
LM_1	29.7	7.1	4.2
LM_3	8.1	1.3	6.2
LM_7	12.6	10.0	1.3

[a] For 20-min incubation, determined by HPLC. [b] Ratio of BP metabolized to (−)-*trans*-7, 8-diol metabolized. [c] $P\text{-}450_{LM4}$ isolated from animals treated with β-naphthoflavone (BNF) or phenobarbital (PB).

TABLE 2. Stereospecificity of Diol Epoxide I and Diol Epoxide II Formation from (−)-*trans*-7, 8-diol by Various Forms by Cytochrome P-450

Metabolites of (−)-*trans*-7, 8-diol	Activity (pmol/min/nmol P-450)[a]					
	LM_2	$LM_{4(BNF)}$[b]	$LM_{4(PB)}$	LM_1	LM_{3b}	LM_7
Diol epoxide I	5.7(20)	117.5(82)	178.8(82)	28.9(27)	7.8(37)	13.5(7)
Diol epoxide II	6.5(22)	10.7(7)	15.7(7)	17.8(17)	4.0(19)	42.8(24)
Ratio, I/II	0.9	11.0	11.4	1.6	2.0	0.3

[a] The numbers in parentheses are the percent of total metabolites extracted into ethyl acetate. [b] $P\text{-}450_{LM4}$ isolated from animals treated with BNF or PB.

TABLE 3. Binding of BP and (−)-*trans*-7, 8-diol Metabolites to DNA

$P\text{-}450_{LM}$ preparation	Activity (pmol/mg DNA)	
	BP	(−)-*trans*-7, 8-diol
LM_2	9.9±3.2	27.3±4.1
$LM_{4(BNF)}$[a]	2.5±0.6	429±4
$LM_{4(PB)}$[a]	4.0±0.2	635±68
Ratio, $LM_{4(BNF)}/LM_2$	0.25	15.7
Ratio, $LM_{4(PB)}/LM_2$	0.4	23.3

[a] $P\text{-}450_{LM4}$ isolated from animals treated with BNF or PB.

Several laboratories have demonstrated that multiple forms of cytochrome P-450 are present in various animal tissues. We have examined the activity of several of the purified forms of cytochrome P-450 from rabbit liver and have found that they are substrate selective and regio- and stereoselective in terms of substrate utilization and product formation (*10*). The data in Tables 1–3 demonstrates this selectivity of the purified cytochrome P-450 forms. The cytochrome P-450s have very different activities, some preferring BP and others preferring (−)-*trans*-7, 8-diol, as the substrates. The activities range from a 3-fold preference for (−)-*trans*-7, 8-diol with $P\text{-}450_{LM4}$ to a 20-fold preference for BP with $P\text{-}450_{LM2}$. Other data in these tables indicate that the different forms of P-450 are stereoselective for substrate, stereoselative for position of oxygenation and for the formation of mutagenic or DNA-binding metabolites.

AHH Activity and BP Metabolism in Man

In order to measure AHH activity and BP metabolism in man, an easily obtainable source of human tissue is required. Tissues meeting this requirment, which we have used, include blood lymphocytes and monocytes. In our laboratory, interindividual and intraindividual variations in the basal and benz(a)anthracene (BA)-induced AHH activity were investigated in cultured monocytes from sets of monozygotic and dizygotic twins. The variation in BA-induced AHH activity among the various individuals show that the mean inducible values differ quite significantly, ranging from 4.26 to 17.69. Statistical analysis of this data suggests that the interindividual variation in AHH activity is largely genetically determined. The data also suggested that the number of genes involved in AHH expression is relatively few in number. Next, we examined the question of the genetics of AHH using the technique of somatic cell hybridization. Hybrids between mouse RAG cells, which had no AHH activity, and human bone marrow cells were examined for the linkage between 22 human isozymes and AHH activity. The results indicated that the expression of AHH activity was associated with the presence of human chromosome 2. The relationship, however, of the total level of AHH activity to an individual's susceptibility to BP or other chemical carcinogen is still the subject of controversy today (*12*). Susceptibility to chemical carcinogens may be more related to the balance between different carcinogen metabolic pathways.

In order to determine whether an individual's mixed-function oxidase activity could be characterized, we analyzed BP metabolite patterns in monocytes and lymphocytes, by using the HPLC technique. The amount of BP metabolism, however, was found to be low in these cells. We modified existing procedures of adding an anti-oxidant to prevent oxidation of minute amounts of metabolites.

Comparison of BP Metabolism in BA-induced and Non-induced Cells

Table 4 summarizes the pattern of BP metabolism in BA-induced and control monocytes from two different individuals. The formation of each BP metabolite is increased about 3-fold in BA-induced monocytes from Individual A as compared to control cells. The fold increase, however, of 7-OH and 9-OH formed in induced cells appears greater than the fold increase observed for other metabolites. Thus, the data suggest that induction preferentially induces enzyme systems forming 9-OH and 7-OH. The amount of 7, 8-diol and 3-OH is increased to a lesser extent in the induced cells than are those of the other phenols. The 9, 10-diol and 4, 5-diol are synthesized in low, barely detectable amounts in both BA-induced and control cells, making any comparison difficult. Formation of quinones varies from sample to sample, and differences in rate of formation in BA-induced and control cells are probably not significant. For Individual B, total BP metabolism is induced 3.6-fold

TABLE 4. Comparison of BP Metabobism in BA-induced and Non-induced Monocytes

BP metabolite	Specific activity (fmol/min/10^6 cells)					
	Individual A			Individual B		
	BA induced	Control	Ratio[a]	BA induced	Control	Ratio[a]
9, 10-Diol	6(0.7)[b]	6(1.4)	1.0(0.5)	13(0.6)	7(1.2)	1.8(0.5)
4, 5-Diol	4(0.5)	0(0.2)	(3.1)	65(3.2)	0(0)	
7, 8-Diol	106(14.1)	47(15.8)	2.3(0.9)	205(10.1)	97(17.2)	2.1(0.6)
9-OH	265(35.2)	70(30.6)	3.8(1.1)	588(29.2)	181(32.1)	3.2(0.9)
7-OH	87(11.8)	20(10.1)	4.4(1.2)	256(12.6)	69(12.2)	3.7(1.0)
3-OH	214(28.6)	76(33.6)	2.8(0.8)	553(27.3)	197(34.8)	2.8(0.8)
1, 6-Quinone	26(3.4)	6(1.5)	4.4(2.3)	108(5.3)	0(0)	
3, 6-Quinone 6, 12-Quinone	44(5.8)	17(8.6)	2.6(0.7)	236(11.6)	14(2.5)	16.9(4.6)
Total	752(100)	242(100)	3.1	2,024(100)	565(100)	3.6
AHH	349	125	2.8	1,216	404	3.0

Duplicate samples of 5 to 6×10^6 frozen BA-induced or non-induced monocytes were incubated with [^{3}H]BP, and the BP metabolites were analyzed by HPLC. For the AHH assay, 4×10^6 monocytes were incubated, and the activity was determined. [a] BA-induced *versus* non-induced cells. [b] Numbers in parentheses, percentage of total metabolites.

TABLE 5. Comparison of BP Metabolism in BA-induced and Non-induced Lymphocytes

BP metabolite	Specific activity (fmol/min/10^6 cells)					
	Individual A			Individual B		
	BA-induced	Control	Ratio[a]	BA-induced	Control	Ratio[a]
9, 10-Diol	1(0.3)[b]	1(0.5)	1.0(0.5)	1(0.4)	0(0.2)	
4, 5-Diol	3(1.2)	0(0.3)	(4.2)	23(6.2)	0(0)	
7, 8-Diol	19(7.4)	13(12.3)	1.4(0.6)	20(5.2)	3(5.8)	6.4(0.9)
9-OH	61(23.3)	20(18.8)	3.0(1.2)	62(16.6)	9(16.6)	6.9(1.0)
7-OH	45(17.4)	11(10.4)	4.1(1.7)	36(9.7)	6(10.6)	6.4(0.9)
3-OH	95(36.4)	43(40.6)	2.2(0.9)	93(25.2)	24(44.1)	3.9(0.6)
1, 6-Quinone	12(4.6)	6(5.5)	2.0(0.8)	43(11.6)	0(0)	
3, 6-Quinone 6, 12-Quinone	25(9.4)	13(11.6)	2.0(0.8)	94(25.3)	12(22.7)	7.6(1.1)
Total	261(100)	107(100)	2.4	372(100)	54(100)	6.9
AHH	142	68	2.1	152	70	2.2

Duplicate samples of 5×10^6 frozen BA-induced or non-induced lymphocytes were incubated with [^{3}H]BP, and the BP metabolites were analyzed by HPLC. For the AHH assay, 3 to 5×10^6 lymphocytes were incubated, and the activity was determined. [a] BA-induced *versus* non-induced cell. [b] Numbers in parentheses, percentage of total metabolites.

by BA with a slight preferential synthesis of 7-OH and a slightly higher formation of 7, 8-diol in control cells.

The summary of BP metabolism in BA-induced and control lymphocytes from the same two individuals is shown in Table 5. In Individual A, BA induction leads to a lower ratio of 7, 8-diol and 3-OH and a higher ratio of 7-OH and 9-OH

TABLE 6. Comparison of BP Metabolism in Monocytes and Lymphocytes from Different Individuals

BP metabolite	Individual A			Specific activity
	Monocyte	Lymphocyte	Ratio[a]	Monocyte
9, 10-Diol	6 (0.7)[b]	1 (0.3)	6.0 (2.7)	13 (0.6)
4, 5-Diol	4 (0.5)	3 (1.2)	1.3 (0.4)	65 (3.2)
7, 8-Diol	106 (14.1)	19 (7.4)	5.6 (1.9)	205 (10.1)
9-OH	265 (35.2)	61 (23.3)	4.3 (1.5)	588 (29.2)
7-OH	87 (11.8)	45 (17.4)	1.9 (0.7)	256 (12.6)
3-OH	214 (28.6)	95 (36.4)	2.3 (0.8)	553 (27.3)
1, 6-Quinone	26 (3.4)	12 (4.6)	2.2 (0.7)	108 (5.3)
3, 6-Quinone	44 (5.8)	25 (9.4)	1.8 (0.6)	236 (11.6)
6, 12-Quinone				
Total	752 (100)	261 (100)	2.9	2,024 (100)

The data for Individuals A and B were taken from Tables 1 and 2 using the values obtained by the

[b] Numbers in parentheses, percentage of total metabolites.

formation. The low amounts of 9, 10-diol and 4, 5-diol and 3-OH formation and preferential increases in 7-OH and 9-OH formation. Thus, in both monocytes and lymphocytes, the induction process appears to induce enzymes favoring 7-OH and 9-OH formation over those enzyme systems, favoring 3-OH and 7, 8-diol formation.

Interindividual Differences in BP Metabolism

Table 6 shows a comparison of BP metabolism by BA-induced monocytes and lymphocytes which have been isolated from the same individual and a comparison of three different individuals. Total BP metabolism in BA-induced monocytes is 2- to 5-fold higher than in BA-induced lymphocytes. The difference may be even greater since the BP metabolic activity in monocytes is more sensitive to freezing and thawing than it is in lymphocytes. One should note that, in this comparison, Individual A has the highest monocyte BP metabolic activity while Individual C has the highest lymphocyte activity. In the comparison of the rate of formation of individual BP metabolites, some common features emerge. The 9, 10-diol and 7, 8-diol are formed at about twice the rate as a percentage of total BP metabolites in monocytes compared to lymphocytes. The percentage of 9-OH formation is 30 to 80% higher in monocytes than in lymphocytes. Conversely, the rate of 3-OH formation is about 20% lower in monocytes than in lymphocytes. Quinone formation is also relatively higher in lymphocytes than in monocytes. In Individual A, the 7, 8-diol: 4, 5-diol ratio is about 26: 1 in monocytes *versus* 6: 1 in lymphocytes with the total diols comprising 15% of total BP metabolites in monocytes and about 9% in lymphocytes. Total phenol synthesis is about 76% in each case. For Individual A, the total phenol: diol ratios are 4.9 for monocytes and 8.7 for lymphocytes. The 7, 8-diol: 4,5-diol ratios in Individual B are 3: 1 in monocytes and about 1: 1 in lymphocytes. The amount of diol formed is 14 and 12% of total BP metabolites,

(fmol/min/10^6 cells)

Individual B		Individual C		
Lymphocyte	Ratio[a)]	Monocyte	Lymphocyte	Ratio[a)]
1 (0.4)	13.0 (1.8)	4 (0.4)	1 (0.2)	4.0 (1.8)
23 (6.2)	2.8 (0.5)	22 (2.2)	9 (2.0)	2.4 (1.1)
20 (5.2)	10.2 (1.9)	105 (10.3)	28 (5.6)	3.8 (1.8)
62 (16.6)	9.5 (1.8)	353 (34.7)	128 (26.0)	2.8 (1.3)
36 (9.7)	7.1 (1.3)	142 (13.9)	80 (16.2)	1.8 (0.9)
93 (25.2)	5.9 (1.1)	350 (34.1)	200 (40.7)	1.7 (0.8)
43 (11.6)	2.5 (0.5)	14 (1.4)	14 (2.8)	1.0 (0.5)
94 (25.3)	2.5 (0.5)	31 (3.0)	33 (6.6)	0.9 (0.5)
372 (100)	5.4	1,021 (100)	493 (100)	2.1

incubation of frozen BA-induced cells. [a)] BA-induced monocytes *versus* BA-induced lymphocytes.

respectively. In this case, phenol formation is 69% of total BP metabolites in monocytes and 52% in lymphocytes with total phenol: diol ratios of 4.9 and 4.4 for monocytes and lymphocytes, respectively. This value for lymphocytes may be abnormally low since oxidation to quinones was higher than normal for these samples. In Individual C, the 7, 8-diol: 4, 5-diol ratios are about 5: 1 and 3: 1 for monocytes and lymphocytes, respectively. Total diols for Individual C represent 13 and 8%, respectively, of total BP metabolites for monocytes and lymphocytes. The phenols represent 83% of total BP metabolites in both these monocytes and lymphocytes. The total phenol: diol ratios are 6.5 and 10.7, respectively, for monocytes and lymphocytes of Individual C.

Quantitatively, the most striking interindividual difference seems to be in the relative amount of 4,5-diol formed by different individuals. This metabolite varied widely, and further work is needed to determine whether this parameter may indicate interindividual differences. In the individual with greater amounts of 4, 5-diol formation, the amount of 9-OH was relatively reduced. In this individual, both the monocytes and lymphocytes formed relatively more 4, 5-diol.

Concluding Remarks

We have presented data which suggest that both genetic and epigenetic mechanisms are important in the regulation of the carcinogen metabolizing enzyme system (*14*). With the use of HPLC, the BP metabolic pattern in cells of all types can be characterized. Our data and others suggest that cells from different animal tissues will show differences in their BP metabolite patterns. Generally, these have been quantitative rather than qualitative differences. Also, the data suggest that induction of enzymes with PAHs may lead to greater ratios of detoxified products than activated products, relative to the ratios formed in control cells. Understanding these different pathways of carcinogen activation and how they may be altered by inhibiting or enhancing the activity of the enzyme system with drugs, pesticides, or natural products occurring in our foods will give us some insight as to how different individuals interact with environmental carcinogens and the factors that govern their response.

REFERENCES*

1. Conney, A. H., Miller, E. C., and Miller, J. A. Substrate-induced synthesis and other properties of benzopyrene hydroxylase in rat liver. J. Biol. Chem., *228*: 753–766, 1957.
2. Gelboin, H. V. Carcinogens, enzyme induction and gene action. Adv. Cancer Res., *10*: 1–81, 1967.
3. Gelboin, H. V., Kinoshita, N., and Wiebel, F. J. Microsomal hydroxylases: Induction and role in polycyclic hydrocarbon carcinogenesis and toxicity. Fed. Proc., *31*: 1298–1302, 1972.
4. Nebert, D. W. and Gelboin, H. V. Substrate-inducible microsomal aryl hydrocarbon hydroxylase in mammalian cell culture. I. Assay and properties of induced enzyme. J. Biol. Chem., *243*: 6242–6249, 1968.
5. Nebert, D. W. and Gelboin, H. V. The *in vivo* and *in vitro* induction of aryl hydrocarbon hydroxylase in mammalian cells of different species, tissues, strains, and developmental and hormonal states. Arch. Biochem. Biophys., *134*: 76–89, 1969.
6. Selkirk, J. K., Cory, R. G., and Gelboin, H. V. High pressure liquid chromatographic separation of 10 benzo(a)pyrene phenols and the identification of 1-phenol and 7-phenol as new metabolites. Cancer Res., *36*: 922–926, 1976.
7. Yang, S. K., Gelboin, H. V., Weber, J. D., Sankaran, V., Fischer, D. L., and Engel, J. F. Resolution of optical isomers by high-pressure liquid chromatography: The separation of benzo(a)pyrene *trans*-diol derivatives. Anal. Biochem., *78*: 520–526, 1977.
8. Yang, S. K., McCourt, D. W., Roller, P. P., and Gelboin, H. V. Enzymatic conversion of benzo(a)pyrene leading predominantly to the diol-epoxide *r*-7-*t*-8-dihydroxy-*t*-9,10-oxy-7,8,9,10-tetrahydrobenzo(a)pyrene through a single enantiomer of *r*-7, *t*-8-dihydroxy-7,8-dehydrobenzo(a)pyrene. Proc. Natl. 'Acad. Sci. U.S., *73*: 2594–2598, 1976.
9. Huberman, E., Sachs, L., Yang, S. K., and Gelboin, H. V. Identification of mutagenic metabolites of benzo(a)pyrene in mammalian cells. Proc. Natl. Acad. Sci. U.S., *73*: 607–611, 1976.
10. Deutsch, J., Leutz, J., Yang, S. K., Gelboin, H. V., Chiang, Y. L., Vatsis, K. P., and Coon, M. J. Regio-and stereoselectivity of various forms of purified cytochrome P-450 in the metabolism of benzo(a)pyrene and (−)-*trans*-7,8-dihydroxy-7,8-dihydrobenzo(a)pyrene as shown by product formation and binding to DNA. Proc. Natl. Acad. Sci. U.S., *75*: 3123–3127, 1978.
11. Okuda, T., Vesell, E. S., Plotkin, V., Tarone, R., Bast, R. C., and Gelboin, H. V. Interindividual and intraindividual variations in aryl hydrocarbon hydroxylase in monocytes from monozygotic and dizygotic twins. Cancer Res., *37*: 3904–3911, 1977.

* With the exception of Ref. *1*, the other references are studies from our laboratory. Reference *15* is a review of this field of research containing references to the numerous studies in this area of research.

12. Gelboin, H. Cancer susceptibility and carcinogen metabolism. N. Engl. J. Med., *297*: 384–387, 1977.
13. Okano, P., Miller, H. N., Robinson, R. C., and Gelboin, H. V. Comparison of benzo(a)pyrene and (−)-*trans*-7,8-dihydroxy-7,8-dihydrobenzo(a)pyrene metabolism in human blood monocytes and lymphocytes. Cancer Res., *39*: 3184–3193, 1979.
14. Okano, P., Whitlock, J. P., and Gelboin, H. V. Aryl hydrocarbon and benzo(a)-pyrene metabolism in rodent liver and human cells. Ann. N.Y. Acad. Sci., in press.
15. Yang S. K., Deutsch, J., and Gelboin, H. V. Benzo(a)pyrene metabolism: Activation and detoxification. *In*; H. V. Gelboin and P.O.P. Ts'o (eds.), Polycyclic Hydrocarbons and Cancer, vol. 1, pp. 205–231, Academic Press, New York, 1978.

GENETIC AND ENVIRONMENTAL FACTORS IN EXPERIMENTAL AND HUMAN CANCER,
H. V. GELBOIN ET AL. (EDS.), JAPAN SCI. SOC. PRESS, TOKYO, PP. 79–87, 1980

Metabolic Activation of Tryptophan Pyrolysates, Trp-P-1 and Trp-P-2 by Hepatic Cytochrome P-450

Ryuichi KATO, Yasushi YAMAZOE, and Tetsuya KAMATAKI
Department of Pharmacology, School of Medicine, Keio University, Tokyo, Japan

Abstract: 1. The metabolic activation of 3-amino-1, 4-dimethyl-5-*H*-pyrido(4, 3-b)indole (Trp-P-1) and 3-amino-1-methyl-5-*H*-pyrido(4, 3-b)indole(Trp-P-2) to mutagenic compounds by hepatic microsomes and purified cytochrome P-450 was assayed by a modified Ames test and activated metabolites were determined by a high performance liquid chromatography (HPLC).
2. The metabolic activation of Trp-P-1 and Trp-P-2 by hepatic microsomes was increased by the treatment of rats with polychlorinated biphenyl mixture (PCB) or 3-methylcholanthrene, but was increased only slightly by phenobarbital.
3. The metabolic activation of purified cytochrome P-448 from PCB- or 3-methylcholanthrene-treated rats was high, but purified cytochrome P-450 from phenobarbital-treated rats showed very low activity.
4. The active metabolites were determined as hydroxylamine and nitroso derivatives of Trp-P-2 by chemical methods and GC-MS after separation by HPLC.
5. The mutagenesis and active metabolite formation by liver microsomes from C57BL/6N mice were increased in parallel with the increase in arylhydrocarbon hydroxylase activity by treatment with 3-methylcholanthrene, whereas these parameters were not affected in DBA/2N mice.
6. Human liver microsomes and purified cytochrome P-450 could activate Trp-P-2 and Trp-P-1 to mutagenic compound(s).

Broiled fish and meat contain substances highly mutagenic to *Salmonella typhimurium*. The mutagenic principles were formed by pyrolysis of protein or amino acid (*1*). Sugimura and coworkers recently isolated two potent mutagens from tryptophan pyrolysate, namely, 3-amino-1, 4-dimethyl-5-*H*-pyrido(4, 3-b)indole (Trp-P-1) and 3-amino-1-methyl-5-*H*-pyrido(4, 3-b)indole(Trp-P-2) (*2*).

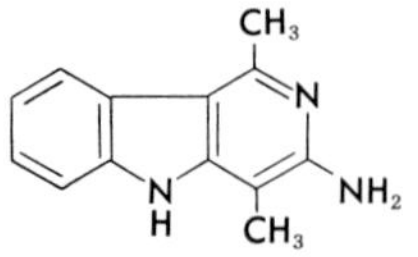

$C_{13}H_{13}N_3$	$C_{12}H_{11}N_3$
Trp-P-1	Trp-P-2

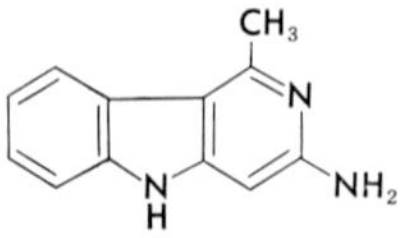

Both compounds exert mutagenic activity after undergoing metabolic activation by liver 9,000×*g* supernatant fraction (*3*) and their carcinogenic natures have been demonstrated (*4*, *5*).

In this communication, the metabolic activation of Trp-P-1 and Trp-P-2 by liver microsomes and purified cytochrome P-450 and the nature of the active metabolites of Trp-P-2 are presented.

MATERIALS AND METHODS

All chemical agents used were purchased from commercial sources except for Trp-P-1 and Trp-P-2 which were donated by Dr. Sugimura, National Cancer Center Research Institute, Tokyo.

Male Sprague-Dawley rats, 8 weeks old, and male C57BL/6N and DBA/2N mice, 7 weeks old, were used. The animals were given intraperitoneally with phenobarbital (80 mg/kg) or 3-methylcholanthrene (20 mg/kg) for 3 days and killed 24 hr later. PCB (polychlorinated biphenyl mixture, Kanechlor 500) was given intraperitoneally and the animals were killed 5 days after injection. The liver microsomes were prepared and cytochrome P-450 was purified and reconstituted as described in a previous paper (*6*).

The human liver microsomes were prepared from autopsy liver and several species of cytochrome P-450 were purified. The detail of the purification will be published elsewhere.

The isolation and quantitation of Trp-P-2 were carried out using a high performance liquid chromatograph (HPLC) (ALC/GPC, Model 204, Water Assoc.) equipped with a UV absorbance detector (Model 440), fluorescence detector (Model 420) and a reverse-phase column, μBondapack C_{18} (4 mmID×30 cm) (*7*).

The mutation assay for Trp-P-1 and Trp-P-2 and metabolites was carried out according to the method of Ames *et al.* using *S. typhimurium* TA98 with some minor modifications (*7*, *8*).

RESULTS AND DISCUSSION

1. *Metabolic activation of Trp-P-1 and Trp-P-2 by hepatic microsomes*

The metabolic activation of Trp-P-1 and Trp-P-2 was markedly stimulated by treatment with 3-methylcholanthrene or PCB, whereas phenobarbital treatment

TABLE 1. Effects of Inducers on Microsomal Activation of Trp-P-1 and Trp-P-2

Inducers	Revertants/μg protein (revertants/pmol P-450)[a]	
	Trp-P-1 (1 μg)	Trp-P-2 (1 μg)
None	3 (5)	5 (8)
Phenobarbital	6 (6)	17 (18)
Methylcholanthrene	184 (217)	466 (548)
PCB	640 (226)	1,133 (400)

[a] The number of revertants per μg protein and per pmol cytochrome P-450 was calculated from the linear curve obtained in the range of 0 to 50 μg protein. The experiment was done in duplicate and the spontaneous revertants were subtracted.

TABLE 2. Requirement of Cofactor and Effects of Inhibitors on the Microsomal Activation of Trp-P-1 and Trp-P-2

Treatment	Revertants/plate	
	Trp-P-1 (1 μg)	Trp-P-2 (1 μg)
Complete[a]	30,000	54,700
−NADPH	126	544
−NADPH-generating system	79	201
+Carbon monoxide	5,100	13,200
+7, 8-Benzoflavone (10 μM)	288	1,049
+*n*-Octylamine (100 μM)	4,900	30,900
+SKF 525-A (250 μM)	9,000	21,400

[a] Liver microsomes (50 μg protein) from PCB-treated rats were used.

resulted in only slight stimulation (Table 1). These results suggest that Trp-P-1 and Trp-P-2 are activated by cytochrome P-448.

The metabolic activation by hepatic microsomes required NADPH and was inhibited by carbon monoxide, 7,8-benzoflavone, *n*-octylamine and SKF 525-A (Table 2). Especially, the inhibition by 7, 8-benzoflavone was very strong, suggesting the involvement of cytochrome P-448 in the metabolic activation.

2. *Metabolic activation of Trp-P-1 and Trp-P-2 by purified cytochrome P-450*

The cytochrome P-448 from PCB-treated rats could activate Trp-P-1 and

TABLE 3. Requirements for the Mutagenic Activation of Trp-P-1 and Trp-P-2

Incubation mixture[a]	Revertants/plate[c]	
	Trp-P-1[b]	Trp-P-2[b]
Complete	13,300 (100%)	45,600 (100%)
−Reductase	260 (2.0%)	1,010 (2.2%)
−P-448	22 (0.2%)	0 (0%)
−NADPH	235 (1.8%)	313 (0.7%)
−Lipid	7,700 (58%)	30,400 (67%)

[a] 0.1 nmol of cytochrome P-448 from PCB-treated rat liver microsomes was used. [b] 3 μg of Trp-P-1 and 1 μg of Trp-P-2 were used. [c] The values are averages of duplicate assays.

TABLE 4. Metabolic Activation of Trp-P-1 and Trp-P-2 to Mutagenic Products by Various Preparations of Cytochrome P-450

Cytochrome P-450 species[a]	Revertants/Plate[c]	
	Trp-P-1[b]	Trp-P-2[b]
PCB P-450	256	3,450
PCB P-448	15,300	40,500
PB P-450	55	29
MC P-448	11,400	34,700

Various species of cytochrome P-450 were purified from liver microsomes from rats treated with PCB, phenobarbital (PB) or methylcholanthrene (MC). [a] 0.1 nmol of each cytochrome P-450 species was used. [b] 3 μg of Trp-P-1 and 1 μg of Trp-P-2 were used. [c] The values are averages of duplicate assays.

Trp-P-2 to mutagenic products in the presence of NADPH-cytochrome P-450 reductase and NADPH. The addition of phosphatidylcholine caused a further increase in the metabolic activation (Table 3).

The metabolic activation of Trp-P-1 and Trp-P-2 by various preparations of cytochrome P-450 was examined. It was clearly demonstrated that the activities

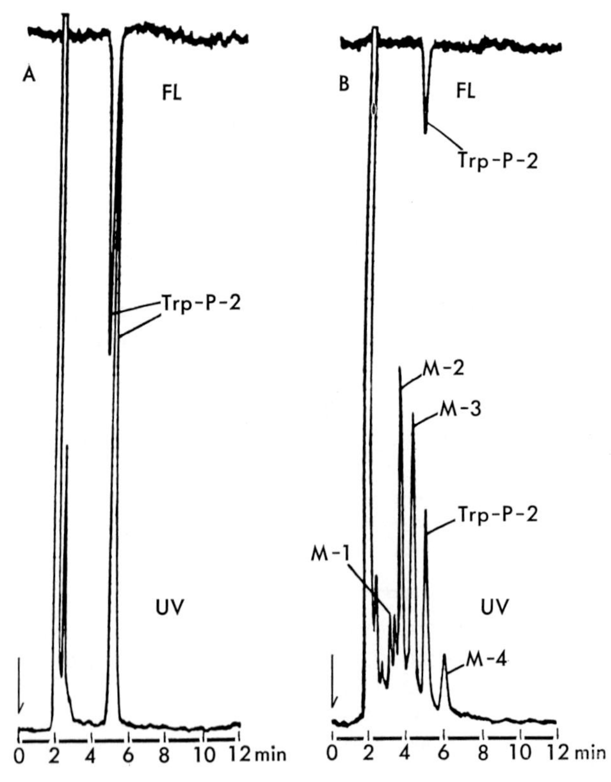

FIG. 1. Reverse-phase liquid chromatograms of Trp-P-2 and its metabolites before (A), and after 30 min of incubation (B). The incubation mixture (20 ml) contained 20 mg microsomal protein from PCB-treated rats, and 4 μmol Trp-P-2 and cofactors.

were low in cytochrome P-450 and high in cytochrome P-448 (Table 4). The activities were especially low in cytochrome P-450 purified from phenobarbital-treated rats.

3. *The metabolism of Trp-P-2 by rat liver microsomes to mutagenic products*

The metabolism of Trp-P-2 by rat liver microsomes was studied by a HPLC. The elution pattern of the metabolites of Trp-P-2 is shown in Fig. 1. Both fluorescence and absorption peaks of Trp-P-2 were decreased by incubation and four metabolites, namely M-1, M-2, M-3, and M-4, were formed. The formations of M-2 and M-3 were increased in proportion to the decrease in Trp-P-2, while the increase in the formation of M-4 started slowly (Fig. 2A).

The time-course of the increase in mutagenic activity ran parallel with that of the increase in the metabolite which forms a complex with pentacyanoammine ferroate (Fig. 2B). These results suggested that M-3 is possibly the metabolite responsible for the mutagenic activity.

Incubation with ascorbic acid prevented the formation of M-4 and increased the formation of M-3, therefore, in further studies ascorbic acid was used for the prevention of further oxidation of M-3.

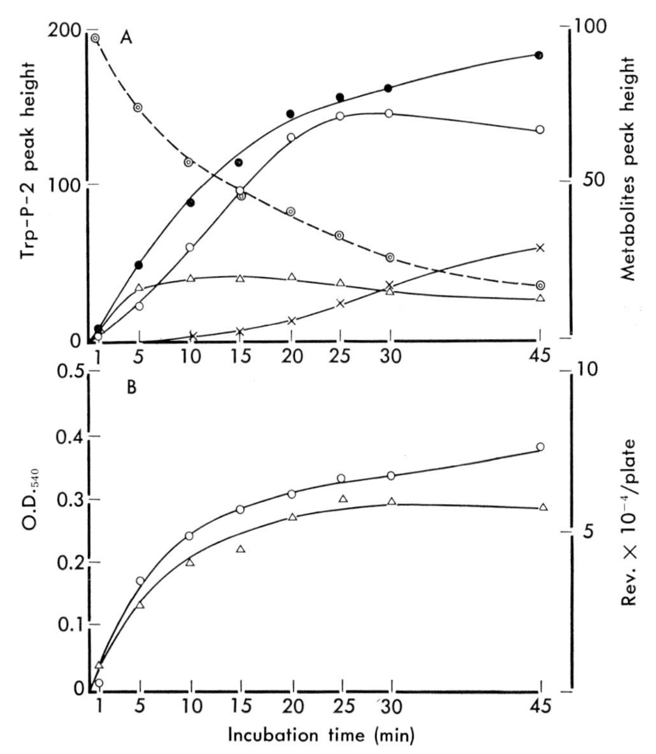

FIG. 2. Disappearance of Trp-P-2 and formation of the metabolites (A), and formation of mutagenic products and pentacyanoammine ferroate complex (B) as a function of incubation time. The liver microsomes from PCB-treated rats were used. A: ⊙ Trp-P-2; △ M-1; ● M-2; ○ M-3; × M-4. B: ○ mutagenic activity; △ ammine ferroate complex.

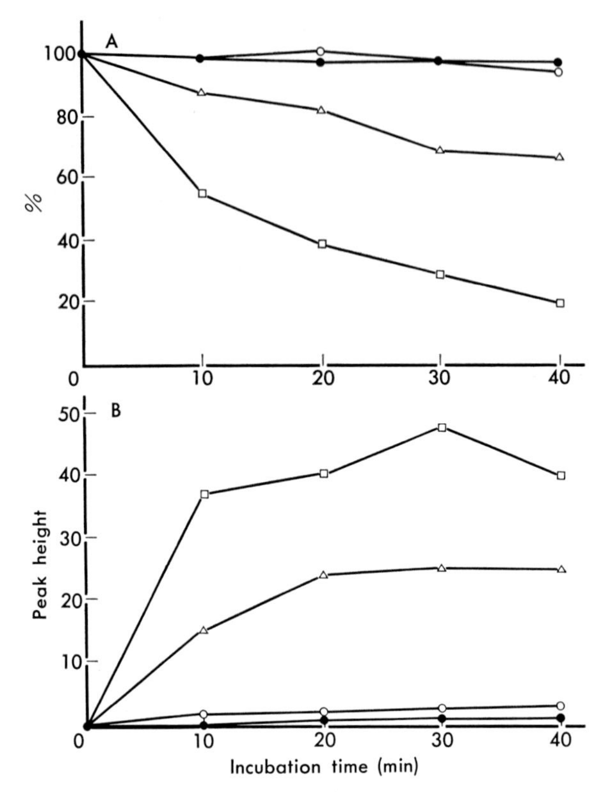

FIG. 3. Metabolism of Trp-P-2 (A) and formation of M-3 (B) by liver microsomes from rats treated with PCB, 3-methylcholanthrene (MC), or phenobarbital (PB). A total of 3 ml of incubation mixture contained 3 mg protein, 0.6 μmol Trp-P-2 and cofactors. ● control; □ PCB; △ MC; ○ PB.

Metabolism of Trp-P-2 by liver microsomes from rats treated with PCB, 3-methylcholanthrene or phenobarbital is shown in Fig. 3A. The metabolism of Trp-P-2 was fast in PCB- or 3-methylcholanthrene-treated rats and slow in phenobarbital-treated or control rats. In accordance with the metabolic rate of Trp-P-2 the formation of M-3 was high in PCB- or 3-methylcholanthrene-treated rats and low in phenobarbital-treated and control rats (Fig. 3B).

It was demonstrated that only M-3 and M-4 had mutagenic activity without metabolic activation by liver microsomes, whereas M-1 and M-2 had no activity.

Formation of the pentacyanoammine ferroate complex with M-3 and M-4 suggested that both metabolites are hydroxylamine and nitroso derivatives (*9*). The oxidation of M-3 to M-4 by potassium ferricyanide or γ-manganese dioxide and the reduction of M-4 and M-3 by titanium trichloride to Trp-P-2 indicated that M-3 and M-4 should be hydroxylamine derivative and nitroso derivative of Trp-P-2, respectively. The fragment patterns in GC-MS of isolated M-3 fraction confirmed this proposal.

As shown in Table 5, the mutagenic activity of M-3 was comparable to that of M-4 and most of the mutagenic activity in the incubated mixture could be accounted for by M-3 formation.

TABLE 5. Mutagenic Activities of Trp-P-2 Metabolites

Sample	Revertants/plate	%
Incubated mixture[a]	42,700	100
M-3 fraction[b]	41,700	97.7
M-4 fraction[c]	39,700	93.0

[a] The incubation mixture (20 ml) containing 1 mg of Trp-P-2 was incubated for 30 min and a portion (15 μl) of the mixture was applied for the mutation assay. [b] After treatment of the incubated mixture (100 μl) with the same volume of acetonitrile, 30 μl of the extract (eqv. to 15 μl of the incubated mixture) was injected into the column, then the fraction containing M-3 was subjected to the mutation assay. [c] The same extract described above but after γ-manganese dioxide treatment was subjected to HPLC and mutation assay.

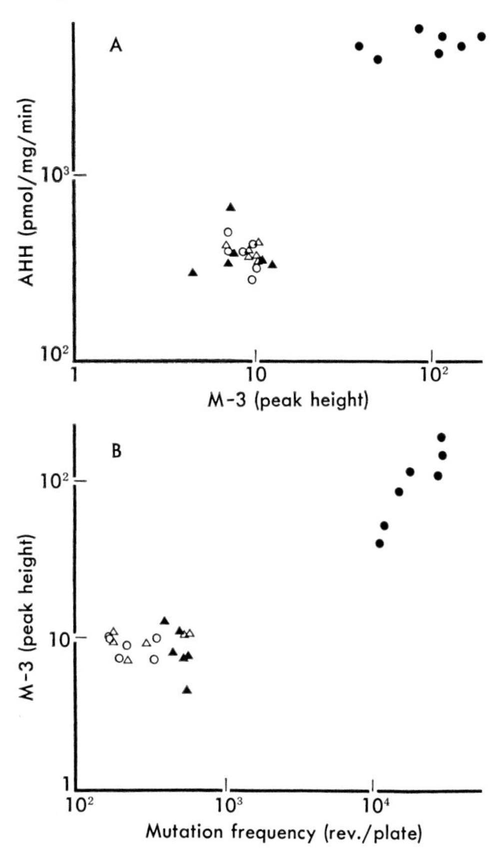

FIG. 4. Metabolic activation and mutagenesis of Trp-P-2 in control and MC-treated C57BL/6N and DBA/2N mice. (A) Relationship between formation of M-3 and AHH. (B) Relationship between mutagenesis of Trp-P-2 and formation of M-3. The mice were treated with 3-methylcholanthrene (MC, 80 mg/kg, i.p.), 40 hr. before sacrifice. ○ C57BL/6N (control); △ DBA/2N (control); ● C57BL/6N (MC); ▲ DBA/2N (MC).

4. Metabolic activation of Trp-P-2 by liver microsomes from 3-methylcholanthrene-responsive and non-responsive mice

To confirm the involvement of cytochrome P-448 in the metabolic activation of Trp-P-2 and that M-3 is a possible active metabolite, we employed C57BL/6N and DBA/2N mice as sources of microsomes for the metabolic activation of Trp-P-2 (*10*).

C57BL/6N mice were responsive to treatment with 3-methylcholanthrene and the activity in arylhydrocarbon hydroxylase (AHH) was increased as shown in Fig. 4A whereas the same treatment did not increase the activity in DBA/2N. The formation of M-3 was markedly increased in the responsive mice by treatment with 3-methylcholanthrene, but not in DBA/2N mice (non-responsive mice).

The mutagenesis of Trp-P-2 was markedly increased by methylcholanthrene treatment only in the responsive mice in accordance with the increase in M-3 formation (Fig. 4B). Similarly, the mutagenesis was markedly increased by methylcholanthrene treatment only in the responsive mice in accordance with the increase in AHH activity. These results indicated the involvement of cytochrome P-448 and M-3 in the mutagenesis of Trp-P-2 in the activation by liver microsomes.

5. The metabolic activation of Trp-P-1 and Trp-P-2 by human liver microsomes

To assess the mutagenic and carcinogenic activities of tryptophan pyrolysates in man, the metabolic activation of Trp-P-1 and Trp-P-2 by human liver microsomes and purified cytochrome P-450 was studied. Human liver microsomes could activate Trp-P-1 and Trp-P-2 to mutagenic compounds, and the activities were comparable with non-treated rat liver microsomes (Table 6). In contrast to rat liver microsomes, 7, 8-benzoflavone stimulated metabolic activation of Trp-P-2 in human liver microsomes.

TABLE 6. Metabolic Activation of Tryptophan Pyrolysis Products by Human Liver Microsomes

Microsomes (mg)	Revertants/plate	
	Trp-P-1 (0.5 μg)	Trp-P-2 (0.5 μg)
0	17	0
0.05	60	720
0.1	199	830
0.2	273	855

TABLE 7. Metabolic Activation of Tryptophan Pyrolysis Products by Cytochrome P-450 Purified from Human Liver Microsomes

Fractions	I-a	I-b	II-a	II-b
	(Revertants/plate, 0.2 nmole P-450)			
Trp-P-1 (1 μg)	15	191	44	44
Trp-P-2 (1 μg)	75	635	70	129
	(Specific content of P-450)			
	5.3	1.2	16.3[a]	7.9

[a] Almost homogeneous on SDS polyacrylamide gel electrophoresis.

Human liver microsomes were solubilized and the cytochromes P-450 were separated into four fractions (I-a, I-b, II-a, and II-b) by column chromatography. These four cytochrome P-450-containing fractions showed different capacities in the activation of Trp-P-1 and Trp-P-2 to mutagenic compounds (Table 7).

These results indicated that Trp-P-1 and Trp-P-2 can be considered as potential mutagens and carcinogens in man.

REFERENCES

1. Sugimura, T., Kawachi, T., Nagao, M., Yahagi, Y., Seino, Y., Okamoto, T., Shudo, K., Kosuge, T., Tsuji, K., Wakabayashi, K., Iitaka, Y., and Itai, A. Mutagenic principles in tryptophan and phenylalanine pyrolysis products. Proc. Jpn. Acad., *58*: 58–61, 1977.
2. Kosuge, T., Tsuji, K., Wakabayashi, K., Okamoto, T., Shudo, K., Iitaka, Y., Itai, A., Sugimura, T., Kawachi, T., Nagao, M., Yahagi, T., and Seino, Y. Isolation and structure studies of mutagenic principles in amino acid pyrolysates. Chem. Pharm. Bull., *26*: 611–619, 1978.
3. Nagao, M., Honda, M., Seino, Y., Yahagi, T., Kawachi, T., and Sugimura, T. Mutagenicities of protein pyrolysates. Cancer Lett., *2*: 355–340, 1977.
4. Takayama, S., Hirakawa, T., and Sugimura, T. Malignant transformation *in vitro* by tryptophan pyrolysis products. Proc. Jpn. Acad., *54*: 418–422, 1978.
5. Ishikawa, T., Takayama, S., Kitagawa, T., Kawachi, T., Kinebuchi, M., Matsukura, N., Uchida, E., and Sugimura, T. *In vivo* experiments on tryptophan pyrolysis products. *In*; E. C. Miller *et al.* (eds.), Naturally Ocurring Carcinogens-Mutagens and Modulators of Carcinogenesis, pp. 159–167, Japan Sci. Soc. Press, Tokyo / University Park Press, Baltimore, 1979.
6. Ishii, K., Ando, M., Kamataki, T., Kato, R., and Nagao, M. Metabolic activation of mutagenic tryptophan pyrolysis products (Trp-P-1 and Trp-P-2) by a purified cytochrome P-450-dependent monooxygenase system. Cancer Lett., *9*: 271–276, 1980.
7. Ishii, K., Yamazoe, Y., Kamataki, T., Kato, R., Metabolic activation of mutagenic pyrolysis products by rat liver microsomes. Cancer Res., in press.
8. Yamazoe, Y., Ishii, K., Kamataki, T., Kato, R., and Sugimura, T. Isolation and characterization of active metabolites of tryptophan-pyrolysate mutagen, Trp-P-2. Chem.-Biol. Interact., *30*: 125–138, 1980.
9. Boyland, E. and Nery, R. Arylhydroxylamines. IV. Their colorimetric determination. Analyst, *89*: 95–102, 1964.
10. Nebert, D. W., Robinson, J. R., Niwa, A., Kumaki, K., and Poland, A. P. Genetic expression of arylhydrocarbon hydroxylase activity in the mouse. J. Cell Physiol., *85*: 393–414, 1975.

PHARMACOGENETICS AND IMMUNOGENETICS

GENETIC AND ENVIRONMENTAL FACTORS IN EXPERIMENTAL AND HUMAN CANCER,
H. V. GELBOIN ET AL. (EDS.), JAPAN SCI. SOC. PRESS, TOKYO, PP. 91–102, 1980

Genetic and Environmental Factors Affecting the Metabolism of Carcinogens

Elliot S. VESELL

Department of Pharmacology and Specialized Cancer Center, Pennsylvania State University College of Medicine, Hershey, Pa. 17033, U.S.A.

Abstract: The metabolism of some drugs and carcinogens in man is highly complex, involving many different enzymes. In addition, multiple factors (Table 1) can alter rates of metabolism, absorption, distribution, and excretion of certain drugs and carcinogens, thereby affecting critical reaction site concentrations of these drugs as well as concentrations of proximate and ultimate carcinogens. Since ethical restrictions prohibit investigation of carcinogens in humans *in vivo*, we employed instead as models of carcinogens safe test compounds, such as antipyrine and tracer doses of ^{14}C-aminopyrine. Changes in the disposition of these model compounds could be quantitated and attributed to specific factors and conditions listed in Table 1. In addition to genetic factors that control large interindividual variations in drug and carcinogen disposition among normal human subjects under near basal conditions, many environmental factors were identified as being capable of altering rates of drug or carcinogen elimination.

In patients a complex, dynamic interaction occurs between certain environmental factors listed in Table 1 and underlying genetic factors that under near basal conditions control large interindividual variations in drug disposition. From the point of view of constructing and interpreting tests for detection of carcinogens, these results suggest existence of extreme differences among human subjects in rates of carcinogen absorption, distribution, metabolism, and excretion. These differences involve genetic traits on which multiple environmental factors impinge episodically. Therefore, the hazard of extrapolating results derived from only a few subjects under a particular set of conditions to many other subjects with different genetic constitutions and environmental backgrounds must be recognized and avoided.

Variability of Human Subjects in Response to Carcinogens

The metabolism of some chemical carcinogens is complex. Numerous factors have been identified that are capable of altering both qualitatively and quantita-

TABLE 1. A Partial List of Variables Affecting Drug Disposition in Experimental Animals

Variables in the external environment	Variables in the internal environment	Pharmacologic variables
Aggregation	Adjuvant arthritis	Drugs
Air exchange and composition	Age	Short *versus* long-term administration, bioavailability, dose, withdrawal, presence of other drugs or food, routes of administration, volume of material injected, tolerance, vehicle, *etc.*
Barometric pressure	Alloxan diabetes	
Cage design, materials	Cardiovascular function	
Cedar and other softwood bedding	Castration and hormone replacement	
Cleanliness	Circadian and seasonal variations	
Coprophagia	Dehydration	
Diet (food and water)	Disease—hepatic, renal, malignant, endocrine (thyroid, adrenal, pituitary)	
Exercise	Estrous cycle	
Gravity	Fever	
Hepatic microsomal enzyme induction or inhibition by insecticides, piperonyl butoxide, heavy metals, detergents, organic solvents, ammonia, vinyl chloride, aerosols containing eucalyptol, *etc.*	Gastrointestinal function, patency and flora	
Handling	Genetic constitution (strain and species differences)	
Humidity	Hepatic blood flow	
Light cycle	Infection	
Migration	Malnutrition, starvation	
Noise level	Pregnancy	
Temperature	Sex	
	Shock (hemorrhagic or endotoxic)	
	Stress	

tively the metabolism of certain chemical carcinogens once they enter a human subject. These factors can affect the absorption, distribution, biotransformation, and excretion of carcinogens, thereby changing their concentrations and duration of action at critical sites. In short, carcinogens in one subject need not be carcinogenic in all subjects. Variability in carcinogenicity can arise from differences among subjects in carcinogen pharmacokinetics.

The factors responsible for this variability in responsiveness of human subjects to many foreign compounds, including carcinogens, are factors that characterize the condition of a particular subject with respect to that subject's age, sex, height, weight, genetic constitution, diet, personal habits such as smoking, as well as ethanol and drug ingestion, chemical environment at work and at home, general health, and functional capacity of critical organ systems. Table 1 presents a partial list of factors capable of altering the disposition of certain drugs in laboratory animals. In this discussion drugs such as antipyrine, aminopyrine, amobarbital, bishydroxycoumarin, diazepam, phenylbutazone, and warfarin will have to serve as "models" for some carcinogens because these so-called "test drugs" can be investigated with very small risk in normal volunteers after single oral doses, whereas carcinogens cannot. The word model applies not to the metabolism of these compounds, because obviously the specific details of the metabolic profile of test drug and carcinogen differ greatly, but rather to the particular effects exerted on their metabolism by the factors in Table 1.

While the validity of this assumption is certainly questionable in many cases, in other instances a generalization may be permissible because the particular model drug and carcinogen are metabolized primarily by the hepatic cytochrome P-450-mediated monooxygenases. Thus, factors listed in Table 1 could alter the metabolism of both test drug and carcinogen in a similar manner and to a similar extent.

Whatever the chemical structure of the particular carcinogen, the generalized mechanism for its activation, deactivation and initiation of carcinogenesis may be similar. Thus, certain factors listed in Table 1 may exert effects on the individual steps by which a particular precarcinogen is converted to proximate and ultimate carcinogens. Therefore, despite these differences in detail of metabolism, the principle may be valid that several factors listed in Table 1 probably influence some reaction rates in the metabolic conversion of some precarcinogens and proximate carcinogens, thereby affecting both the concentration and duration of action of certain ultimate carcinogens. The uniqueness of each subject with respect to the factors (Table 1) that influence the absorption, distribution, biotransformation, and excretion of a chemical compound in that subject leads to the prediction that human subjects might exhibit different susceptibilities to developing cancer after exposure to a particular carcinogen. Several epidemiologic studies confirmed this prediction, showing certain subjects to be more susceptible to cancer than others under conditions of equal exposure to the same precarcinogen, in the same dose for the same period of time. From the point of view of developing and interpreting tests for carcinogenic effects of chemical compounds, these observations and facts suggest limitations of tests performed entirely *in vitro* or based on nonhuman systems and the need to remember that what is carcinogenic for one subject need not be carcinogenic for other subjects. This difference among humans in responsiveness to exogenous chemicals finds expression in the old adage that one man's food is another man's poison.

Genetic Factors That Cause Interindividual Variation in Drug and Carcinogen Metabolism

In different mouse strains the gene that affects inducibility of the P-448-mediated enzyme aryl hydrocarbon hydroxylase (AHH) is directly correlated to the carcinogenic index ($r = 0.90$) (*1*). Mice possessing the genetically transmitted trait for high AHH inducibility are more susceptible than mice with the gene for low AHH inducibility to fibrosarcomas initiated by subcutaneously administered 3-methylcholanthrene. This genetically controlled difference in drug responsiveness had been demonstrated previously for many other drugs and is illustrated in Table 2 for hexobarbital (*2*). In human subjects, large interindividual variations in AHH inducibility in cultures of peripheral lymphocytes occur, and genetic factors also appear to account for a major portion of this variability (*3–5*). However, these heritable variations in AHH inducibility cannot yet be used in diagnostic tests for cancer susceptibility because of the high degree of day-to-day variability in control

TABLE 2. Sleeping Time, Hexobarbital Oxidase Activity and Brain Hexobarbital Level on Awakening in Various Strains of Male Mice[a)]

Strain		Sleeping time ±SD (min)	Hexobarbital oxidase (μmol/g liver/15 min)	Brain hexobarbital level (γ/g)
AL/N	(24)[b)]	86±16	0.20±0.01	45±6
BALB/C	(24)	46±6	0.34±0.02	46±4
C3H/HeN	(24)	41±7	0.34±0.02	53±5
C57BL/6N	(24)	73±12	0.23±0.01	52±5
DBA/2N	(24)	85±17	0.21±0.01	48±4
STR/N	(24)	47±6	0.36±0.01	59±4
CAF_1	(24)	72±10	0.24±0.02	46±4
CdF_1	(24)	58±9	0.28±0.02	54±5
NIH	(24)	37±5	0.32±0.03	48±4
CFW/N	(24)	40±7	0.37±0.02	46±3
GP	(24)	42±8	0.37±0.03	52±5

[a)] Differences in sleeping time and liver hexobarbital oxidase activity between AL/N, DBA/2N, C57BL/6N, CAF_1 and CdF_1 mice and other strains are significant ($P<0.05$). [b)] The numbers in parenthesis indicate the number of animals used.

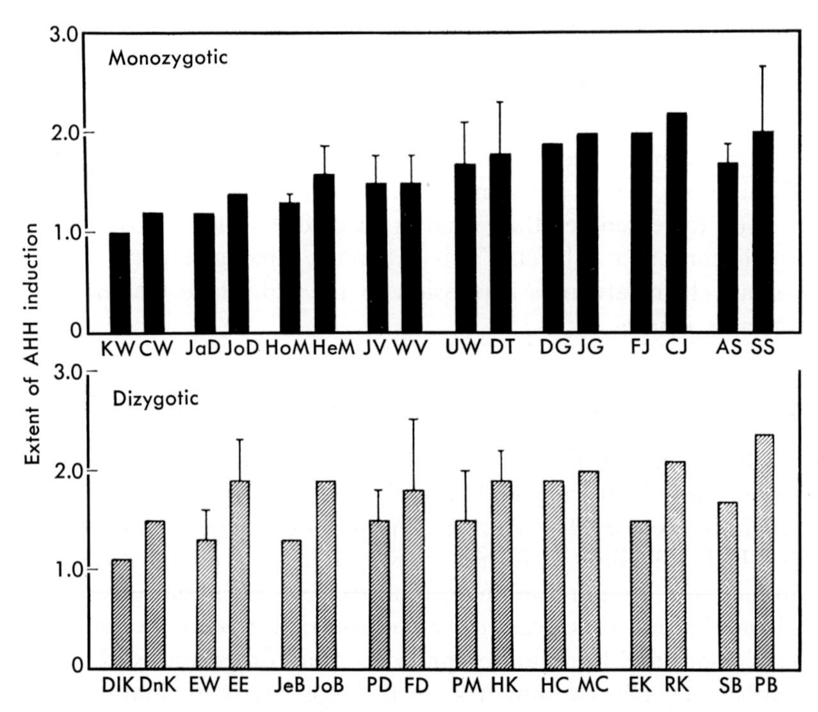

FIG. 1. Variation in the extent of AHH induction between monozygotic and dizygotic twins. The values shown are the means of 2 to 5 determinations on each individual, in some cases using different lots of fetal calf serum. The height of each vertical bar denoted the mean induction ratio; brackets, SD when 3 or more determinations were performed. Both members of each twinship were studied simultaneously on all occasions. Data from studies by Atlas *et al.* (*4*).

AHH activity of a given subject due to presently undefined environmental variables that influence the culture system (*4*, *5*). If lymphocytes from monozygotic (MZ) and dizygotic (DZ) twins are cultured simultaneously on the same day, these day-to-day variations are overcome, and the underlying genetic control of interindividual variations in AHH inducibility can be demonstrated (Fig. 1) (*4*). Genetic factors cause variations to be larger with DZ than within MZ twinships (Fig. 1) (*4*).

However, until technical advances in the AHH assay either successfully overcome day-to-day variations in measurement of control AHH activity or obscure these variations by elevating the fold inducibility of AHH beyond its present range, AHH inducibility in cultured mitogen-activated lymphocytes or in other similar systems cannot serve reliably as a biochemical marker for determining susceptibility to bronchogenic carcinoma.

The role of genetic factors in controlling large interindividual differences in rates of metabolism of various chemicals in man is so important that other examples will be cited. Table 3 shows the results of an experiment revealing large variations among normal subjects in the inducibility of antipyrine metabolism in normal twins after 2 weeks of daily phenobarbital administration (*6*). The extent of inducibility was regulated by genetic factors and predictable from initial rate of antipyrine elimination before exposure to phenobarbital (Fig. 2) (*6*). Figure 3 shows that in the uninduced and uninhibited state normal subjects exhibit large interindividual differences in antipyrine and bishydroxycoumarin elimination rates; the magnitude of these interindividual differences in rates of drug elimination reached

TABLE 3. Response of Plasma Antipyrine Half-life to Phenobarbital Administration in 16 Twins

Twin	Age, sex	Plasma antipyrine half-life		Decrease in half-life produced by phenobarbital (%)	Percentage difference between sibs in response to phenobarbital
		Before phenobarbital (hr)	After phenobarbital (hr)		
Identical twins					
Dan. E.	22, M	13.6	9.6	29.4	0
Dav. E.	22, M	13.6	9.6	29.4	
A. M.	35, F	8.0	6.3	21.2	0
B. Z.	35, F	8.0	6.3	21.2	
Bar. J.	23, F	18.2	8.4	53.8	0
Bev. J.	23, F	18.2	8.4	53.8	
B. F.	26, F	10.8	7.3	32.4	2.6
B. J.	26, F	11.4	7.4	35.0	
Fraternal twins					
F. D.	49, M	12.0	10.3	14.2	
P. D.	49, M	9.3	9.3	0	14.2
C. K.	49, M	17.5	5.5	68.6	
N. R.	49, F	14.5	5.8	60.0	8.6
H. H.	47, F	12.3	9.2	25.2	
P. M.	47, F	6.5	5.5	15.4	9.8
E. W.	54, F	15.0	6.9	54.0	
E. E.	54, F	9.0	6.9	23.3	30.7

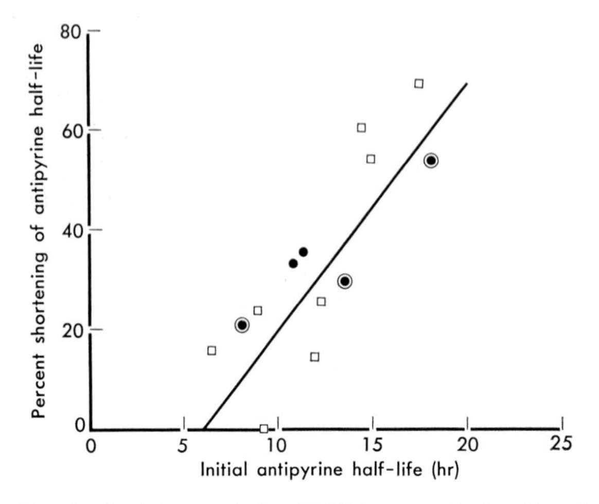

FIG. 2. Positive correlation (0.84) between the initial antipyrine half-life in plasma and the phenobarbital-induced shortening of antipyrine half-life. Data from studies by Vesell and Page (*6*). ● identical twin; □ fraternal twin; ⊙ identical twins with identical values.

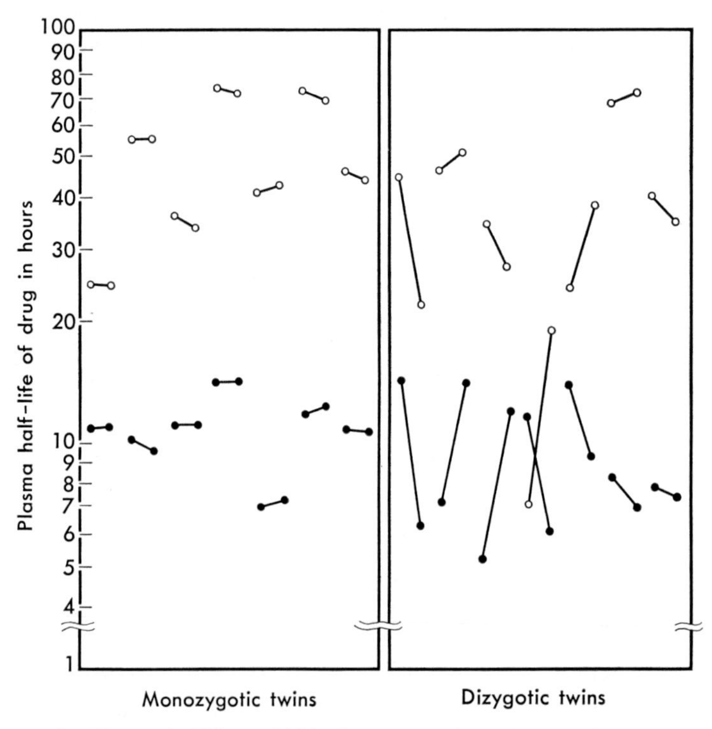

FIG. 3. Plasma half-lives of bishydroxycoumarin and antipyrine were measured separately at an interval of more than 6 months in healthy MZ and DZ twins. A solid line joins the values for each set of twins for each drug. Note that intratwin differences in the plasma half-life of both bishydroxycoumarin and antipyrine are smaller in MZ and in DZ twins. Data are from studies by Vesell and Page (*13*, *14*). ● antipyrine; ○ bishydroxycoumarin.

6-fold and 10-fold for antipyrine and bishydroxycoumarin, respectively. The differences vanished in MZ twins but were preserved to a large extent in most sets of DZ twins (Fig. 3). MZ twins have identical genetic constitution, whereas DZ twins share approximately 50% of their genes, but differ in approximately 50% of their genes. Twin and family studies also identified genetic factors as the main cause of large interindividual variations in the clearance from plasma of amobarbital, nortriptyline, phenylbutazone, phenytoin, and salicylates. It must be stressed that these twin and family studies were mainly performed under carefully controlled environmental conditions in normal subjects not taking other drugs, not smoking or exposing themselves to other compounds known to alter rates of drug elimination. Therefore, one wonders how imposition of some of the environmental factors listed in Table 1 affects the genetically controlled near basal rate of drug elimination.

Before leaving the topic of genetic constitutions as a factor affecting interindividual differences in rates of drug metabolism, it should be emphasized that the histocompatibility locus (HLA) system provides fresh opportunities for exploring associations between genetic constitution, carcinogen metabolism and susceptibility to certain forms of cancer. It has already been demonstrated that certain HLA genotypes are more frequently associated with Hodgkin's disease (A1, B5, B8, B18) and acute lymphatic leukemia (A2, B8, B12) but not with mammary carcinoma or malignant melanoma (*7*). In general, an association between certain HLA genotypes and a disease or condition such as a peculiar type or rate of carcinogen metabolism suggests an infectious or autoimmune etiology.

Environmental Factors That Can Affect Carcinogen Disposition

The critical role played by many of the environmental factors listed in Table 1 in the disposition of carcinogens or other foreign compounds can be illustrated by considering the consequences of coadministration of an agent that entirely binds or inactivates the carcinogen at its site of administration. Binding would prevent the carcinogen from progressing beyond its site of entry to its site of metabolic conversion and toxicity. Thus, harmful effects of the carcinogen could be avoided. In pharmacology several examples exist of inactivation of one drug by presence of another or by the particular environmental conditions prevailing at sites of drug administration. Well-known examples of this binding and inactivation are the interaction between tetracyclines and antacids containing multivalent cations and the destruction of an oral dose of insulin by the low pH in the stomach, respectively. While these examples are crude, finer aspects of how multiple variables both individually and collectively can affect carcinogen absorption, distribution, and excretion are now largely unknown and do not seem to be attracting much investigative interest. Nonetheless, potential effects exerted by these variables present problems in the interpretation of *in vitro* tests for carcinogenicity or of tests employing only a single subject under a single set of conditions. Results of such tests may not apply to many other subjects. Crucial questions concern not only the precise number of other subjects to whom the results obtained in one or a few people may be accurately extrapolated, but also how certain common conditions and factors may change the results of these tests even in the same subject.

From the point of view of carcinogen metabolism, the complexities of biotransformation of many carcinogens are such that we have not yet progressed to a consideration of how factors in Table 1 can affect the individual rate of each pathway. Additional complexity with respect to carcinogen metabolism arises from immunologic and catalytic evidence for multiple discrete forms of cytochrome P-450 in rat liver (*8*). Presumably, different carcinogens and drugs require different forms of cytochrome P-450 for metabolic activation, thereby raising the question of which environmental factors influence which multiple forms. All forms of P-450 are not equally susceptible to induction by exogenously administered chemicals, some chemicals enhancing all forms, others elevating concentrations of only one or two forms. The specificity of effects on carcinogen metabolism of the factors listed in Table 1, including the inducing or inhibiting actions of other exogenous chemicals administered before or during carcinogen exposure, must be considered. Simultaneously a heterogeneous effect may also occur at critical sites other than metabolism; for example, a certain factor or chemical may have an inductive effect on hepatic P-450-mediated monooxygenases but it may also simultaneously alter carcinogen absorption, plasma or tissue binding, excretion or some combination of these processes in ways that might either be additive or counterdirectional to the metabolic effect. Such multiplicity of drug effects is well recognized in pharmacology and explains the dictum that a drug rarely exerts only a single action. Heterogeneity of cytochrome P-450 raises questions concerning genetic control of these multiple forms and potential relationships between genetic differences among subjects in this regard and susceptibility to cancer through increased activation of carcinogens. Since certain cytochrome P-450's are present in various tissues, drug metabolism occurs in several extrahepatic sites. Some drugs and carcinogens are metabolized by bacteria in the gut. Because the bacterial flora changes according to diet, dietary differences among people and countries can influence the metabolism of some drugs and carcinogens. Recently, Kappas *et al.* (*9*) demonstrated that even within the same normal subject certain dietary changes can markedly alter rates of drug metabolism.

Interindividual Variations in Response of Normal Human Subjects to Various Factors and Conditions That Affect Drug Disposition

Over the past decade we developed and used the antipyrine and aminopyrine tests to identify and quantitate effects of factors listed in Table 1 on the capacity of human subjects to eliminate various compounds (*6*, *10*, *11*). The antipyrine and aminopyrine tests consist of measuring elimination rates of each of these drugs in normal subjects under near basal, environmentally controlled conditions. Under such conditions plasma and saliva clearances of each drug are highly reproducible in a given subject; then these values are remeasured in each subject after imposition of a single environmental change, such as chronic phenobarbital administration. The magnitude of the alteration in clearance of the test drug quantitatively indicates the effect of the particular factor under study on the hepatic metabolism of the test compound. The virtue of this form of the test is that each subject serves as control, thereby eliminating entirely the contribution of genetic factors to large interindividual differences in rates of elimination of a drug such as antipyrine or aminopyrine. In addition, by using each subject as control, many environmental factors can be either entirely eliminated or greatly reduced as sources of interindividual variation. Thus, a much increased sensitivity of examination of each of the factors listed in Table 1 can be attained. Dose-response curves can be constructed and the quantitative impact of the factors can be compared under various conditions. Over the past decade approximately 50 such studies have been performed on different environmental factors that can affect antipyrine and aminopyrine disposition. These studies identified many diverse factors capable of altering drug disposition. Specific results, virtues, and limitations of this test are described elsewhere (*11*).

Under some circumstances it may not always be possible to obtain multiple elimination rates of antipyrine in the same subject. Rather it may prove feasible to obtain only a single rate in a patient and then depend on comparisons of antipyrine clearances in different patient groups. This approach limits the sensitivity of the antipyrine test so that some factors may not be detected by it in this form, whereas they would if each subject served as his or her own control. In patients with various diseases there is a very large range in the magnitude of alteration in

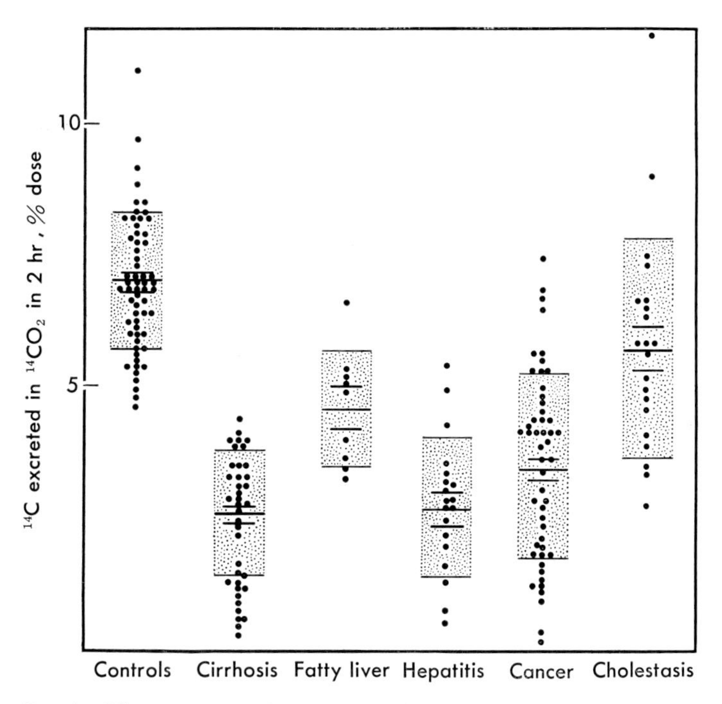

FIG. 4. The percentage of administered ^{14}C excreted as $^{14}CO_2$ in breath 2 hr after oral administration of [^{14}C]aminopyrine. Transverse lines represent mean ± SEM; hatched areas represent SD. Data from studies by Hepner and Vesell (*15*).

TABLE 4. Percentage Deviation of Each Drug Test from the Mean of the Value in Control Subjects (Mean±SD)

	Aminopyrine breath test	Diazepam breath test	Antipyrine metabolic clearance rate	Indocyanine green
Controls	−0.3±21.0	+0.5±21.0	3.0±36.0	−2.0±31.0
Anticonvulsant drugs	+56.0±46.0	+97.0±53.0	+122.0±74.0	+6.0±18.0
Hepatocellular disease	−62.0±21.0	−44.0±18.0	−55.0±27.0	−56.0±24.0
Hepatic neoplasm	−43.0±22.0	−31.0±35.0	−38.0±37.0	−34.0±24.0
Cholestasis	+10.0±34.0	+20.0±42.0	−36.0±14.0	−30.0±23.0

drug metabolism and disposition dependent on the course and severity of the disease process, as well as on the presence of additional factors such as drug ingestion and functional status of the liver, heart, and kidneys. This large range in the extent of impairment of drug metabolism is illustrated by our study on aminopyrine metabolism in cancer patients with hepatic metastases (Fig. 4).

The question arises of how closely derangement in the disposition of a drug such as antipyrine, aminopyrine, diazepam, or indocyanine green can serve to predict in a single patient the magnitude of alteration of the other 3 drugs. Table 4 discloses that in hepatocellular diseases the extent of alteration in the disposition of one of these 4 test compounds closely reflects the degree of impairment in the disposition of the other 3 compounds. However, Table 4 also reveals that in patients taking inducing drugs to control seizures the magnitude of enhancement in elimination rate of each test compound differed greatly from that of the others. Thus, in induction one test compound could not be used to predict the extent of change in disposition of others (*12*). This example is cited to show how hazardous it may be to attempt to extrapolate an effect from one drug to another.

Finally, large interindividual differences occur in the magnitude of change in antipyrine clearance produced by a given factor (Table 1). Some factors, such as diurnal variation and fever, exerted only a small influence on the mean value of antipyrine clearance, with some subjects showing no alteration whatever. Other factors, such as chronic phenobarbital or disulfiram administration, produced much larger mean effects. Nevertheless, even for these latter factors, if the data are examined for individual subjects, a broad range of response is generally observed; large interindividual variations occur despite the fact that subjects are under apparently homogeneous environmental conditions. These large interindividual variations in response to a particular factor are non-random because when each subject is retested under similar conditions, individual values are generally reproducible. One implication of this individuality of response to a particular factor is that the concept of "extrapolation" of results from one subject to another, as well as from one test drug to another, becomes hazardous, if not illogical. If, as has been repeatedly observed, regardless of the magnitude of the *mean* quantitative change, individual subjects exhibit large differences and deviations from a "mean" response, then the concept of extrapolation between subjects or drugs is tenuous.

Complex Dynamic Interactions between Environmental and Genetic Factors Capable of Affecting the Disposition of Drugs and Carcinogens

The present view of the role of the factors listed in Table 1 with respect to their effects on drug and carcinogen disposition in human subjects is that the genetic and environmental factors listed in Table 1, rather than being in conflict or opposition, are complementary and coordinated, interacting at multiple sites in a complex, dynamic manner to affect rates of drug absorption, distribution, metabolism, and excretion. A rapidly developing area of clinical pharmacology concerns these multiple and changing interactions between genes in man that control rates of drug metabolism and numerous factors in the environment of each subject. Despite underlying genetic control of large interindividual differences in hepatic drug-metabolizing capacity in humans under near basal, environmentally controlled conditions, the enzymes by which the liver biotransforms drugs and the genes regulating them are particularly susceptible to perturbation by many environmental factors. Because so many environmental factors can alter rates of drug elimination in patients and because the relative role that each of these factors plays in influencing rates of drug metabolism changes even in the same subject with time and with numerous other variables, it is very difficult to identify even at a particular time which factors are operating and what contribution each factor makes to a subject's total drug-metabolizing capacity. For this reason, the role of each of these factors is generally investigated in normal, non-medicated subjects under environmentally controlled conditions.

There has been a tendency to partition phenotypic variation in the disposition of drugs and carcinogens into neat, separate genetic and environmental packages treated as though they never interacted. This tendency needs to be replaced by the concept that in practice both genetic constitution and environmental factors affecting drug and carcinogen disposition interact often and at several levels (*10*, *11*). Although it has been convenient in twin studies to separate genetic from environmental contributions as though they were discrete, unrelated entities, this circumstance based on the twin techniques used should not obscure the fact that in reality both the transcriptional and translational mechanisms by which genetic information is expressed require environmental participation. Conversely, some environmental factors that alter rates of drug and carcinogen disposition in man do so by affecting genetic mechanisms (*3*, *4*, *6*). Environmental chemicals, such as DDT, polychlorinated biphenyls, and polycyclic hydrocarbons, can alter, through induction, near basal, genetically controlled, hepatic drug-metabolizing enzyme activity. Disease states can change markedly a subject's genetically controlled rate of drug elimination. Assurance of an environmentally controlled, uninduced or uninhibited state can never be complete but is partially attainable through repeated measurements of clearance of a particular test drug, such as antipyrine, and through a careful history of exposures at work and at home to compounds or conditions capable of altering a subject's rate of drug elimination. Simultaneously acting environmental factors, each with a different capacity in different subjects for changing the underlying genetically controlled rates of drug and carcinogen metabolism, make it exceedingly difficult to attribute quantitatively different portions of the total interindividual variation to specific single environmental factors.

REFERENCES

1. Nebert, D. W. and Felton, J. S. Importance of genetic factors influencing the metabolism of foreign compounds. Fed. Proc., *35*: 1133–1141, 1976.
2. Vesell, E. S. Genetic and environmental factors affecting hexobarbital metabolism in mice. Ann. N.Y. Acad. Sci., *151*: 900–920, 1968.
3. Kellermann, G., Luyten-Kellermann, M., and Shaw, C. R. Genetic variation of aryl hydrocarbon hydroxylase in human lympohcytes. Am. J. Human Genet., *25*: 327–331, 1973.
4. Atlas, S. A., Vesell, E. S., and Nebert, D. W. Genetic control of interindividual variations in the inducibility of aryl hydrocarbon hydroxylase in cultured human lymphocytes. Cancer Res., *36*: 4619–4630, 1976.
5. Okuda, T., Vesell, E. S., Plotkin, E., Tarone, R., Bast, R. C., and Gelboin, H. V. Interindividual and intraindividual variations in aryl hydrocarbon hydroxylase in monocytes from monozygotic and dizygotic twins. Cancer Res., *37*: 3904–3911, 1977.
6. Vesell, E. S. and Page, J. G. Genetic control of the phenobarbital-induced shortening of plasma antipyrine half-lives in man. J. Clin. Invest., *48*: 2202–2209, 1969.
7. Svejgaard, A., Hauge, M., Jersild, C., Platz, P., Ryder, L. P., Nielsen, L. S., and Thomsen, M. The HLA system: An introductory survey. *In*; L. Beckman and M. Hauge (eds.), Monographs in Human Genetics, vol. 7, pp. 1–103, S. Karger, Basel, 1975.
8. Nebert, D. W. and Atlas, S. A. The A*h* locus: Aromatic hydrocarbon responsiveness of mice and men. Human Genet., *1*: 149–160, 1978.
9. Kappas, A., Anderson, K. E., Conney, A. H., and Alvares, A. P. Influence of dietary protein and carbohydrate on antipyrine and theophylline metabolism in man. Clin. Pharmacol. Ther., *20*: 643–653, 1976.
10. Vesell, E. S. Pharmacogenetics: Multiple interactions between genes and environment as determinants of drug response. Am. J. Med., *66*: 183–187, 1979.
11. Vesell, E. S. Commentary: The antipyrine test in clinical pharmacology: Conceptions and misconceptions. Clin. Pharmacol. Ther., *26*: 275–286, 1979.
12. Hepner, G. W., Vesell, E. S., Lipton, A., Harvey, H. A., Wilkinson, G. R., and Schenker, S. Disposition of aminopyrine, antipyrine, diazepam, and indocyanine green in patients with liver disease or on anticonvulsant drug therapy: Diazepam breath test and correlations in drug elimination. J. Lab. Clin. Med., *90*: 440–456, 1977.
13. Vesell, E. S. and Page, J. G. Genetic control of dicumarol levels in man. J. Clin. Invest., *47*: 2657–2663, 1968.
14. Vesell, E. S. and Page, J. G. Genetic control of drug levels in man: Antipyrine. Science, *161*: 72–73, 1968.
15. Hepner, G. W. and Vesell, E. S. Quantitative assessment of hepatic function by breath analysis after oral administration of (^{14}C)aminopyrine. Ann. Intern. Med., *83*: 632–638, 1975.

GENETIC AND ENVIRONMENTAL FACTORS IN EXPERIMENTAL AND HUMAN CANCER,
H. V. GELBOIN ET AL. (EDS.), JAPAN SCI. SOC. PRESS, TOKYO, PP. 103–109, 1980

The Role of the Immune System in Ultraviolet Carcinogenesis: A Review

Margaret L. KRIPKE

Cancer Biology Program, NCI Frederick Cancer Research Program, Frederick, Md. 21701, U.S.A.

Abstract: UV carcinogenesis in inbred mice has been used as a model to investigate the interaction between the immune system and developing skin cancers. This tumor system is particularly useful for these investigations because UV-induced skin cancers of mice are highly antigenic, and many are immunologically rejected upon transplantation to mice of the same inbred strain. In spite of their strong antigenicity, these tumors grow progressively in the primary host and rarely undergo regression *in situ*. This suggests that the immune system of the primary host is altered in a way that favors, rather than impedes, tumor growth.

In experiments addressing the question of why the primary host fails to reject these antigenic tumors, we found that UV irradiation of mice induces suppressor lymphocytes. These suppressor cells prevent the induction of an immune reaction against tumors induced by UV radiation. This seems to be the mechanism whereby UV-induced skin cancers escape immunologic destruction during their development in UV-irradiated mice. Such studies provide evidence that the immune system plays a significant role in carcinogenesis. In addition, they emphasize the importance of the regulatory pathways of the immune system in controlling tumor growth.

It has been known since the early 1900's that UV radiation from sunlight can induce skin cancers in man and in experimental animals (*1*). However, only recently has it been recognized that the immune system may play an important role in the pathogenesis of this type of skin cancer. Evidence supporting this contention comes from human patients that have undergone renal transplant. One adverse consequence of renal transplantation, which involves longterm immunosuppression to prevent rejection of the grafted kidney, is a high risk of developing UV-radiation-induced skin cancers early in life (*2–4*). Although this association between the immune system and skin carcinogenesis in humans is tenuous, the information from experimental animals on this relationship is quite extensive (*5*). In fact, the system of UV carcinogenesis in inbred mice provides a model that illustrates quite clearly how the immune response to tumor antigens is regulated, and the importance of immunologic regulatory mechanisms in determining the outcome of this neoplastic disease.

Immunogenicity of UV-induced Murine Skin Cancers

The reason that this tumor system is particularly useful for studies of immunologic regulation is because the UV-induced skin cancers are extremely antigenic (*6*). In fact, in some strains of mice, more than 70% of the primary tumors are immunologically rejected upon transplantation to syngeneic hosts (*7, 8*). However, these skin cancers, which can be either fibrosarcomas or squamous carcinomas depending upon the strain of mice (*8*) and the conditions of induction (*9*), grow progressively in immunosuppressed mice (*7, 8*). The transplantation immunity induced by these tumors is like that of chemically-induced fibroasarcomas, in that each tumor is antigenically distinct (*7, 10*). Recent studies by Spellman and Daynes (*11*) suggest that there also may be minor cross-reacting antigens, although these may represent viral or fetal antigens rather than tumor-specific antigens. In general, *in vitro* assays of both humoral (*12*) and cell-mediated immunity (*13*) to these tumors also demonstrate specificity for individual tumors.

The finding that many primary tumors, particularly in C3H mice, were immunologically rejected by normal syngeneic recipients raised the question of how these tumors were able to survive and grow progressively in their original host. It seemed paradoxical that these antigenic skin cancers were not eliminated by the immune system. Therefore, a series of experiments was initiated in an attempt to determine the basis for this apparent contradiction (*14*). Primary skin cancers were removed from C3H mice, cut into small fragments, and transplanted to groups of normal and immunosuppressed syngeneic mice. One tumor fragment was reimplanted subcutaneously (s.c.) in the original host on the ventral side, in an area not exposed to UV radiation. Although most of the tumor fragments were rejected by the normal recipients, they grew progressively in the immunosuppressed mice, and all implants grew in the original host. This demonstrated that the failure of the primary host to reject its developing tumor was a systemic one, and was not due to the location of the tumor in a protected local environment. Futhermore, other primary tumors from UV-irradiated animals could be transplanted similarly into mice with primary UV-induced tumors, indicating that the systemic failure to reject these tumors was not specific for an individual tumor.

Immunologic Alterations in UV-irradiated Mice

By testing animals at various times during the course of UV treatment to induce skin cancers, we found that even precancerous mice were unable to reject transplanted UV-induced tumors (*14*). Very early in the induction period, before any primary skin cancers were discernible, tumors that were rejected by age-matched, unirradiated mice, grew progressively in the UV-irradiated animals. Recent studies have shown that this susceptibility of UV-treated mice to transplanted UV-induced tumors is systemic, since it is also manifested when the tumor cells are injected intravenously (i.v.). Some UV-induced tumors that form few or no pulmonary tumor colonies when injected i.v. into normal syngeneic mice produce many pulmonary and extrapulmonary metastases in UV-irradiated recipients (*15*).

Careful dose-response experiments in BALB/c mice have demonstrated that the susceptibility of UV-irradiated mice to tumor challenge is directly proportional to the dose of radiation (*16*). Furthermore, the effectiveness of a single dose of UV-radiation in inducing tumor susceptibility is not increased by fractionation into multiple smaller doses. This implies that there is a fundamental difference in mechanism between the induction of susceptibility to tumor challenge and UV carcinogenesis, because in skin cancer induction, fractionated doses of UV radiation are much more effective than a single large dose.

In analyzing the basis for the susceptibility of UV-irradiated mice to challenge with highly antigenic UV-induced tumors, we first considered whether it could be due to nonspecific immunosuppression. This did not appear to be the case, since UV-irradiated mice could respond normally to a variety of exogenous antigens (*14*, *17–20*). Furthermore, UV-irradiated mice do not appear to have altered responsiveness to other types of tumors, including several types of primary and transplanted tumors of viral or chemical etiology (*20*). One important exception to this generalization is that UV-treated mice respond poorly to contact allergens (*17*, *21*). However, generalized immunosuppression did not seem to account for the dramatic susceptibility of UV-treated mice to UV-induced syngeneic tumors.

The immunologic basis for the lack of tumor rejection in UV-irradiated mice was demonstrated in both *in vivo* and *in vitro* tests. Lymphocytes from UV-irradiated mice immunized with a syngeneic UV-induced fibrosarcoma did not kill these tumor cells in an *in vitro* microcytotoxicity test. In contrast, lymphocytes from unirradiated mice immunized in the same way were specifically cytotoxic *in vitro* to the immunizing tumor (*22*). *In vivo* studies demonstrated that the unresponsiveness of the UV-irradiated host could be passively transferred by reconstituting lethally X-irradiated syngeneic mice with lymphoid cells taken from UV-irradiated donors. This unresponsiveness is due, at least in part, to the presence of suppressor cells in the lymphoid organs of UV-treated animals. Mixing spleen and/or lymph node cells from UV-irradiated mice with normal lymphoid cells prior to their injection into lethally X-irradiated recipients renders the normal lymphocytes unreactive against tumor challenge (*23*). This suppressive activity also can be demonstrated by i.v. injection of lymphocytes from UV-treated mice directly into normal syngeneic recipients; such recipients are unable to reject UV-induced tumors (*24*, *25*).

Characteristics of UV-induced Suppressor Cells

The discovery of suppressor cells in the lymphoid tissue of UV-irradiated animals has provided an opportunity to study how the immune response against syngeneic tumors is regulated. Aspects of this problem that are currently under investigation are the origin, specificity, and activity of the UV-induced suppressor cells. These cells are T lymphocytes, based on their ability to be deleted from a spleen cell suspension by treatment with anti-theta serum and complement, and by their enrichment after the incubation of spleen cells on a nylon wool column (*24–26*). They act selectively to prevent the rejection of UV-induced tumors, but not tumors induced by the chemical carcinogen, methylcholanthrene (*25*). This type of restricted reactivity is unusual, in that other tumor-related suppressor cells are either specific for an individual tumor (*27*) or lack specificity altogether (*28*). With the UV-induced suppressors, we have an intermediate situation in which the suppressor cells seem to control immune reactivity against a set of non-cross-reacting tumor antigens.

Recent experiments indicated that these suppressor cells do not inhibit *in vitro* cytotoxic responses, and all attempts to establish an *in vitro* assay for the detection of these cells have been unsuccessful so far (*13*; Thorn, Fisher, and Kripke, submitted for publication). The cells do function in a local *in vivo* assay, similar in design to a Winn test. In this assay, tumor cells are injected s.c. into highly immunosuppressed recipients, and progressive tumor growth ensues. When normal lymphocytes are mixed with the tumor cells prior to injection, tumor growth is inhibited. This local tumor rejection is prevented by including UV-induced suppressor T cells in the mixture.

The suppressor cells do not appear to reside in the spleen, since splenectomy of UV-irradiated mice does not render them resistant to tumor challenge (*25*). Furthermore, neither the spleen nor the thymus is required for the generation of the suppressor cells, since thymectomizing or splenectomizing mice before UV treatment does not prevent their appearance (Thorn, Fisher, and Kripke, submitted for publication). Since suppressor cells are present in the spleen (*21*, *24*, *25*), lymph nodes (*29*), and possibly the thymus (*24*) of intact UV-irradiated animals, it seems most likely that these cells originate in the recirculating pool of peripheral T lymphocytes, from which they are distributed to the lymphoid organs.

Induction of Suppressor Cells by UV Radiation

One of the most interesting questions raised by these findings is how UV irradiation of the skin leads to such profound immunologic alterations. Although we do not yet know the precise mechanisms involved, some of our recent experiments have identified a pathway connecting UV treatment and suppressor cell induction.

In our earlier studies on the immunologic competence of UV-irradiated animals, we noted that the reactivity of mice to dinitrochlorobenzene (DNCB), a contact-sensitizing agent, was depressed after a few hours of UV irradiation (*17*). The basis for this unresponsiveness has been analyzed in detail (*21*), and we found that there is no deficiency in the ability of lymphocytes from UV-treated animals to respond effectively to DNCB if they are removed from their UV-irradiated host. This implied that the deficiency resided in the earlier, antigen-processing steps of the immune response and suggested that macrophages, rather than lymphocytes, were the target in this effect of UV radiation. Direct evidence supporting this suggestion was provided by recent experiments performed in collaboration with Drs. Greene, Sy, and Benacerraf at Harvard Medical School (*30*). In these studies, splenic adherent cells from UV-irradiated mice were found to be incapable of presenting contact-sensitizing antigens to normal lymphocytes in a way that triggers the lymphocytes to respond to the antigen. Thus, UV irradiation of mice produces a defect in effective antigen presentation. Furthermore, this defective antigen presentation not only prevents the induction of a normal immune response, but also stimulates the production of suppressor T lymphocytes that are specific for the "improperly" presented antigen.

This phenomenon may explain the presence of the suppressor cells in UV-irradiated animals that prevent the rejection of UV-induced tumors. Following UV irradiation, new antigens, related in specificity to tumor antigens, appear in the skin. At the same time, the defect in antigen processing occurs, resulting in the inappropriate presentation of these new antigens to the lymphoid system. This alteration in antigen presentation would then lead to the production of suppressor cells with specificity for the tumor antigens present on UV-induced skin cancers. Although we have no direct proof of a connection between the antigen-presenting defect and the UV-induced suppressor cells, this working hypothesis is consistent with all of the information that is presently available.

CONCLUSIONS

This tumor system provides an opportunity to investigate in detail some of the ways in which immune responses to syngeneic and autochthonous tumors can be regulated. In addition, it provides an illustration of the intimate relationship between the immune system and primary cancers. It is clear that an immune response is subject to both positive (helper) and negative (suppressor) regulation. In the UV carcinogenesis system, the immune response to developing tumors is dominated by suppressive forces, and this system probably represents an extreme case of such negative regulation. In other murine tumor systems, for example those associated with the Moloney sarcoma and polyoma viruses, the balance of regulation is toward facilitating an immune response against neoplastic cells. Although most experimental cancers probably fall in between these examples, the extremes serve to illustrate the potential of the immune system in influencing carcinogenic processes.

ACKNOWLEDGMENT

This research was sponsored by the National Cancer Institute under Contract No. NO1-CO-75380 with Litton Bionetics, Inc.

REFERENCES

1. Blum, H. F. Carcinogenesis by Ultraviolet Light, Princeton University Press, Princeton, N.J., 1959.
2. Penn, I. Malignant tumors in organ transplant recipients. Recent Results Cancer Res., *35*: 1–51, 1970.
3. Koranda, F. C., Dehmel, E. M., Kahn, G., and Penn, I. Cutaneous complications

in immunosuppressed renal homograft recipients. J. Am. Med. Assoc., *229*: 419–424, 1974.

4. Marshall, V. Premalignant and malignant skin tumours in immunosuppressed patients. Transplantation, *17*: 272–275, 1974.
5. Kripke, M. L. Immunologic effects of UV radiation and their role in photocarcinogenesis. *In*; K. Smith (ed.), Photochemical and Photobiological Reviews, vol. 5, Plenum Press, New York, pp. 257–292, 1980.
6. Graffi, A., Pasternak, G., and Horn, K.-H. Die Erzeugung von Resistenz gegen isologe Transplantate UV-induzierter Sarkome der Maus. Acta Biol. Med. Ger., *12*: 726–728, 1964.
7. Kripke, M. L. Antigenicity of murine skin tumors induced by ultraviolet light. J. Natl. Cancer Inst., *53*: 1333–1336, 1974.
8. Kripke, M. L. Latency, histology, and antigenicity of tumors induced by ultraviolet light in three inbred mouse strains. Cancer Res., *37*: 1395–1400, 1977.
9. Spikes, J. D., Kripke, M. L., Connor, R. J., and Eichwald, E. J. Time of appearance and histology of tumors induced in the dorsal skin of C3H female mice by ultraviolet radiation from a mercury arc lamp. J. Natl. Cancer Inst., *59*: 1637–1643, 1977.
10. Pasternak G., Graffi, A., and Horn, K.-H. Der Nachweis individualspezifischer Antigenitat bei UV-induzierten Sarkomen der Maus. Acta Biol. Med. Ger., *13*: 276–279, 1964.
11. Spellman, C. W. and Daynes, R. A. Ultraviolet light induced murine suppressor lymphocytes dictate specificity of anti-ultraviolet tumor immune responses. Cell. Immunol., *38*: 25–34, 1978.
12. DeLuca, D., Kripke, M. L., and Marchalonis, J. J. Induction and specificity of antisera from mice immunized with syngeneic UV-induced tumors. J. Immunol., *123*: 2696–2703, 1979.
13. Thorn, R. M. Specific inhibition of cytotoxic memory cells produced against UV-induced tumors in UV-irradiated mice. J. Immunol., *121*: 1920–1926, 1978.
14. Kripke, M. L. and Fisher, M. S. Immunologic parameters of ultraviolet carcinogenesis. J. Natl. Cancer Inst., *57*: 211–215, 1976.
15. Kripke, M. L. and Fidler, I. J. Enhanced experimental metastasis of UV-induced fibrosarcomas in UV-irradiated syngeneic mice. Cancer Res., *40*: 625–629, 1980.
16. DeFabo, E. and Kripke, M. L. Dose-response characteristics of immunologic unresponsiveness to UV-induced tumors produced by UV irradiation of mice. Photochem. Photobiol., *30*: 385–390, 1979.
17. Kripke, M. L., Lofgreen, J. S., Beard J., Jessup, J. M., and Fisher, M. S. *In vivo* immune responses of mice during carcinogenesis by ultraviolet irradiation. J. Natl. Cancer Inst., *59*: 1227–1230, 1977.
18. Norbury, K. C., Kripke, M. L., and Budmen, M. B. *In vitro* reactivity of macrophages and lymphocytes from UV-irradiated mice. J. Natl. Cancer Inst., *59*: 1231–1235, 1977.
19. Spellman, C. W., Woodward, J. G., and Daynes, R. A. Modification of immunological potential by ultraviolet radiation. I. Immune status of short-term UV-irradiated mice. Transplantation, *24*: 112–119, 1977.
20. Kripke, M. L., Thorn, R. M., Lill, P. H., Civin, C. I., Pazmino, N. H., and Fisher, M. S. Further characterization of immunological unresponsiveness induced in mice by ultraviolet radiation. Transplantation, *28*: 212–217, 1979.
21. Jessup, J. M., Hanna, N., Palaszynski, E., and Kripke, M. L. Mechanisms of de-

pressed reactivity to dinitrochlorobenzene and ultraviolet-induced tumors during ultraviolet carcinogenesis in BALB/c mice. Cell. Immunol., *38*: 105–115, 1978.
22. Fortner, G. W. and Kripke, M. L. *In vitro* reactivity of splenic lymphocytes from normal and UV-irradiated mice against syngeneic UV-induced tumors. J. Immunol., *118*: 1483–1487, 1977.
23. Fisher, M. S. and Kripke, M. L. Systemic alteration induced in mice by ultraviolet light irradiation and its relationship to ultraviolet carcinogenesis. Proc. Natl. Acad. Sci. U.S., *74*: 1688–1692, 1977.
24. Spellman, C. W. and Daynes, R. A. Modification of immunological potential by ultraviolet radiation. II. Generation of suppressor cells in short-term UV-irradiated mice. Transplantation, *24*: 120–126, 1977.
25. Fisher, M. S. and Kripke, M. L. Further studies on the tumor-specific suppressor cells induced by ultraviolet radiation. J. Immunol., *121*: 1139–1144, 1978.
26. Spellman, C. W. and Daynes, R. A. Properties of ultraviolet light-induced suppressor lymphocytes within a syngeneic tumor system. Cell. Immunol., *36*: 383–387, 1978.
27. Fujimoto, S., Greene, M. I., and Sehon, A. H. Immunosuppressor T cells and their factors in tumor-bearing hosts. *In*; S. K. Singhal and N. R. Sinclair (eds.), Suppressor Cells in Immunity, pp. 136–147, University of Western Ontario Press, London, 1975.
28. Gorczynski, R. M. Immunity to murine sarcoma virus-induced tumors. II. Suppression of T cell-mediated immunity by cells from progressor animals. J. Immunol., *112*: 1815–1825, 1974.
29. Daynes, R. A., Schmitt, M. K., Roberts, L. K., and Spellman, C. W. Phenotypic and physical characteristics of the lymphoid cells involved in the immunity to syngeneic UV-induced tumors. J. Immunol., *122*: 2458–2464, 1979.
30. Greene, M. I., Sy, M. S., Kripke, M. L., and Benacerraf, B. Impairment of antigen-presenting cell function by ultraviolet radiation. Proc. Natl. Acad. Sci. U.S., *76*: 6591–6595, 1979.

GENETIC AND ENVIRONMENTAL FACTORS IN EXPERIMENTAL AND HUMAN CANCER,
H. V. GELBOIN ET AL. (EDS.), JAPAN SCI. SOC. PRESS, TOKYO, PP. 111–126, 1980

Immunodeficiency and Malignancy

J. H. KERSEY, A. H. FILIPOVICH, and B. D. SPECTOR

Departments of Laboratory Medicine and Pathology and Pediatrics, University of Minnesota, Minneapolis, Minn. 55455, U.S.A.

Abstract: In recent years, significant interest has centered on the possible role of the immune system in the prevention of malignancy through the process of recognition and destruction of newly formed malignant cells. This "surveillance" nature, while widely popularized, has generally not been well supported by available data as will be reviewed in this paper. In fact, some authors have suggested that the immune system is capable of "stimulation" of malignant cells under appropriate conditions.

The major conclusion to be discussed in the present manuscript is that individuals who are immunodeficient have a greatly increased risk of development of malignancy of the lymphoreticular system. While the precise pathogenetic mechanisms responsible for the increase is not clear, the observations raise certain questions that are currently undergoing experimental exploration.

The Study of Tumors in Immunodeficient Individuals

Reports of tumors arising in patients with genetic disorders associated with immunodeficiency appeared in increasing numbers in the late 1950s and early 1960s (*1–3*). During the succeeding decade attention was drawn to the significant number of *de novo* malignancies, particularly lymphomas occurring in organ allograft recipients (*4*, *5*), and in patients who had received chronic immunosuppressive treatment for nonmalignant disorders (*6*). Individuals who had been treated with immunosuppressive antimetabolites for primary malignancies ranging from ovarian cancer to Hodgkin's disease were noted to develop acute myeloid leukemia as a second malignancy (*7*).

Persistent interest in the link between immunodeficiency and human cancer is reflected in the ongoing activities of two special cancer case registries which have been established within the U.S.A. for the retrospective collection of data regarding tumors diagnosed in immunocompromised persons.

The Immunodeficiency-Cancer Registry (ICR) was established at the Uni-

TABLE 1. Summary of Immunodeficiency-Cancer Registry (ICR, May 1979) (N=303 Cases[a])

Immunodeficiency disease	Lymphomas				
	Histiocytic	Lymphocytic	Hodgkin's	Others	Subtotal
A-T (2.0–31) 9.0[b]	13	16	11	21	61
WAS (1.5–22) 6.0	12	3	4	22	41
CVI (2.0–69) 46.0	7	8	5	18	38
↓ IgA (3.0–62) 30.0	2	1	1	2	6
X-LA (1.1–20) 8.0	1	1	1	2	5
SCID (0.1–2.25) 0.9	4	2	1	4	11
↓ IgM (2.5–41) 11.0	3	1	1	1	6
Nl./Elv. Igs (7.0–10) 9.0	1	1	0	1	3
Others (1.5–71) 18.0	1	1	2	0	4
Totals	44	34	26	71	175 (58%)

Abbreviations: A-T=ataxia-telangiectasia; WAS=Wiskott-Aldrich syndrome; CVI=common variable SCID=severe combined immunodeficiency; ↓ IgM=selective IgM deficiency; Nl./Elv. Igs=immuno- X-linked hyperimmunoglobulin M (2 cases), hyperimmunoglobulin E syndrome (1 cases), episodic lympho- infancy (1 case); NOS=not otherwise specified. [a] An additional 11 malignancies were observed in seven

versity of Minnesota in Minneapolis in 1973 (*8*). This registry has collected 303 cases of malignancies (Table 1) which developed in individuals with 1 of 14 different categories of naturally occurring immunodeficiency (classified as Primary Specific Immunodeficiencies by the WHO Committee on Immunodeficiency) (*9*). One hundred seventy-six or 63% of such ICR cases with known age at diagnosis of malignancy occurred in children less than 15 years of age.

Data base of malignancies among patients who had received immunosuppressive or cytotoxic treatment has been operated by Dr. Israel Penn at the Veteran's Hospital in Denver, Colorado since 1968. This Denver Transplant Tumor Registry (DTTR) contains 733 tumors (Table 2) which occurred *de novo* in patients who had received allogeneic organ transplants for reasons other than prior malignancy (*10*).

Analysis of data collected on malignancy in immunodeficiency is, however, complicated by the heterogeneity of the immunodeficiency states and malignancy types represented in these series. It is clear that patients with naturally occurring immunodeficiencies (NOIDs) represent a variety of distinct underlying defects. The 14 NOIDs in the ICR differ from one another as to age of onset and severity of immunodeficiency symptoms, scope and degree of cellular abnormalities (which frequently affect several organ systems), prognosis, and malignancy patterns. Patients with NOID typically suffer frequent, prolonged, and repeated infections with bacterial and viral pathogens, the sources of chronic antigenic stimulation.

Leukemias				Epithelial tumors	Other tumors	Total
Lymph	Myel	NOS	Subtotal			
22	0	4	26	18	1	106
0	4	4	8	1	3	53
3	1	0	4	35	0	77
1	0	0	1	8	1	16
3	2	2	7	2	1	15
1	3	0	4	0	0	15
0	0	0	0	2	0	8
0	0	0	0	0	0	3
1	0	0	1	4	1	10
31	10	10	51 (17%)	70 (23%)	7 (2%)	303 (100%)

immunodeficiency; ↓ IgA=selective IgA deficiency; X-LA=X-linked (Bruton's) agammaglobulinemia; deficiency with normal or elevated immunoglobulins; others=immunodeficiency with thymona (4 cases), penia and lymphocytotoxin (1 case), DiGeorge syndrome (1 case), transient hypergammaglobulinemia of patients. [b] Age range and median time of diagnosis of malignancy.

TABLE 2. Summary from Denver Transplant Tumor Registry (DTTR, March 1979) (*N*=733 malignancies[a])

	Lymphomas					Leukemias	Epithelial tumors		Other tumors	Total
	Histiocytic	Lymphocytic	Hodgkin's	Other	Subtotal		Skin and lips	Other		
De novo tumors in allogeneic transplant recipients	93	10	3	44	150 (20.5%)	17 (2.3%)	277 (37.8%)	211 (28.8%)	78 (10.6%)	733 (100%)

Age at transplant: 5–70 years; mean: 40 years. Malignancy developed 1–158 months posttransplant; mean: 38 months. [a] Forty patients had more than one variety of malignancy. From Ref. *10*.

Malignant events generally become apparent early in life and without exposure to immunosuppressive drugs, although many affected patients have received one or more courses of immunostimulatory or immunoreconstituting treatment (*11–15*).

In human allograft recipients, an immunodeficient state is induced shortly before surgery and sustained indefinitely by immunosuppressive or cytotoxic drugs. The allogeneic graft, as well as opportunistic or reactivated infectious agents chronically stimulate the purposefully compromised immune system in these patients. Corticosteroids and azathiopurine may have carcinogenic effects apart from their immunosuppressive potential (*10*). Other factors that may compound the susceptibility of the transplant recipient to tumorigenesis include the nature of the predisposing disease (*16*) and chronic metabolic or hormonal abnormalities. The latency period to malignancy varies with tumor histology.

Little information is currently available which would permit comparison of variables of immunologic function among these patients in the several "susceptible" populations who do or do not eventually develop tumors.

Although there are serious limitations inherent in the analysis of data collected retrospectively, the descriptions of malignancy patterns can be used to generate hypotheses to be tested in future prospective clinical studies or in the laboratory.

Features of Tumors in Immunodeficient Individuals

1. Incidence

The cancer incidence for all NOIDs has been estimated at 4.0, 3.1, and 1.7 (*8, 17, 18*). In the ICR, age at diagnosis ranges from 1 month to 69 years with a male to female predominance of 2: 1. Twenty-five percent of the tumors reported occurred in syndromes felt to have an X-linked recessive inheritance. Thirteen family groupings of NOID have been collected, of which nine are concordant for histologic type and site of malignancy (*8*).

For three of these disorders (*i.e.*, ataxia-telangiectasia (A-T), Wiskott-Aldrich syndrome (WAS), selective IgA deficiency), similar cancer incidence rates were observed in registry-generated surveys (*8, 17*). The two syndromes with the highest incidence figures, WAS and A-T, represent disorders which include phenotypic abnormalities outside of the immune system. WAS had the highest cancer rate, 12% for each of the series, an 118-fold age-adjusted excess risk compared to population based figures (*19*). Patients with this disease experience thrombocytopenia, eczema, variable immune dysfunction with a variety of pathogens and autoimmune reactions, and survive to a mean age of 6.5 years. Forty-one out of 53 or 76.9% of tumors in WAS originated in cells of lymphoid origin.

The percentages of patients with A-T who develop cancer stand at 10.3 and 11.7, respectively (*8, 17*). This syndrome has an autosomal recessive inheritance presenting in early childhood with cerebellar ataxia and bulbar telangiectasia. Variable immunodeficiency of B and T cells is generally diagnosed beyond infancy and patients frequently survive into the second and third decades. Insulin-resistant diabetes and tumors may develop in later years (*20*). Eighty-three out of 106 or *78%* of such neoplasms were lymphoid.

Comparatively low incidence rates of 1.2 and 2.4% were found among the most common primary specific immunodeficiency, selective IgA deficiency (*8, 17*). Epithelial tumors comprise the majority in patients with selective IgA deficiency, and are twice as common when cases associated by diagnoses with IgA deficiency as part of their disorder (hypogammaglobulinemia, dysgammaglobulinemia, A-T, and common variable immunodeficiency) are compared with those cases in the ICR assumed to have had normal or elevated levels of circulating IgA.

Lymphoproliferative disorders predominate in severe combined immunodeficiency (SCID), an etiologically diverse disorder. Acute leukemia was the most common diagnosis associated with Bruton's X-linked agammaglobulinemia, presumed to be an intrinsic disorder of B cells. The largest group of epithelial tumors, especially gastric carcinoma, was reported in common variable immunodeficiency, a heterogenous disease category characterized by dysgammaglobulinemia and variable T cell dysfunction, with onset of infectious complications and autoimmune phenomena frequently occurring in the second to fifth decades of life.

Several years ago, we estimated that the cancer mortality rate in immunodeficient children was to be 100 times greater than for the general pediatric population (*21*). Refined cancer incidence and mortality rates for NOIDs await more accurate estimates of the populations at risk.

Six percent of renal transplant recipients have developed *de novo* malignancy, an incidence rate which has been reported to be 100 times greater than in the general population (*22*). Twenty-four percent of patients surviving 5 years after successful transplant have developed a malignant neoplasm (*23*). Tumors of the skin and lips make up 37.8% of the DTTR total (277/733), and represented more than one-half of all epithelial cancers recorded; 20.5% of *de novo* tumors were submitted as lymphomas (*10*). To date, all *de novo* lymphomas in which the genetic source of malignant cells could be identified were of recipient origin (*24*). Overall cancer incidence in recipients of cardiac allografts is 8%, and 54% of these are lymphomas (*16*).

2. *Characteristics of tumors*

a) *Malignant lymphomas and lymphoproliferative disorders*

General associations. The striking predominance of lymphomas in immunocompromised individuals clearly indicates a strong association between a defective immune system and a host that is susceptible to this class of tumors. There were 175 lymphomas among 303 patients in the ICR (Table 1) and 150 among 733 malignancies recorded in the DTTR (Table 2). These represent 58 and 20.5%, respectively, of reported tumor diagnoses. Age-specific cancer rates were available for two groups of immunocompromised patients: (1) the relative risk of nodal malignancies in WAS was computed at 350 times the expected rate (*25*), and (2) the risk of developing lymphoma increased 35-fold following renal transplantation (*5*). In both the ICR and DTTR, lymphomas showed a shorter latency period than did other tumors (*5, 26*), and were diagnosed more often in children and relatively young adults (*8, 16*). Among cardiac allograft recipients studied at Stanford University in California, the six lymphomas were reported exclusively in individuals with the same pretransplant disease, idiopathic cardiomyopathy (*16*). Malignant lymphomas have been associated with several autoimmune diseases (*6*), such as Sjogren's syndrome, systemic lupus erythematosus, autoimmune hemolytic anemia, which may represent inherited or acquired diseases of immunoregulation and possibly activation of endogenous viruses. The latent susceptibility to development of lymphoma may, in these latter diseases, be exacerbated by additional immunosuppressive treatment (*6*) or splenectomy.

Whereas exposure to benzene and other hydrocarbons has been perceived for a long time as a risk factor in carcinogenesis, it is only recently that a significant excess of deaths attributable to lymphomas following occupational benzene exposure has been reported (*27*). Increased relative risks were noted for reticulum cell sarcomas, lymphosarcomas, and Hodgkin's disease in persons 45 years or older. The authors speculate that chromosomal damage, lymphopenia, and other immunologic aberrations, which have been shown to follow chronic benzene exposure in some humans, may contribute to this observed excess (*27*).

It has also been shown that progressive cell-mediated immunodeficiency, such as that seen in late stage and disseminated non-Hodgkin's lymphoma as well as Stage III and IV Hodgkin's disease occurring in the general population, portends a worse prognosis (*28*). Data regarding the course of malignant lymphomas, response to therapy, and survival post diagnosis in patients with NOIDs are currently under analysis.

Site and histology. Involvement of the central nervous system by lymphoma is an unusual characteristic seen in immunocompromised hosts. Eighty-seven percent of the 54 such cases in the DTTR (42% of all lymphomas) originated in and remained confined to the central nervous system. Among 40 WAS patients with lymphoma in the ICR, 10 or 25% (mean age 6.6 years) developed a primary lymphoma of the brain; 80% carried a diagnosis compatible with the histiocytic lymphoma of Rappaport (*29*). Less than 2% of the lymphomas from patients in the general population arise in the central nervous system (*10*); the mean age of such patients in one large study was 52 years and only 15% of all primaries of the brain were of the "histiocytic" type (*30*). It is unclear why lymphomas in immunocompromised individuals should favor the central nervous system. Absence of a lymphatic drainage system, and decreased effectiveness of extent immunity within the central nervous system have been proposed to explain the seemingly "immunologically privileged" nature of this site (*31*). Furthermore, neurotropic viruses (*e.g.*, herpes group) may serve as antigenic stimuli at this location. A hemagglutinating strain of papovavirus was isolated from one such "reticulum cell sarcoma" of the brain as well as from the urine of a patient with WAS at the time of diagnosis (*32*).

The 350-fold excess of the diffuse histiocytic type (reticulum cell sarcoma) determines almost entirely the overall increased risk of lymphoma following renal transplantation (*5*). Application of the Lukes and Collins classification for the non-Hodgkin's lymphomas showed that some of the posttransplant diffuse histiocytic tumors, and similar tumors from patients with WAS, were classifiable as B immunoblastic sarcomas (*33*). This tumor type, presumed to involve transformed committed lymphocytes, is not only rare in large series of lymphomas, but has previously been reported only in adults (*34*, *35*). Recently, a monoclonal immunoblastic sarcoma of donor origin was reported as the first example of a lymphoma arising in a human with documented graft *versus* host disease following allogeneic bone marrow transplantation (*36*).

Twenty cases of Kaposi's sarcoma, accounting for 13% of posttransplant lymphomas, have been recorded in the DTTR (*37*). It has been postulated that this tumor of skin and visceral organs, histologically comprised of several cell types, represents the result of an immunologic reaction between normal and antigenically altered lymphocytes (*38*). A herpes-type virus has been isolated from such tumors (*39*). This tumor is endemic to malaria-infested regions of Africa, but in the United States and Europe it is most prevalent among patients receiving immunosuppressive thereapy (*40*). Alteration of immunosuppression in some of the 20 renal transplant patients resulted in regression of lesions in both benign and malignant forms (*37*). We are unaware of Kaposi's sarcoma occurring in patients with NOID diseases.

Hodgkin's disease, the most frequently occurring tumor of lymphoid tissue in the general population, may represent a primary neoplasm of the monocyte-macrophage series. Activation of the immune system to eradicate the tumor appears to be a critical prognostic variable (*41*). Twenty-five cases of Hodgkin's disease have been collected in the ICR with the greatest number of cases, 11 of these, associated with A-T. Pathologic material from eight of the 26 cases of Hodgkin's disease has recently been submitted to an independent morphologic evaluation. Results showed 50% to be of the rare lymphocytic depletion type (*42*). This type, extremely uncommon under 10 years of age, comprises a small minority of the Hodgkin's disease cases in the population at large (*43*), and is associated with poor prognosis due to lowered resistance to the disease and intercurrent infection (*41*). Three of four of this type were observed in patients with A-T who were 4, 8, and 10 years old at the time of diagnosis (42).

Tumors such as histiocytic lymphoma and lymphocytic-depletion type Hodgkin's disease are most frequently diagnosed in the older age group characterized by the most consistent evidence of "immunologic involution" (*44*).

Lymphoproliferative processes associated with Epstein-Barr virus (EBV). An interesting facet of the link between immunodeficiency and lymphoma is the number and types of such neoplasms arising in patients with the X-linked lymphoproliferative syndrome (XL-P), a condition that probably is the result of ineffective response to EBV. Dr. D. Purtilo of the University of Massachusetts at Worcester has studied the cases of 50 males in two kindreds thought to represent this inherited immunodeficiency. He reports that 40% of the patients with this diagnosis have developed malignant lymphomas including American Burkitt's, histiocytic lymphoma, and B immunoblastic sarcoma as well as plasmacytoma (*45*). Variable phenotypic expression within a given family, findings of subtle immune defects to several pathogens, as well as an increased incidence of cardiac and neurologic congenital anomalies in relatives (*46*) has led to speculation that this syndrome may involve several organ systems.

In other cases of immunoblastic proliferation and/or hypergammaglobulinemia associated with EBV infection, findings of associated humoral and cell-mediated immunodeficiency (*47*) or failure to generate immune interferon (*48*) in response to the organism have been implicated.

EBV has been recovered from patients with renal allografts, or from others receiving immunosuppressive treatment with two to three times the frequency of isolates from the normal population (*49*). A polyclonal "immunoblastic sarcoma" clinically simulating malignant lymphoma has been reported in a renal transplant recipient who had been exposed to infectious mononucleosis shortly prior to surgery

(*50*). The extent to which other examples of atypical infectious mononucleosis or other viral disorders, and aggressive "benign" polyclonal, or "premalignant" but fatal lymphoproliferative processes contribute to the overall number of lymphomas occurring in immunodeficient patients has not yet been determined.

Other lymphoproliferative processes. Another four of six posttransplant lymphomas examined at the University of Minnesota were found to be polyclonal by the determination of intracytoplasmic immunoglobulin by the immunoperoxidase technique (*33*). The remaining two were negative for immunoglobulins of any class. A fatal lymphoproliferative disorder and concomitant polyclonal gammopathy have developed in four patients with SCID following transfer factor (*13*) therapy or intraperitoneal cultured thymic epithelial grafts (*14*). Many of the lymphoid tumors reported to the ICR are submitted as "unclassifiable" malignancies. In addition, the behavior of some "lymphomas" in NOIDs may be unusual. For example, we recently observed a patient with common variable immunodeficiency whose "malignant" follicular center cell lymphoma (diagnosed incidentally three years prior to death) had spontaneously disappeared at the time of autopsy (Snovar, D. and Filipovich, A. H., unpublished observations).

b) *Leukemias*

The proportion of leukemias is noticeably greater than the 3.3% of the general population (*51*) for the following two groups: patients with naturally occurring immunodeficiency diseases (17%, 51/303, Table 1), and patients with various autoimmune diseases who had received some form of immunosuppressive therapy (18%, 15/83 patients) (*52*). In another review, 58 patients treated with alkylating agents (and in most cases corticosteroids) for nonneoplastic diseases who developed variants of acute myelogenous leukemia were reported (*53*). The rising number of such cases apparent in the 1970s are being attributed to the broader application of cytotoxic drugs. Overall, patients treated for primary tumors with a multiplicity of chemotherapeutic agents appear to carry a 50-fold increased long-term risk of developing acute nonlymphocytic leukemias. For patients who have received treatment for multiple myeloma, the risk rises to 100-fold (*7*).

Seven of 10 lymphocytic leukemias from patients with A-T which have been studied with immunologic markers have been shown to be of T lymphocyte origin; 3/7 cases are known to be associated with abnormalities of the 14 q chromosome. This small series of leukemias in one NOID evaluated by functional markers suggests that the proportion of T phenotypes is somewhat higher than that seen in other series of lymphoid malignancies, in presumably nonimmunodeficient patients (*54*, *55*).

Acute myelogenous leukemia was associated with WAS and SCID.

c) *Carcinomas*

Skin and lip carcinomas are the most frequently reported tumors of epithelial origin in transplant recipients (277/488 in the DTTR, Table 2). In one study where the comparison population was drawn from the same geographic area as the cases, the risk of skin cancer in renal transplant recipients was 7.1 times the expected rate, due primarily to squamous cell carcinoma (relative risk, 36.4) (*56*). Although skin cancers in general are inconsistently reported, and are not even reported by the U.S. Third National Cancer Survey, it is known that in the public at large basal cell carcinomas out number squamous cell carcinomas 2: 1. Multiple skin tumors were found more often in the DTTR than in the general population (43%, 22%, respectively) and occurred in much younger individuals (*57*). "Premalignant" hyperkeratoses associated with exposure to solar irradiation are a frequent finding in transplant recipients.

Carcinomas of the gastrointestinal tract in the ICR have affected only patients presumed to have IgA deficiency as part of the immunodeficiency disease. Of these, gastric carcinomas were the most common epithelial tumor reported to the ICR overall (45%, 26/55); 72% (18/26) of these malignancies arose in patients with common variable immunodeficiency at a median age of 54 years (range 15–67 years).

d) *Other features*

Except for the excess number of skin and lip cancers, cervical cancer, *in situ*, in the DTTR, is the only other common malignancy of the general population (*e.g.*, lung, colon, breast) that figures predominantly in either cancer case registry. Multiple primaries have not been reported in excess in patients reported to the ICR (*8*).

Hypotheses Relating Immunodeficiency and Malignancy

Theories regarding the etiologies of malignancies in the heterogenous populations of immunodeficient patients must take into account the various findings of these cancer registries. In general terms, immunodeficiency is associated with unusual sites and histologies of tumors, many of which can be viewed as malignancies of the immune system itself.

Two alternative hypotheses may explain the incidence of lymphoreticular malignancies in these patients. One hypothesis suggests that lymphoid cells from these patients are intrinsically abnormal due to an inherited defect of cellular physiology. Such an abnormal cell would be expected to both behave abnormally (and be "immunodeficient") and at the same time be at greater risk for malignant transformation. In this hypothesis, the malignancy and the immunodeficiency are viewed as "parallel" rather than "series" events. The alternative "series" hypothesis suggests that these malignancies arise when immunocytes proliferate due to failure of immunoregulation. Thus, selected diffuse histiocytic lymphomas which occur in posttransplant patients and some NOID diseases (especially those lymphomas identified as B-immunoblastic sarcomas) might represent a subtype of lymphoma that reflects such underlying defects of immunoregulation. Immunoregulation is a normal function of the immune system which operates in conjunction with that intact system's critical role of protecting the rest of the organism from extracellular infectious agents. It is generally recognized that the immunoregulatory function rests primarily among T lymphocytes. Two distinct populations of "helper" and "suppressor" T cells comodulate the expansion of subpopulations of B and T cells which are stimulated, and in many cases, overstimulated by a variety of antigens. Resident lymphocytes in the immunocompromised host undergo an initial response to pathogens and/or an allograft. This process may result in polyclonal proliferation which is, in a sense, promoted by a lack of normal suppressive signals. Such a state of polyclonal proliferation may progress to a polyclonal tumor mass or give rise to a monoclonal malignancy if a new irreversible cytogenetic defect occurs.

This sequence of events may be analogous to that proposed for the evolution of frankly malignant immunoblastic lymphomas in α-chain disease (*58*). In this primary disorder of immunoglobulin synthesis, there is an early stage characterized by proliferation of plasma cells, typically infiltrating the small intestine, which produces a population of abnormal α-heavy chains. It is sometimes possible to reverse this phase with antibiotic therapy, which may clear the intestinal infection which has stimulated replication of α-chain secreting plasma cells. In the second stage, there is progressive deterioration of the patient's general condition associated with the development of a "malignant" immunoblastic tumor which is thought to arise from the same B cell clone responsible for the plasmacytosis. Dissemination of this tumor outside of the abdomen is rare, and occasionally the malignant cells no longer appear to have the capacity to produce α-chain disease protein. There is some evidence that certain cases of α-chain disease represent progression from a benign monoclonal state to a true malignancy (*58*).

Dysfunction of immunoregulation has also been postulated and to some extent demonstrated in human graft *versus* host disease (*59*, *60*) and systemic lupus erythematosus (*61*). Except for a documented absence of concanavalin A suppressor function in three patients with SCID who subsequently developed lymphoproliferative disorders, no comprehensive prospective studies of the integration of enhancing and suppressing lymphocyte populations have yet been carried out on patients with the various NOIDs.

Another example of an immunoregulatory defect, that of suppressor cell dysfunction prior to transplantation and chemical immunosuppression, has been observed in patients with idiopathic cardiomyopathy, the same disease associated exclusively with postcardiac transplant lymphomas (*16*, *62*). Immunosuppressive therapy of a host chronically stimulated by his allograft could affect immunoregulation by depressing overall immune function.

A related issue in the discussion of immunologic defense against cancer is its role in limiting proliferation of endogenous and exogenous oncogenic viruses. It has been proposed that one of the evolutionary imperatives in the refinement of immunoregulation was the advantage to the host of preventing expression of virogenes (RNA tumor viruses) carried in the lymphocytes. Experimentally, activation of these oncogenic viruses occurs with activation and proliferation of lymphocytes, particularly in conjunction with murine graft *versus* host disease (*49*). A similar process has been suggested to account for the greatly increased risk of lymphoma in allograft recipients (host stimulated by graft) as compared to patients receiving similar immunosuppressive therapy for other indications (*49*).

On the other hand, failure to eradicate cells infected with exogenous or reactivated DNA viruses has been documented in association with several aggressive lymphoproliferative disorders in transplant patients (EBV) (*50*) and XL-P (EBV)

(*45*), and in two patients with A-T (influenza virus) who were not harboring a malignancy (*63*). A systematic study of the abilities of immunosuppressed and immunodeficient patients to mount cytotoxic T cell responses to virus-bearing cells may prove useful in ultimately evaluating the role of this specific defect in the susceptibility to malignancy.

The once dominant idea of immunosurveillance, the process of eliminating small foci of malignantly transformed cells which express "foreign" surface antigens, stands to make at the least a brief theoretical recovery with the current interest in natural killer cells (NK). These "null" cells, possibly a variety of pre-T cells, have been isolated from spleens and peripheral blood of humans. They have the capacity to lyse a wide variety of cultured and malignant cells without presensitization (*64*). They show preferential activity against human tumor and virus-infected cells (*65*), and their efficiency is increased following viral infectious and interferon production. While a role for the NK cell *in vivo* remains to be defined, lack of *in vitro* activity has been demonstrated in one patient with SCID (*66*), one patient with systemic lupus erythematosus (*65*), as compared to normal cytotoxic activity in patients with common variable immunodeficiency, Bruton's agammaglobulinemia (*66*), and in persons undergoing intensive immunosuppressive therapy (*65*).

Amplification of a genetically restricted, cytotoxic response against presumed tumor associated antigens (acute myelogenous leukemia) has been accomplished *in vitro*, using cells obtained from patients during remission (*67*). This manipulation holds some promise for future immunotherapeutic intervention, but the potency of this T cell-mediated function *in vivo* during primary escalation of the leukemia must still be questioned.

Consideration should be given to the possibility that immunodeficient states and their associated malignancies are both manifestations of the same predisposing event. For the purpose of exposition, we will use the example of A-T. With this disease, ineffective repair of damaged DNA has been demonstrated *in vitro*. Immunodeficiency is rarely apparent early in life and is an inconsistent clinical feature. Chromosomal defects, including a "characteristic" break at the 14 q12, are present in many cells and have been detected in subpopulations of nonmalignant T cells. The extent to which these dictate "acquired" or progressive T cell dysfunction in some patients has not yet been shown. Fourteen q translocations have been associated with a variety of lymphoid malignancies, including the three T cell leukemias in A-T mentioned previously, and in Burkitt's lymphoma. It is of interest that three out of five Burkitt lymphomas reported to the ICR occurred in patients with A-T.

Bloom's syndrome is also marked by frequent chromosomal rearrangements and has findings of immunodeficiency as well as a marked predisposition to lymphoid malignancy and gastric carcinoma (*68*), a pattern similar to that seen in A-T.

It is also possible the metabolic errors such as adenosine deaminase deficiency associated with SCID, may independently result in both immunodeficiency and malignancy.

It should be kept in mind that within a given immunodeficiency syndrome, excess cancer risk for differing tumors may be due to varying proportions of several contributing factors.

Future Studies

Several directions can be pursued in the study of the role of immunodeficiency in human malignancy. The first, a careful retrospective examination of clinical and histologic case material is currently being conducted by the ICR. As homogenous groupings and potential risk factors are identified among the heterogenous immunodeficiency populations, case/control studies may present feasible approaches to the examination of such variables as infection, *in vitro* correlates of immunoregulatory function and premalignant conditions and lesions. Immunoreconstitution, including bone marrow transplantation, a method of permanently correcting a variety of inherited defects resulting in immunodeficiency, should be investigated as an approach to the "prevention" of malignancies peculiar to immunodeficiency states. Of interest in this regard are the current observations indicating that while patients with SCID develop lymphoreticular malignancies when they are untreated or receive a variety of immunostimulating agents, lymphomas apparently do not develop if successful bone marrow transplantation is accomplished.

ACKNOWLEDGMENT
This work was supported in part by a contract from the United States Public Health Service (CP-43384).

REFERENCES

1. Good, R., Kelly, W., and Gabrielson, A. E. Studies of the immunologic deficiency diseases agammaglobulinemia, Hodgkin's disease and sarcoidosis. *In*; P. Babor and P. Miescher (eds.), IInd International Symposium on Immunopathology Brook Lodge (Michigan, U.S.A.), pp. 353–384, Benno Schwabe and Co., Basel, 1962.
2. Page, A. R., Hansen, A. E., and Good, R. A. Occurrence of leukemia and lymphoma in patients with agammaglobulinemia. Blood, *21*: 197–206, 1963.
3. Boder, E. and Sedgwick, R. Ataxia-telangiectasia: A review of 101 cases. *In*; G. Walsh (ed.), Cerebellum Posture and Cerebral Palsy, vol. 8, pp. 110–118, Little Club Clinics Developmental Medicine, National Spastics Society, Medical Education and Information Unit, London, 1963.
4. Penn, I. (ed.). Malignant Tumors in Organ Transplant Recipients, Springer-Verlag, New York, 1970.
5. Hoover, R. and Fraumeni, J. F., Jr. Risk of cancer in renal transplant recipients. Lancet, *ii*: 55–57, 1973.
6. Louie, S. and Schwartz, R. S. Immunodeficiency and the pathogenesis of lymphoma and leukemia. Sem. Hematol., *15*: 117–138, 1978.
7. Casciato, D. A. and Scott, J. L. Actue leukemia following prolonged cytotoxic agent therapy. Medicine, *58*: 32–47, 1979.
8. Spector, B. D., Perry G. S., III, and Kersey, J. H. Genetically determnied immunodeficiency diseases (GDID) and malignancy: Report from the Immunodeficiency-Cancer Registery. Clin. Immunol. Immunopathol., *11*: 12–29, 1978.

9. WHO Report: Immunodeficiency. Report of a WHO Scientific Group. Clin. Immunol. Immunopathol., *13*: 296–359, 1979.
10. Penn, I. Tumor incidence in human allograft recipients. Transplant. Proc., *11*: 1047–1051, 1979.
11. Spitler, L. E. Transfer factor therapy in the Wiskott-Aldrich syndrome. Results of long-term followup in 32 patients. Am. J. Med., *67*: 59–66, 1979.
12. Stoop, J. W., Eijsvoogel, V. P., Zegers, B.J.M., Blok-Schut, B., van Bekkum, D. W., and Ballieux, R. E. Selective severe cellular immunodeficiency. Effect of thymus transplantation and transfer factor administration. Clin. Immunol. Immunopathol., *6*: 289–298, 1976.
13. Gelfand, E. W., Baumal, R., Huber, J., Crookston, M. C., and Schumak, K. H. Polyclonal gammopathy and lymphoproliferation following transfer factor in severe combined immunodeficiency disease. N. Engl. J. Med., *289*: 1385–1389, 1973.
14. Borzy, M. S., Hong, R., Horowitz, S. D., Gilbert, E., Kaufman, D., DeMendonca, W., Oxelius, V.-A., Dictor, M., and Pachman, L. Fatal lymphoma after transplantation of cultured thymus in children with combined immunodeficiency disease. N. Engl. J. Med., *301*: 565–568, 1979.
15. Dutau, G., Corberand, J., Abbal, M., Blanc, M., Claverie, P., and Rochiccioli, P. Ataxie-telangiectasie avec deficit immunitaire mixte et formes apparentees. J. Genet. Hum., *23*: 281–299, 1975.
16. Anderson, J. L., Bieber, C. P., Fowles, R. E., and Stinson, E. B. Idiopathic cardiomyopathy, age, and suppressor-cell dysfunction as risk determinants of lymphoma after cardiac transplantation. Lancet, *ii*: 1174–1177, 1978.
17. Aiuti, F., Giunchi, G., Bardare, M., Baroni, C., Bonomo, L., Dammacco, F., Tursi, A., Borrone, C., Burgio, G. R., Ugazio, R., Businco, L., Rezza, E., Calvani, M., Bricarelli, F. D., Moscatelli, P., Fontana, L., Franceschi, C., Masi, P., Paolucci, G., Cavagni, G., Imperato, C., Carapella, E., Massimo, L., Nicola, P., Panizon, G., Agosti, F., Ricci, G., Segni, P., Vaccaro, G., Vierucci, A., and Zanussi, C. Le immunodeficienze primitive in Italia. Folia Allergol. Immunol. Clin., *25*: 7–15, 1978.
18. Hayakawa, H., Iizuba, N., Yata, J., Yamada, K., and Kobayashi, N. Analysis of registered cases of immunodeficiency in Japan. *In*; Japan Medical Research Foundation (ed.), Immunodeficiency, Its Nature and Etiological Significance in Human Disease, vol. 5, pp. 271–281, University of Tokyo Press, Tokyo, 1978.
19. Perry G. S., III, Spector, B. D., Schuman, L. M., Mandel, J. S., Anderson, V. E., McHugh, R. B., Hanson, M. R., Fahlstrom, S. M., Krivit, W., and Kersey, J. H. The Wiskott-Aldrich syndrome in the United States and Canada (1892–1979). J. Peds., in press, May, 1980.
20. Levin, S. and Perlov, S. Ataxia-telangiectasia in Israel with observations on its relationship to malignant disease. Israel J. Med. Sci., *7*: 1535–1541, 1971.
21. Kersey, J. H., Spector, B. D., and Good, R. A. Cancer in children with primary immunodeficiency diseases. J. Peds., *84*: 263–264, 1974.
22. Penn, I. Immunosuppression and malignant disease. *In*; J. J. Twomey and R. A. Good (eds.), Comprehensive Immunology. The Immunopathology of Lymphoreticular Neoplasms, No. 4, pp. 223–237, Plenum Medical Book Co., New York, 1978.
23. Sheil, A. G. Cancer in renal allograft recipients in Australia and New Zealand. Transplant. Proc., *9*: 1133–1136, 1977.
24. Penn, I. Tumors in allograft recipients. N. Engl. J. Med., *301*: 385, 1979.

25. Meruelo, D. and McDevitt, H. O. Recent studies on the role of the immune response in resistance to virus-induced leukemias and lymphomas. Sem. Hematol., *15*: 399–419, 1978.
26. Spector, B. D., Perry, G. S. III, Good, R. A., and Kersey, J. H. Immunodeficiency diseases and malignancy. *In*; J. J. Twomey and R. A. Good (eds.), Comprehensive Immunology. The Immunopathology of Lymphoreticular Neoplasms, No. 4, pp. 203–222, Plenum Medical Book Co., New York, 1978.
27. Vianna, N. J. and Polan, A. Lymphomas and occupational benzene exposure. Lancet *i*: 1394–1395, 1979.
28. Gupta, S. and Good, R. A. Immunodeficiencies associated with chronic lymphocytic leukemia and non-Hodgkin's lymphomas. *In*; J. J. Twomey and R. A. Good (eds.), Comprehensive Immunology. The Immunopathology of Lymphoreticular Neoplasms, No. 4, pp. 565–583, Plenum Medical Book Co., New York, 1978.
29. Rappaport, H. Tumors of the hematopoietic system. Atlas of Tumor Pathology: Section II, Fasc 8, Armed Forces Institute of Pathology, Washington, D. C., 1966.
30. Henry, J. M., Heffner, R. R., Dillard, S. H., Earle, K. M., and Davis, R. L. Primary malignant lymphomas of the central nervous system. Cancer, *34*: 1293–1302, 1974.
31. Barker, C. F. and Billingham, R. E. Immunologically privileged sites. *In*; H. G. Kunkel and F. J. Dixon (eds.), Advances in Immunology, vol. 25, pp. 1–54, Academic Press, New York, 1977.
32. Takemoto, K. K., Rabson, A. S., Mullarkey, M. F., Blaese, R. M., Garon, C. F., and Nelson, D. Isolation of papovavirus from brain tumor and urine of a patient with Wiskott-Aldrich syndrome. J. Natl. Cancer Inst., *53*: 1205–1207, 1974.
33. Frizzera, G. Malignant lymphomas in transplant patients. Presented at the Symposium on Iatrogenic Cancer, Joint ASCP-CAP Meeting, New Orleans, March 1979.
34. Lichtenstein, A., Levine, A. M., Lukes, R. J., Cramer, A. D., Taylor, C. R., Lincoln, T. L., and Feinstein, D. I. Immunoblastic sarcoma. A clinical description. Cancer, *43*: 343–352, 1979.
35. Lukes, R. G. and Collins, R. D. Lukes-Collins classification and its significance. Cancer Treat. Rep., *61*: 971–979, 1977.
36. Gossett, T. C., Gale, R. P., Fleischman, H., Austin, G. E., Sparkes, R. S., and Taylor, C. R. Immunoblastic sarcoma in donor cells after bone-marrow transplantation. N. Engl. J. Med., *300*: 904–907, 1979.
37. Penn, I. Kaposi's sarcoma in organ transplant recipients. Report of 20 cases. Transplantation, *27*: 8–11, 1979.
38. Warner, T.F.C.S. and O'Loughlin, S. Kaposi's sarcoma: A by-product of tumor rejection. Lancet, *ii*: 687–688, 1975.
39. Giraldo, G., Beth, E., Haguenau, F., Noury, G., Puissant, A., and Huraux, J.-M. Sarcome de Kaposi. Études des cultures de tissu. Études virologiques et immunologiques. Ann. de Derm., *100*: 283–284, 1973.
40. Editorial: Disentangling Kaposi's sarcoma. Br. Med. J., 1044, October 14, 1978.
41. Twomey, J. J., Good, R. A., and Case, D. C., Jr. Immunological changes with Hodgkin's disease. *In*; J. J. Twomey and R. A. Good (eds.), Comprehensive Immunology. The Immunopathology of Lymphoreticular Neoplasms, No. 4, pp. 585–608, Plenum Medical Book Co., New York, 1978.
42. Frizzera, G., Rosai, J., Dehner, L. P., Spector, B. D., and Kersey, J. H. Malignancies in primary immunodeficiency (ID) diseases, with special emphasis on lymphoreticular neoplasia. Fed. Proc., abstr. No. 5529, 1979.

43. Rosenberg, S. A. and Kaplan, H. S. Hodgkin's disease and other malignant lymphomas. Calif. Med., *113*: 23–28, 1970.
44. Yunis, E. J., Fernandes, G., and Good, R. A. Aging and involution of the immunologic apparatus. *In*; J. J. Twomey and R. A. Good (eds.), Comprehensive Immunology. The Immunopathology of Lymphoreticular Neoplasms, No. 4, pp. 53–80, Plenum Medical Book Co., New York, 1978.
45. Purtilo, D. T. Opportunistic non-Hodgkin's lymphomas in X-linked recessive immunodeficiency and lymphoproliferative syndromes. Sem. Oncol., *4*: 335–343, 1977.
46. Purtilo, D. T., Sullivan, J. L., and Paquin, L. A. Biomarkers in immunodeficiency syndromes predisposing to cancer. *In*; H. T. Linch and H. Guirgis (eds.), Biological Markers in Cancer, in press, 1980.
47. Britton, S., Andersson-Anvret, M., Gergely, P., Henle, W., Jondal, M., Klein, G., Sandstedt, B., and Svedmyr, E. Epstein-Barr-virus immunity and tissue distribution in a fatal case of infectious mononucleosis. N. Engl. J. Med., *298*: 89–91, 1978.
48. Virelizier, J. L., Lenoir, G., and Griscelli, C. Persistent Epstein-Barr virus infection in a child with hypergammaglobulinemia and immunoblastic proliferation associated with a selective defect in immune interferon secretion. Lancet, *ii*: 231–234, 1972.
49. Schwartz, R. S. Immunoregulation, oncogenic viruses, and malignant lymphomas. Lancet, *i*: 1266–1269, 1972.
50. Houlihan, K. and Ascher, N. Unpublished observations, University of Minnesota, Minneapolis, Minn.
51. Third National Cancer Survey: Incidence Data. National Cancer Institute Monograph 41, DHEW Publ. No. (NIH) 75–787, Bethesda, Md., 1975.
52. Penn, I. Malignancies associated with immunosuppressive or cytotoxic therapy. Surgery, *83*: 492–502, 1978.
53. Grunwald, H. W. and Rosner, F. Acute leukemia and immunosuppressive drug use. A review of patients undergoing immunosuppressive therapy for nonneoplastic diseases. Arch. Intern. Med., *139*: 461–466, 1979.
54. Spector, B. D., Perry, G. S. III, Gajl-Peczalska, K. J., Coccia, P., Nesbit, M. E., and Kersey, J. H. Malignancy in children with and without genetically determined immunodeficiencies. *In*; R. L. Summitt and D. Bergsma (eds.), Birth Defects: Original Article Series, vol. XIV, No. 6A, pp. 85–89, Alan R. Liss, Inc., New York, 1978.
55. Bloomfield, C. D., Gajl-Peczalska, K. J., Frizzera, G., Kersey, J. H., and Goldman, A. I. Clinical utility of lymphocyte surface markers combined with the Lukes-Collins histologic classification in adult lymphoma. N. Engl. J. Med., *301*: 512–518, 1979.
56. Hoxtell, E. O., Mandel, J. S., Murray, S. S., Schuman, L. M., and Goltz, R. W. Incidence of skin carcinoma after renal transplantation. Arch. Dermatol., *113*: 436–438, 1977.
57. Penn, I. Tumors arising in organ transplant recipients. *In*; G. Klein and S. Weinhouse (eds.), Advances in Cancer Research, vol. 28, pp. 31–61, Academic Press, New York, 1978.
58. Seligmann, M. and Rambaud, J. C. Alpha-chain disease: A possible model for the pathogenesis of human lymphomas. *In*; J. J. Twomey and R. A. Good (eds.), Comprehensive Immunology. The Immunopathology of Lymphoreticular Neoplasms, No. 4, pp. 425–447, Plenum Medical Book Co., New York, 1978.
59. Reinherz, E. L., Parkman, R., Rappaport, J., Rosen, F. S., and Schlossman, S. F. Aberrations of suppressor T cells in human graft-*versus*-host disease. N. Engl. J. Med., *300*: 1061–1068, 1979.

60. Graze, R. R. and Gale, R. P. Chronic GVHD: A syndrome of disordered immunity. Am. J. Med., *66*: 611–620, 1979.
61. Horowitz, S., Borcherding, W., Moorthy, A., Chesney, R., Schulte-Wissermann, H., and Hong, R. Induction of suppressor T cells in systemic lupus erythematosus by thymosin and cultured thymic epithelium. Science, *197*: 999–1001, 1977.
62. Fowles, R. E., Bieber, C. P., and Stinson, E. B. Defective *in vitro* suppressor cell function in idiopathic congestive cardiomyopathy. Circulation, *59*: 483–491, 1979.
63. Nelson, D. L., Biddison, W. E., and Shaw, S. Analysis of influenza virus specific cytotoxic T-cell responses in immunodeficiency patients. Ped. Res., *13*: No. 4: abstr. No. 758, 1979.
64. Herbermann, R. B. and Holden, H. T. Natural cell-mediated immunity. *In*; G. Klein and S. Weinhouse (eds.), Advances in Cancer Research, Vol. 27, pp. 305–377, Academic Press, New York, 1978.
65. Santoli, D. and Koprowski, H. Mechanisms of activation of human natural killer cells against tumor and virus-infected cells. Immunol. Rev., *44*: 125–163, 1979.
66. Koren, H. S., Amos, D. B., and Buckley, R. H. Natural killing in immunodeficient patients. J. Immunol., *120*: 796–799, 1978.
67. Zarling, J. M. and Bach, F. H. Continuous culture of T cells cytotoxic for autologous human leukaemia cells. Nature, *280*: 685–687, 1979.
68. German, J. The association of chromosome instability defective DNA repair, and cancer in some rare human genetic diseases. *In*; S. Arnendares and R. Lisker (eds.), Human Genetics, pp. 64–68, Excerpta Medica, Amsterdam, 1977.

GENETIC AND ENVIRONMENTAL FACTORS IN EXPERIMENTAL AND HUMAN CANCER,
H. V. GELBOIN ET AL. (EDS.) JAPAN SCI. SOC. PRESS, TOKYO, PP. 127–137, 1980

Immunological Features of Patients at High Risk for Leukemogenesis

Takehiko SASAZUKI,*[1] Yasuharu NISHIMURA,*[1] Akira TONOMURA,*[2] and Takehiko KURITA*[3]

Department of Human Genetics[1] and Department of Cytogenetics,*[2] Medical Research Institute, Tokyo Medical and Dental University, Tokyo, Japan, and Department of Pediatrics, Kohnodai Hospital, Ichikawa, Japan*[3]*

Abstract: In order to have insight into the immunogenetic factors for the leukemogenesis in the patients with Down's syndrome who have high risk for leukemia as well as for severe recurrent infectious diseases, immunological functions of 162 patients with Down's syndrome were investigated. Natural killer cell activity against human T cell leukemia line was rather increased. Antigen specific immune response measured by the T cell proliferation *in vitro* against tetanus toxoid and streptococcal cell wall antigen was not abberent at all. This observation may indicate that antigen recognistion by T cells, antigen presentation by macrophages and the T cell macrophage interaction are not affected by the trisomic condition of chromosome 21. Responsiveness to phytohemagglutinin (PHA), on the other hand, significantly diminished in the patient group. It is difficult to explain the fact that monoclonal or oligoclonal T cell response to the specific antigens was not affected, whereas polyclonal T cell response to PHA was significantly affected in the patient group.

Furthermore, T cell responsiveness to alloantigens in mixed lymphocyte culture (MLC) reaction of the patient group was significantly enhanced. Sine the cytotoxic autoantibody present in the young patients with Down's syndrome had a strong effect on the enhancement of the T cell responsiveness in MLC reaction, but had no effect on the antigen specific T cell response *in vitro*, this cytotoxic autoantibodies might specifically kill the suppressor T cells in MLC reaction in the patients. Subsequently this oligoclonal lymphocyte activation observed in the patients might attribute to the high risk of the patients with Down's syndrome to leukemogenesis.

The profiles of T cells from the patients with Down's syndrome analyzed by fluorescence activated cell sorter using two anti T cell antibodies revealed great abnormalities. This might represent the functional abnormalities found in the patient group observed here.

Patients with Down's syndrome, 21 trisomy, are highly susceptible to leukemia as well as to recurrent severe infectious diseases. Immunological abnormalities to explain this predisposition to leukemia and infectious diseases have been investigated in the patient group, however, data thus far have been controversial (*1–3*).

Immunological surveillance, on the other hand, has been reported to have a crucial role in resistance or susceptibility to leukemia or carcinoma in experimental animals (*4*, 5).

Antigen recognition of T (thymus-derived) cells and macrophages, and their cell-cell interactions are known to be the crucial step that leads B cells to produce antibodies as well as to generate killer T cells, both of which play important roles in the immune surveillance system (*4*). This antigen recognition by T cells and macrophages and their cell-cell interaction are under the control of antigen-specific immune response genes (Ir-genes) in mice, which are linked to the murine major histocompatibility complex, H-2 complex (*6*). The T cell-deficient mouse, the nude mouse, did not show any genetic predisposition to leukemogenesis or carcinogenesis, though the nude mouse is expected to lack the T cell-dependent immune surveillance system (*4*). This host resistance to leukemogenesis and carcinogenesis in the nude mouse is partially explained by the existence of natural killer cells which eliminate malignant cells, and which do not belong to the T cell population (*7*, *8*).

In this paper we report the characteristic features of the immune system in patients with Down's syndrome, who have a genetic predisposition to leukemia as well as infectious diseases.

MATERIALS AND METHODS

1. Patients with Down's syndrome

One hundred and sixty-two patients with Down's syndrome (98 males and 64 females) were the subjects of this study. Their age and sex distribution is shown in Fig. 1. Out of 162 patients, 153 were shown to be standard 21 trisomy, 6 were D/G translocation and 3 were mosaicism.

2. Lymphocyte preparation

Peripheral lymphocytes were separated on a Ficoll-Conray density gradient (s.g.=1.077), washed with RPMI-1640 medium, and resuspended in RPMI-1640 with 2 mM L-glutamine, 100 U/ml penicillin, 100 μg/ml streptomycin, and 10% heat-inactivated pooled human male sera.

3. Response of the peripheral lymphocytes to specific antigens or PHA in vitro

Immune response of patients with Down's syndrome and of the age-matched healthy controls were measured by antigen-specific T cell proliferation *in vitro*. Peripheral lymphocytes (2×10^5) were cultured with 1 μg of tetanus toxoid, or 1 μg of streptococcal cell wall antigen in 0.2 ml of RPMI-1640 with the supplements in a flat-bottomed microtiter plate (Nunc) at 37°C at 5% CO_2 in humidified air for 7 days. One μCi of tritiated thymidine was added to each well 16 hr before harvest. Cells were harvested on glass microfibre paper (Whatman GF/C). Incor-

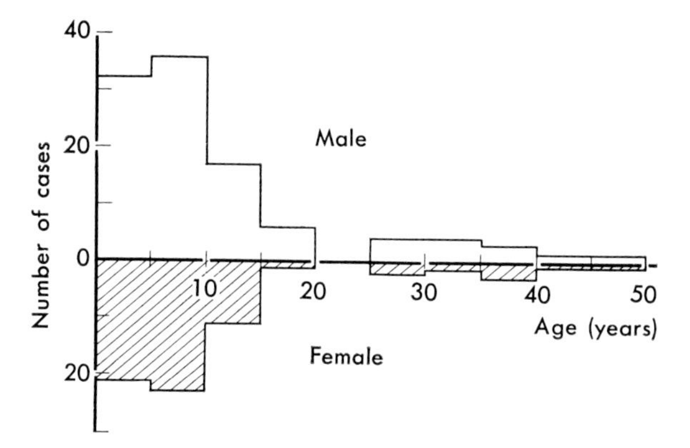

FIG. 1. Age and sex distribution of 162 patients with Down's syndrome including 153 standard trisomy; 6 translocation and 3 mosaicism.

poration of tritiated thymidine into DNA was measured in a liquid scintillation counter.

Responsiveness of the peripheral lymphocytes to phytohemagglutinin (PHA) was determined by the same method as mentioned above. Cells were harvested on day 4 instead of day 7.

4. Mixed lymphocyte culture reaction

One-way mixed lymphocyte culture (MLC) reaction was performed by the method previously described (9) with minor modifications. In brief, 5×10^4 lymphocytes from both patients and controls were used as responding cells and the same number of mitomycin-treated lymphocytes (stimulating cells) from three human leucocyte antigen system A (HLA) nonidentical individuals were cultured with each responder in 0.2 ml of RPMI-1640 with the supplements for 6 days followed by further incubation with 1 μCi of ^{3}H-thymidine for 16 hr. MLC reaction was determined by the incorporation of ^{3}H-thymidine measured in a liquid scintillation counter.

5. Natural killer cell activity

Natural killer cell activity of the patients with Down's syndrome was measured by the chromium-release assay using MOLT 3 cells (human T cell leukemia line) as a target. The target cells were prepared by incubation of 5×10^6 cells in 0.5 ml medium with 50 μCi of $Na_2\,^{51}CrO_4$ for 45 min at 37°C in a water bath. After three washes they were resuspended at 2×10^5 cells/ml and were distributed in 0.1 ml aliquots to U-bottomed microtiter plates (Nunc). Effector cell suspension (5×10^5/0.1 ml) of peripheral blood lymphocytes was added to each well. After a 6 hr incubation at 37°C in 5% CO_2, 0.1 ml of supernatant was collected by centrifugation and counted in an auto-gamma counter.

Spontaneous ^{51}Cr release was determined by the incubation of target cells alone. Maximum ^{51}Cr release was determined by three freezings and thawings of the target cells. The percent lysis was determined by the following formula:

$$\frac{\text{Experimental } ^{51}\text{Cr release} - \text{Spontaneous } ^{51}\text{Cr release}}{\text{Maximum } ^{51}\text{Cr release} - \text{Spontaneous } ^{51}\text{Cr release}} \times 100$$

6. Cytotoxicity test

Presence of cytotoxic antibody in the patients with Down's syndrome was tested by the microcytotoxicity test using peripheral lymphocytes from 3 HLA non-identical healthy individuals. Peripheral lymphocytes (1×10^5) were incubated with 50 μl of patient's sera at 15°C for 30 min, followed by a further 3 hr incubation with 50 μl of rabbit complement. Cytotoxicity was determined by the trypan blue staining method. Killing of lymphocytes at more than 20% was determined to be positive cytotoxicity.

7. Profile of the peripheral T lymphocytes

Nonadherent cells of peripheral lymphocytes to goat anti-human Ig-coated plastic Petri dishes were used as T lymphocytes. Profile of peripheral T lymphocytes was investigated by a fluorescence-activated cell sorter (FACS-II) using the rabbit antisera against brain-associated T cell antigen (BAT antigen) or natural thymocytotoxic autoantibody (NTA) from patients with systemic lupus erythematosus, followed by staining with fluorescence conjugated goat anti-rabbit Ig or goat anti-human Ig antibody.

RESULTS

1. Response of the patients with Down's syndrome to soluble antigens or PHA

Immune responsiveness of the patients with Down's syndrome to the purified tetanus toxoid measured by antigen-specific T cell proliferation *in vitro* is shown in Fig. 2. The immune responsiveness to tetanus toxoid was vigorous both in the patient and age-matched normal control groups. No anomalous immune response

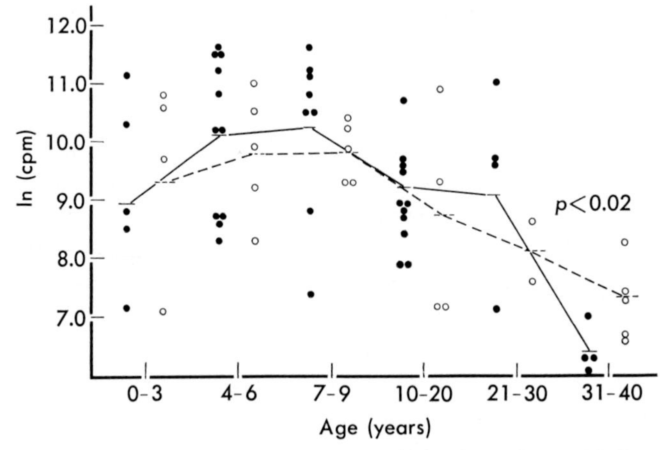

FIG. 2. Immune response to tetanus toxoid in the patients with Down's syndrome. 2×10^5 cells/well; 1μg antigen/well. ● Down's syndrome; ○ control. Intensity of the immune response was expressed by natural log count per min (ln pm).

was observed in the patient group when compared to the age-matched control group except in the patients between 31 and 40 years who showed a lower response to tetanus toxoid than the age-matched control group ($P<0.02$).

Immune response of the patients with Down's syndrome to streptococcal cell wall antigen measured by the antigen-specific T cell proliferation *in vitro* was higher in the young patients (younger than 20 years) and slightly lower in the adult patients (older than 21 years) than the age-matched normal controls (Fig. 3). These differences in immune response to streptococcal cell wall antigen between the patients and the age-matched controls are not statistically significant except that in the patients 7 to 9 years old from the immune response was significantly greater than that of the age-matched controls ($P<0.01$).

Responsiveness of the peripheral lymphocytes from the patients with Down's syndrome to PHA was slightly, but significantly, lower than that of the age-matched normal control groups (Fig. 4). There was no difference in responsiveness to PHA between the young and the aged in patients with Down's syndrome.

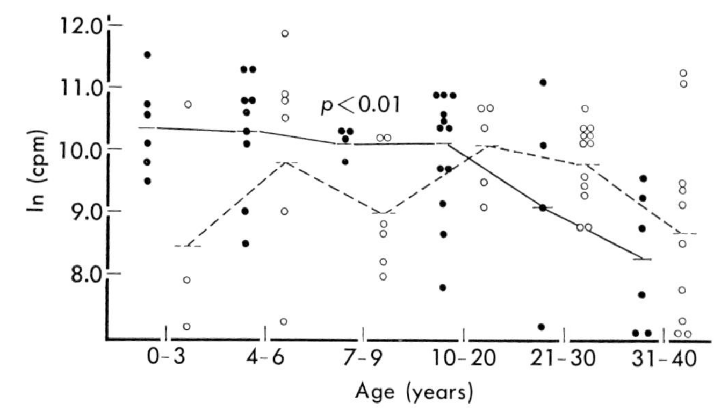

FIG. 3. Immune response to streptococcal cell wall antigen in the patients with Down's syndrome. 2×10^5 cells/well; 1 μg antigen/well. ● Down's syndrome; ○ control.

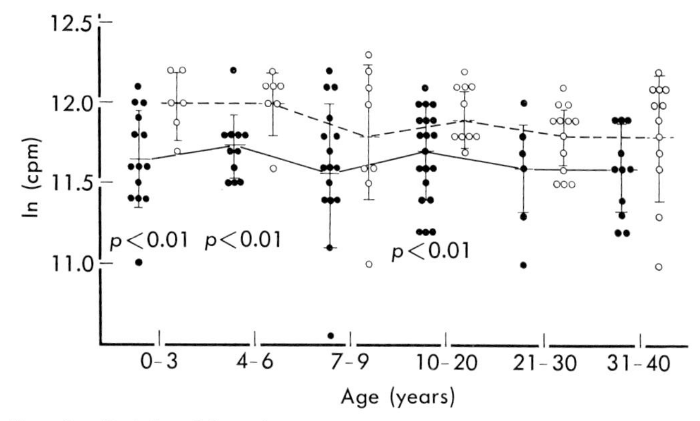

FIG. 4. Peripheral lymphocyte response to PHA in the patients with Down's syndrome. 2×10^5 cells/well; PHA $50\times$, 20 μl/well. ● Down's syndrome; ○ control.

2. MLC reaction of the patients with Down's syndrome

The responsiveness of lymphocytes from the patients with Down's syndrome in the MLC reaction against the stimulating lymphocytes from healthy individuals was greater than that of lymphocytes from the age-matched healthy controls (Fig. 5). These different responses in the MLC reaction are statistically significant in the young patients (younger than 6 years old) and in the aged patients (older than 31 years old).

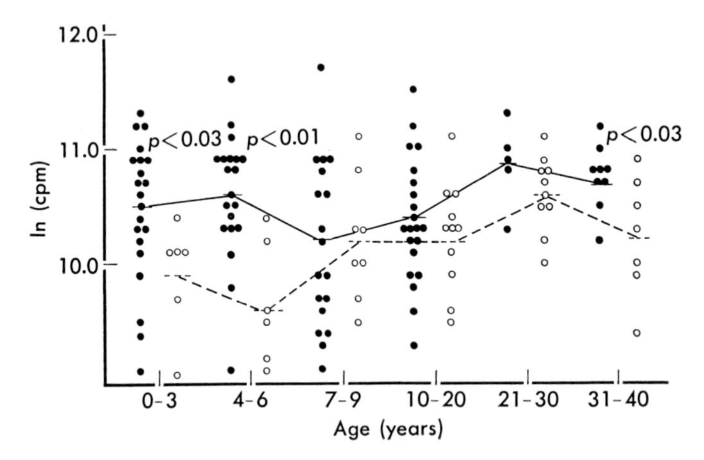

FIG. 5. Mixed lymphocytic culture reaction in the patients with Down's syndrome. Stimulator (mitomycin C-treated), 5×10^4 cells/well; responder, 5×10^4 cells/well. ● Down's syndrome; ○ control.

3. Natural killer cell activity of the patients with Down's syndrome

Natural killer cell activity of the patients with Down's syndrome against the human T cell leukemia line (MOLT 3) was slightly, but significantly, higher in the infant patients (younger than 3 years) ($P<0.01$) and in the patients between 10 and 20 years old ($P<0.0001$) than that in the age-matched control groups (Fig. 6).

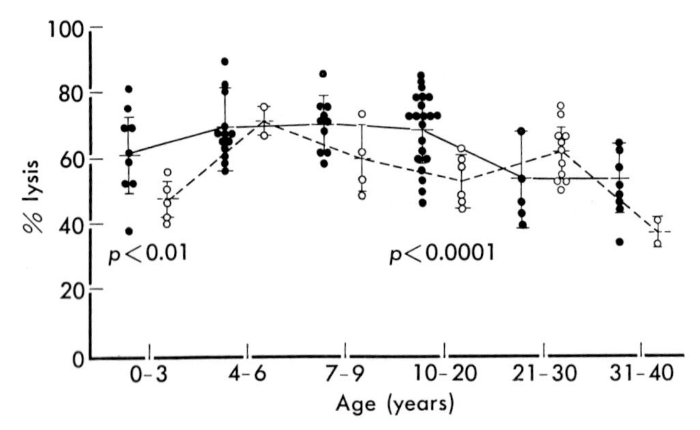

FIG. 6. Natural killer cell activity in the patients with Down's syndrome. Target: MOLT 3 (human T cell leukemia line). Target: effector=1: 25. ● Down's syndrome; ○ control.

4. Presence of cytotoxic antibodies to peripheral lymphocytes in the patients with Down's syndrome

Cytotoxic antibodies against the peripheral lymphocytes from HLA non-identical individuals were present in 8 of 33 young patients with Down's syndrome (younger than 9 years old) but were not present in 21 patients over 10 years of age (Fig. 7).

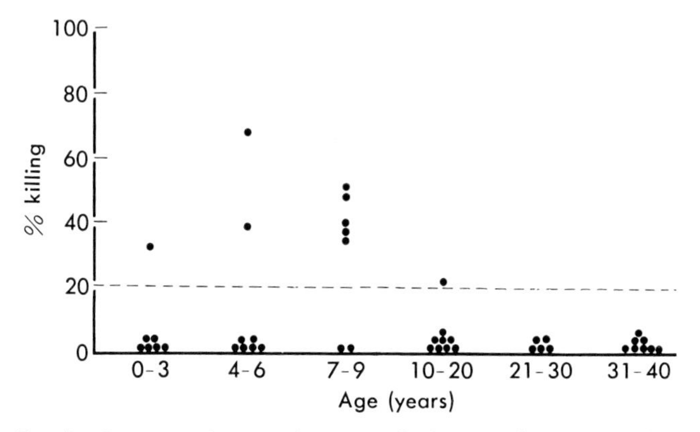

FIG. 7. Presence of cytotoxic autoantibody to peripheral lymphocytes in the patients with Down's syndrome.

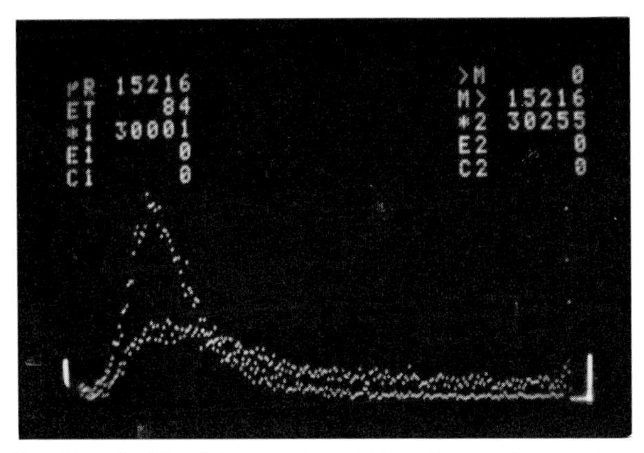

FIG. 8a. Profile of the peripheral T lymphocytes from patients with Down's syndrome stained with anti-BAT antisera.

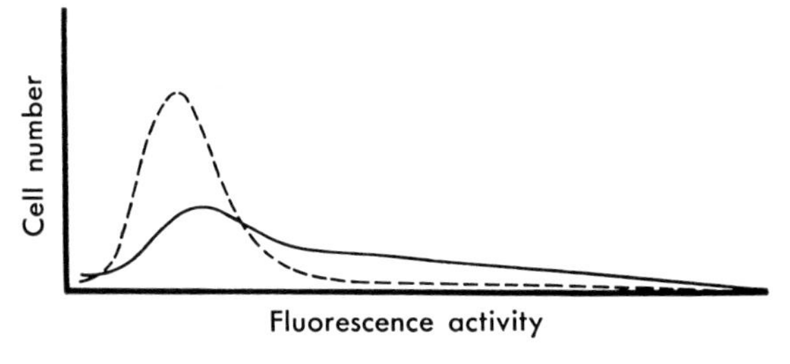

FIG. 8b. Schematic profile of the peripheral T lymphocytes from the patients with Down's syndrome stained with anti-BAT antisera. --- control; —— patient.

5. Cell profile of the peripheral T lymphocytes from the patients with Down's syndrome

The peripheral T lymphocytes brightly stained with anti-BAT antisera were greatly increased and those dully stained with the same antisera were markedly decreased in the patients with Down's syndrome (Fig. 8). The peripheral T lym-

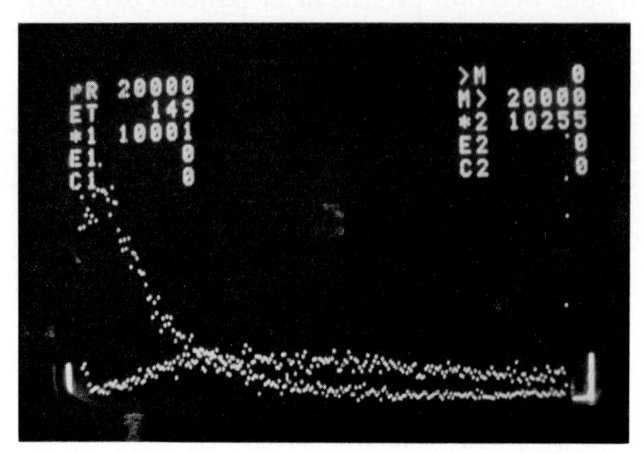

FIG. 9a. Profile of the peripheral T lymphocytes from the patients with Down's syndrome stained with NTA.

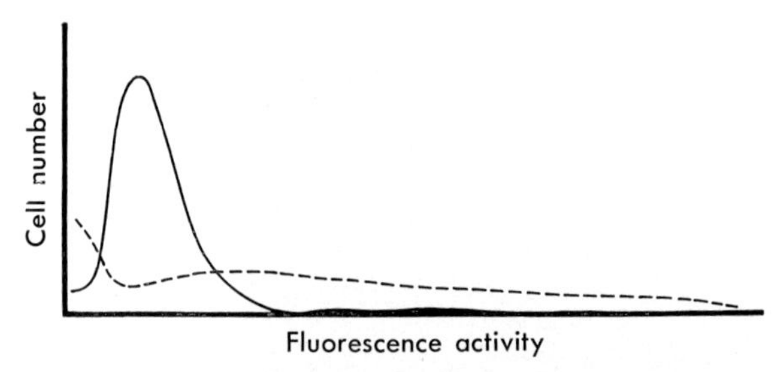

FIG. 9b. Schematic profile of the peripheral T lymphocytes from patients with Down's syndrome stained with NTA. --- control; —— patient.

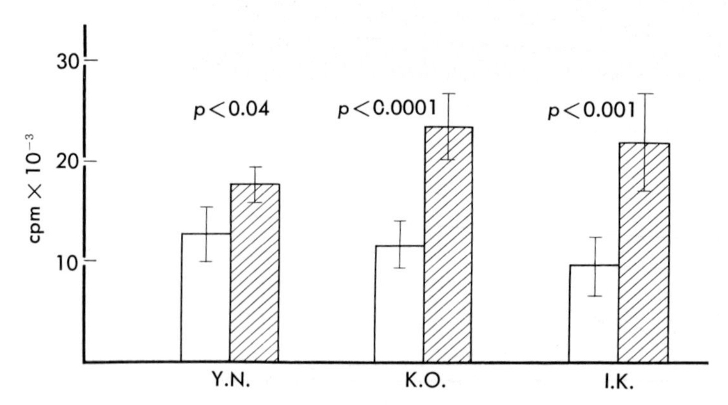

FIG. 10. Effect of the cytotoxic autoantibody from the patients with Down's syndrome on the MLC reaction. ▨ cultivated with 20 μl of patient's plasma; □ cultivated with 20 μl of normal human plasma.

phocytes brightly stained with human NTA were markedly decreased in the patients with Down's syndrome (Fig. 9).

These characteristic profiles of the peripheral T lymphocytes were observed in both young and old patients with Down's syndrome.

6. Effect of cytotoxic antibodies in the sera from the patients with Down's syndrome on the MLC reaction

Twenty μl of serum from a patient with Down's syndrome markedly enhanced the responsiveness of peripheral lymphocytes from healthy individuals in the MLC reaction. This enhancement of responsiveness in the MLC reaction was almost twice as great as the controls (Fig. 10).

DISCUSSION

Holland *et al.* (*10*) reported that the incidence of death by leukemia in the patients with Down's syndrome was 18 times higher than expected. Tonomura *et al.* (*11*) reported that in the Japanese population of 1,000 patients with Down's syndrome 7 died from acute leukemia, and they estimated that in the age group from newborn to 4 years the incidence of leukemia in the patients with Down's syndrome was almost 100 times greater than that in the age-matched normal controls. For these reasons we mainly selected young patients with Down's syndrome (younger than 20 years) as well as aged patients, as shown in Fig. 1. Immune response of patients with Down's syndrome to both tetanus toxoid and streptococcal cell wall antigen was not abberent in young patients with Down's syndrome. This may indicate that antigen recognition by T cells, antigen presentation by macrophages (antigen-presenting cells) and the T cell-macrophage interaction are not affected by the trisomic condition of chromosome 21. However, the responsiveness of aged patients (older than 31 years) was slightly lower than that of the age-matched control group. This fact might partially be explained by the rapid ageing of patients with Down's syndrome (*12*).

Responsiveness to PHA, on the other hand, significantly diminished in the patients with Down's syndrome. Burgio *et al.* (*13*) also reported that responsiveness to PHA of patients with Down's syndrome was greatly decreased in the aged group. Since T cells and macrophages are involved in the response to PHA, it may be that either T cells or macrophages decreased in number in the patients, or either T cells or macrophages had functional abnormalities due to the 21 trisomic condition. It is very difficult, however, to explain the fact that monoclonal or oligoclonal T cell response to the specific antigens were not affected, whereas polyclonal T cell response to PHA was significantly affected.

Moreover, the oligoclonal T cell response of the patients with Down's syndrome to the HLA-D antigens on allogenic lymphocytes in the MLC reaction was significantly enhanced compared to that of age-matched controls. Since cytotoxic antibodies present in the patients with Down's syndrome had a strong effect on the enhancement of the T cell response in the MLC reaction, the increased responsiveness of the patients in the MLC reaction might well be due to these cytotoxic autoantibodies, which seem to kill the suppressor T cells in the MLC reaction. These cytotoxic autoantibodies, however, did not have any effect on the antigen-specific T cell response *in vitro*. From these results it seems that the patients with Down's syndrome might have cytotoxic autoantibodies which are highly specific for the suppressor T cells in the MLC reaction. Oligoclonal lymphocyte activation generated by the lack of suppressor T cells due to cytotoxic autoantibodies seems to explain the malignant lymphoid transformation in patients with Down's syndrome.

Since natural killer cell activity and T cell recognition of foreign antigens were not decreased in the patient group, it is unlikely that the high susceptibility to leukemia of patients with Down's syndrome is due to the decreased ability to eliminate transformed malignant cells through the dysfunction of the immune surveillance system.

The abnormal profiles found in the peripheral lymphocytes of the patients with Down's syndrome may represent the functional abnormalities observed in the patient group such as a lack of suppressor T cells. However, it was difficult at this stage to show a clear relationship between the abnormal cell profiles and immunological abnormalities in the patients with Down's syndrome.

In summary, patients with Down's syndrome did not seem to have any deficiencies in immune surveillance, including T cell recognition and natural killer activity. The patients, however, showed significantly enhanced responsiveness in the MLC reaction due to the presence of cytotoxic autoantibodies which seemed to kill the suppressor T cells in this reaction.

This oligoclonal lymphocyte activation may be attributed to the high risk of the patients with Down's syndrome to leukemogenesis. Cell profiles of the peripheral lymphocytes from the patients with Down's syndrome differed markedly from those of normal individuals, which may represent the functional abnormalities found in the patient group such as a lack of suppressor T cells in the MLC reaction.

ACKNOWLEDGMENTS

The authors are grateful to Dr. T. Yoshiki, Sapporo City Hospital for his help in determining cytotoxic autoantibodies in the patient group. We sincerely thank Dr. T. Hirano, Department of Immunology, School of Medicine, Tokyo University and Y. Shiratori, Department of Internal Medicine, School of Medicine, Tokyo University, for checking cell profiles of the T lymphocytes from the patient group.

The authors are deeply indebted to Mr. T. Koishi, President of the association for the parents of patients with Down's syndrome (Kobato-Kai) for his kind cooperation. We are also indebted to Dr. T. Iwamoto, Yamayuri-en, Kanagawa, for his help in obtaining specimens from aged patients with Down's syndrome. Thanks are aslo due to Ms. S. Takeda, Kohnodai Hospital, Chiba, for her patient cooperation.

This work was supported in part by a grant (1979) for cancer research from the Ministry of Education, Science and Culture of Japan.

REFERENCES

1. Siegel, M. Susceptibility of mongoloids to infection. I. Incidence of pneumonia, influenza A and *Shigella dysenteriae* (Sonne). Am. J. Hyg., *48*: 53–62, 1948.

2. Adinolfi, M., Gardner, B., and Martin, W. Observations on the levels of γG, γA, and γM globulins, anti-A and anti-B agglutinins, and antibodies to *Escherichia coli* in Down's anomaly. J. Clin. Pathol., *20*: 860–864, 1967.
3. Gregory, L., Williams, R., and Thompson, E. Leukocyte function in Down's syndrome and acute leukaemia. Lancet, *i*: 1359–1361, 1972.
4. Möller, G. and Möller, E. The concept of immunological surveillance. Transplant. Rev., *28*: 3–16, 1976.
5. Burnet, F. M. Immunological Surveilance, Pergamon Press, Oxford, 1970.
6. Benacerraf, B. and McDevitt, H. O. Histocompatibility-linked immune response genes. Science, *175*: 273–279, 1972.
7. Kiessling, R., Klein, E., Pross, H., and Wigzell, H. "Natural" killer cells in the mouse. II. Cytotoxic cells with specificity for mouse Moloney leukemia cells. Characteristics of the killer cell. Eur. J. Immunol., *5*: 117–121, 1975.
8. Herberman, R. B., Nunn, M. E., and Lavrin, D. H. Natural cytotoxic reactivity of mouse lymphoid cells against syngeneic and allogeneic tumors. I. Distribution of reactivity and specificity. Int. J. Cancer, *16*: 216–229, 1975.
9. Sasazuki, T., McMichael, A., Radvany, R., Payne, R., and McDevitt, H. Use of high dose X-irradiation to block back stimulation in the MLC reaction. Tissue Antigens, *7*: 91–96, 1976.
10. Holland, W. W., Doll, R., and Carter, C. O. The mortality from leukaemia and other cancers among patients with Down's syndrome (Mongols) and among their parents. Br. J. Cancer, *XVI* (2): 177–186, 1962.
11. Tonomura, A., Yamada, K., Aoki, H., Oishi, H., and Kurita, T. Cytogenetic studies of 1,000 cases with Down's syndrome. Jpn. J. Human Genet., *16*: 84–85, 1971.
12. Øster, J., Mikkelsen, M., and Nielsen, A. Mortality and life-table in Down's syndrome. Acta Pediatr. Scand., *64*: 322–326, 1975.
13. Burgio, G. R., Ugazio, A. G., Nespoli, L., Marcioni, A. F., Bottelli, A. M., and Pasquali, F. Derangements of immunoglobulin levels, phytohemagglutinin responsiveness and T and B cell markers in Down's syndrome at different ages. Eur. J. Immunol., *5*: 600–603, 1975.

GENETIC AND ENVIRONMENTAL FACTORS IN EXPERIMENTAL AND HUMAN CANCER,
H. V. GELBOIN ET AL. (EDS.), JAPAN SCI. SOC. PRESS, TOKYO, PP. 139–148, 1980

Experimental Studies on the Carcinogenicity of Fungus-contaminated Food from Linxian County

Min-Hsin LI, Shih-Hsin LU, Chuan JI, Yinglin WANG, Mingyao WANG, Shujun CHENG, and Guizhen TIAN

Cancer Institute, Chinese Academy of Medical Sciences, Beijing, The People's Republic of China

Abstract: Some foodstuffs in Linxian County, a high-risk area for esophageal cancer in China, are frequently tainted by fungi. Our previous work showed the formation of dimethylnitrosamine, diethylnitrosamine, methylbenzylnitrosamine, and a new nitrosamine, *N*-3-methylbutyl-*N*-1-methylacetonylnitrosamine, in cornbread following inoculation of *Fusarium moniliforme* or other fungus, with an addition of a small amount of $NaNO_2$ after incubation. Recently, papilloma and carcinoma developed in the forestomach of rats fed *F. moniliforme*-inoculated cornbread, with an addition of $NaNO_2$ after incubation, for 238–883 days. Early carcinomas of the forestomach were also found in rats fed the fungus-inoculated cornbread alone for 621 and 701 days. Mammary tumors developed in 8 females in these groups. These results give evidence for the presence of carcinogens, *i.e.*, nitrosamines and/or mycotoxins, in the fungus-contaminated food. Furthermore, we demonstrated that some species of fungi would not only reduce nitrates to nitrites, but also increased the amount of secondary amines in contaminated food that provide a favorable condition for the synthesis of carcinogenic nitrosamines from these precursors. Pickled vegetables, commonly consumed in Linxian County, are heavily contaminated by *Geotrichum candidum*. Adenocarcinoma of the stomach and other tumors developed in rats subsequent to prolonged feeding of an extract of pickles. Papillomas of the forestomach were observed in mice intubated with a concentrated fluid of pickles. Roussin red methyl ester (di-μ-methanethiolato-tetranitrosodiiron) was identified in the extract of pickles which showed, however, weak mutagenic and tumorigenic actions.

A nationwide epidemiological survey on the mortality rates of cancer for the period 1973–1975 in China showed that Linxian County, Henan Province, ranked first in the average age-sex-adjusted mortality rate of esophageal cancer with 161.33/100,000 in men and 102.88/100,000 in women. During the past 35 years the trends in mortality rates in this county have shown little change, and the incidence and mortality rates of carcinoma of the esophagus have increased gradually with a definite rise for each age-group (*1*, *2*). Epidemiological evidence indicates the importance of environmental factors in the causation of this malignant disease.

Nitrosamines and Mycotoxins in Fungus-contaminated Cornbread

Our previous work showed that a relatively high content of the precursors of nitrosamine, *i.e.*, nitrates, nitrites, and secondary amines, and possibly some preformed *N*-nitroso compounds were found in staple foods collected from Linxian County (*2*). Furthermore, some foodstuffs in the same area are frequently tainted by *Fusarium moniliforme* Sheld., *Geotrichum candidum* Link, *Penicillium cyclopium* Westl.,

TABLE 1. Formation of Nitrosamines in Cornbread Inoculated with Fungus[a)]

Species of fungus	TLC			GC/MS[c)]				
	No. exp.	Positive reaction	Negative reaction	No. exp.	DMNA	DENA	MBNA	MAMBNA
Fusarium moniliforme Sheld.	39	34	5	27	1	0	2	24
Aspergillus flavus Link	9	6	3	3	1	2	0	0
Aspergillus terreus	4	4	0	2	0	0	0	2
Aspergillus niger v. Tiegh	2	2	0	2	1	1	0	0
Geotricum candidum Link	1	1	0	1	0	0	0	1
Mixed fungi[b)]	4	4	0	4	2	2	0	0
Control	10	1	9	1	1	0	0	0

[a)] To each sample 400 mg $NaNO_2$ was added after an 8-day incubation. [b)] Mixed fungi: *Fusarium moniliforme* Sheld, *Aspergillus flavus* Link, *Aspergillus niger* v. Tiegh, *Penicillium cyclopium* Westl, and *Penicillium oxalicum* Currie et Thom. [c)] DMNA, dimethylnitrosamine; DENA, diethylnitrosamine; MBNA, methylbenzylnitrosamine; MAMBNA, *N*-3-methylbutyl-*N*-1-methylacetonylnitrosamine.

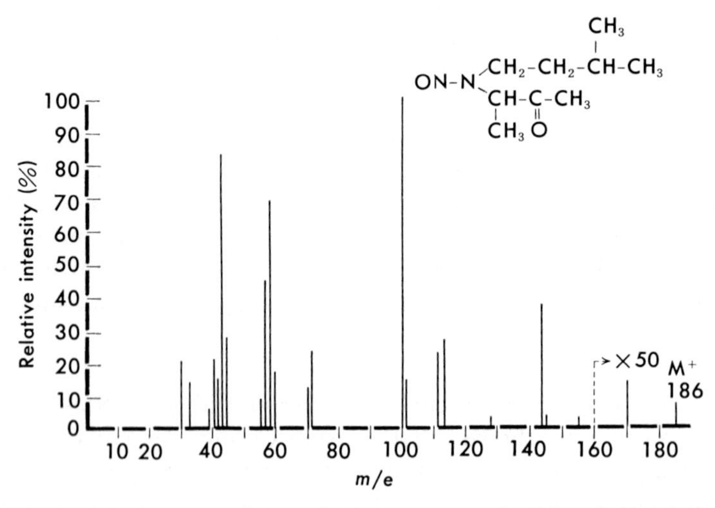

FIG. 1. Mass spectrum of a new *N*-nitroso compound, *N*-3-methylbutyl-*N*-1-methylacetonylnitrosamine (MAMBNA), formed in the cornbread inoculated with *Fusarium moniliforme* with $NaNO_2$ added later.

Aspergillus flavus Link, *Penicillium chrysogenum* Thom, and other fungi (*3*). Recently, we demonstrated the formation of dimethylnitrosamine, diethylnitrosamine, methylbenzylnitrosamine, and a new *N*-nitroso compound, *N*-3-methylbutyl-*N*-1-methylacetonylnitrosamine (MAMBNA) in cornbread following inoculation of *F. moniliforme* or other fungi, with an addition of a small amount of $NaNO_2$ after an 8-day incubation (Table 1, Fig. 1) (*4–6*). Ames test had revealed the mutagenic action of this new nitrosamine and papillomas were induced in the forestomach of mice treated with the secondary amine of MAMBNA plus $NaNO_2$ for 5 months.

Experimental investigation on the carcinogenic action of mouldy cornbread was carried out in 2 groups of rats (Table 2). In the first group, female rats of the Wistar strain were fed with *F. moniliforme*-inoculated cornbread and water *ad libitum*. The cornbread was disinfected before inoculation of the fungus and to each 400 g sample 400 mg $NaNO_2$ was added after an 8-day incubation. The presence of

TABLE 2. Carcinoma of the Forestomach in Rats Fed with *Fusarium moniliforme*-contaminated Cornbread[a)]

Food	No. of rats	Lesions in the forestomach		Mammary tumor
		Papilloma	Carcinoma	
F. moniliforme-inoculated cornbread	31 ♀	2 (554, 596 days)	2 (621, 701 days)	4 (454, 774 days)
F. moniliforme-inoculated cornbread plus $NaNO_2$	35 ♀	10 (238–768 days)	5 (664–885 days)	4 (360–838 days)
Uninoculated cornbread	10 ♀	—	— (330–700 days)	—

a) Epithelial hyperplasia was observed in the lower portion of the esophagus and in the glandular stomach of some experimental rats.

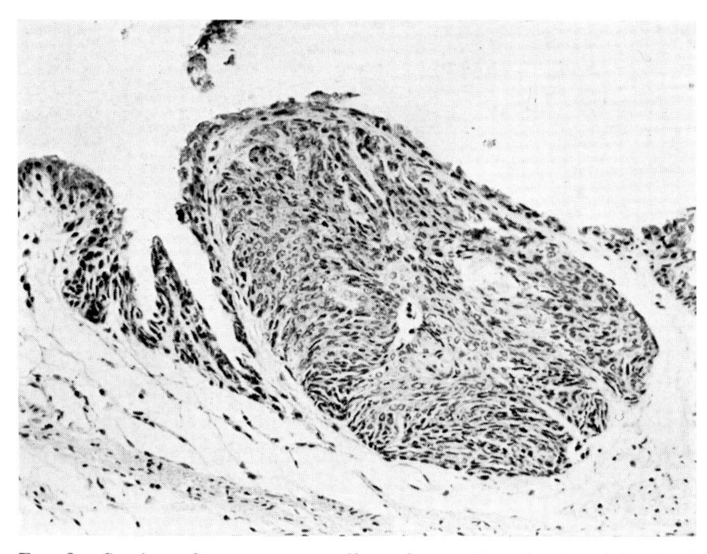

FIG. 2. Section of a squamous cell carcinoma that developed in the forestomach of a rat (RIII-37) fed with *F. moniliforme*-contaminated cornbread plus $NaNO_2$ for 838 days. ×200.

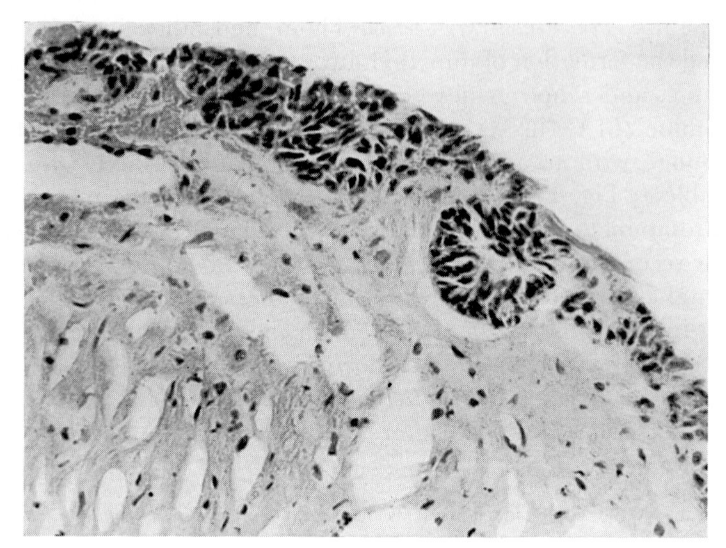

FIG. 3. Carcinoma *in situ* formed in the forestomach of a rat (RIII-38) fed with *F. moniliforme*-contaminated cornbread plus $NaNO_2$ for 840 days. ×400.

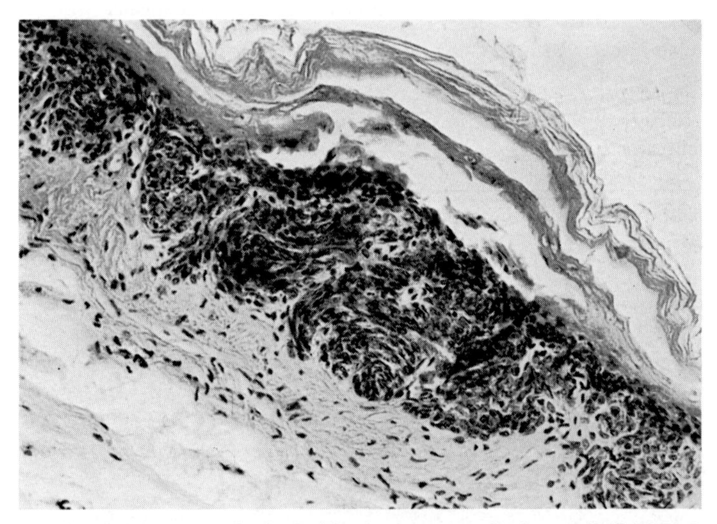

FIG. 4. Carcinoma *in situ* formed in the forestomach of a rat (RII-20) fed with *F. moniliforme*-contaminated cornbread alone for 621 days. ×200.

MAMBNA in cornbread was determined regularly by thin-layer chromatography during the course of the experiment. Ten papillomas, 4 early carcinomas and 1 squamous cell carcinoma developed in the forestomach of 35 rats fed the mouldy food for 238–883 days (Figs. 2, 3, and 5). Most carcinoma lesions were observed after 775 days of treatment. Besides papillomas, there was papillary hyperplastic growth and other types of epithelial hyperplasia occurred in the forestomach of treated animals. In some experimental rats simple epithelial hyperplasia was observed in the lower portion of the esophagus and in the glandular stomach. In

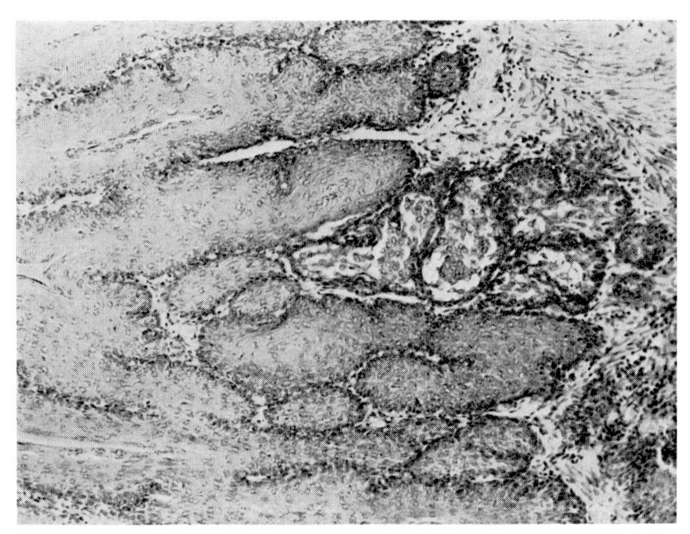

FIG. 5. Early carcinoma that developed from a papillomatous growth in the forestomach of a rat (RIII-25) fed with *F. moniliforme*-contaminated cornbread plus $NaNO_2$ for 664 days. ×100.

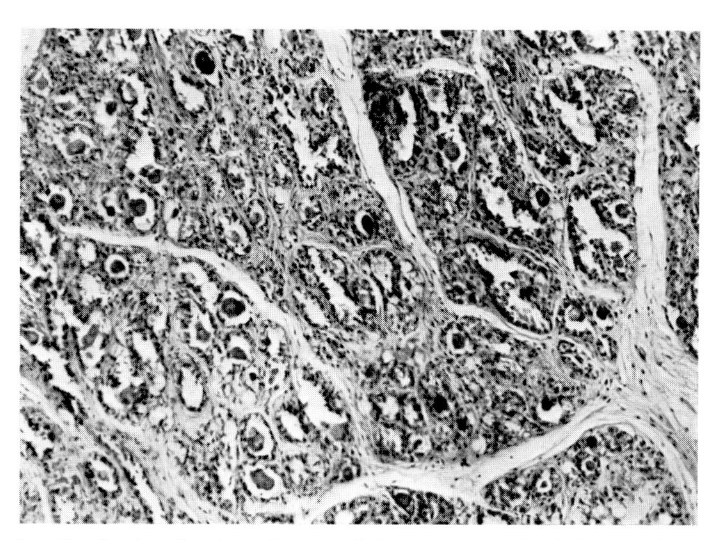

FIG. 6. Section showing adenoma of the mammary gland that developed in a rat (RII-30) fed with *F. moniliforme*-contaminated cornbread alone for 758 days. ×100.

addition, 1 adenoma and 3 fibroadenomas of the mammary gland were noted in 4 females. A second group of female rats was fed the fungus-inoculated cornbread alone. Two papillomas and 2 early carcinomas developed in the forestomach of 31 rats after 554–701 days of the feeding experiment (Fig. 4). Epithelial hyperplasia of the esophagus, forestomach and glandular stomach was observed in some animals of this group. They also developed 2 fibroadenomas and 2 adenomas of the mammary gland (Fig. 6). It is known that certain mycotoxins, such as zearalenone, produced by *Fusarium* fungi can exert an estrogenic activity in target organs (*7*).

Another group of 10 female rats served as the control and were fed with conventional cornbread. No epithelial lesions were noted in the forestomach of these animals. The present experimental results give evidence for the presence of carcinogens, *i.e.*, nitrosamines and/or mycotoxins, in the fungus-contaminated food, and the findings of chemical carcinogens, beside mycotoxins, formed in mouldy foodstuffs opens a new research field in cancer etiology.

Fungi Enhanced the Formation of Nitrosamines in Food

Our further investigations showed that some species of fungi such as *F. moniliforme*, *Aspergillus versicolor* (Vuill.) Tirab., *Penicillium brevi-compactum* Dierckx and *Penicillium lividum* Westl. can reduce nitrates to nitrites, as has been observed with certain bacteria, and their ability to reduce nitrates varied with the species (Fig. 7). Fungi contained in pickled vegetables can also reduce nitrates to nitrites, reaching high levels after a 4–5 day incubation (Table 3). However, this reducing action is lost following steam disinfection of the pickled vegetables indicating that fungi and other microorganisms are responsible for the nitrate reduction.

Fungus contamination also raised the amount of secondary amines in food and promoted the synthesis of nitrosamines from their precursors. When cornbread inoculated with a common species of fungus encountered in Linxian, such as *F.*

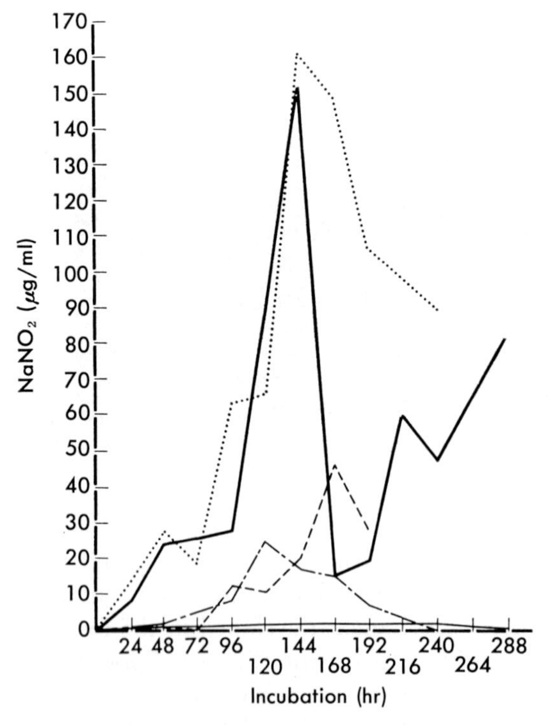

FIG. 7. Reduction *in vitro* of $NaNO_3$ (170 mg) by fungi showing changes in content of $NaNO_2$ during incubation in Sabouraud's glucose agar. —— control; **——** *Fusarium moniliforme*; ...... *Aspergillus versicolor*; —-— *Penicillium brevi-copamctum*; - - - *Penicillium lividum*.

TABLE 3. Reduction of Nitrates by Fungi in the Pickled Vegetables from Linxian

Group[a]	$NaNO_3$ (mg)	Content of $NaNO_2$ (μg/ml)							
		24	48	72	96	120	144	168	192 hr
Exp. A	1,000	0.3	23.1	44.5	79.4	82.0	33.4	14.7	0.6
Control	0	0.2	0.1	0.2	0.4	0.4	0.7	1.2	1.0
Exp. B	1,000	16.4	40.2	92.0	104.0	87.0	18.7	—	—
Control	0	0.4	0.8	0.8	0.8	0.7	0.7	—	—

[a] To each 100 g sample 200 ml distilled water was added and placed at room temperature (summer).

TABLE 4. Effect of Fungus on the Content of Secondary Amines in Cornbread

Fungus	Content of secondary amines (mg/kg)[a]				
	2 days	4 days	5 days	6 days	7 days
Control	1.375	1.187	1.562	1.562	—
Cladosporium herbarum	2.375	4.194	4.952	4.970	5.723
Penicillium chrysogenum	2.824	5.382	6.083	6.695	6.474
Aspergillus flavus	3.136	5.962	4.737	4.613	4.034
Geotrichum candidum	2.326	2.715	3.998	3.794	5.212
Fusarium moniliforme	5.933	21.132	22.907	26.547	27.855

[a] An average of 5 samples.

TABLE 5. Effect of *Geotrichum candidum* on the Formation of Nitrosamines

Group	No. exp.	Day of incu-bation	Diethyl-amine (mg)	Methyl-phenyl-amine (mg)	$NaNO_2$ (mg)	$NaNO_3$ (mg)	pH[a]	TLC[b]	
								DENA	MPNA
G. candidum	2	4-6	55.0-71.5	—	35.0-46.5	—	5.1	(+)	—
Control	2	4-6	55.0-71.5	—	35.0-46.5	—	6.1	(−)	—
G. candidum	3	1-7	—	1.6-2.7	—	8.5-170	4.1-5.1	—	(+)-(+++)
Control	3	1-7	—	1.6-2.7	—	8.5-170	6.7	—	(−)

[a] At the end of experiment. [b] DENA, diethylnitrosamine; MPNA, methylphenylnitrosamine.

moniliforme, *P. chrysogenum* or *Cladosporium herbarum* (Pers.) Link, a marked increase in content of secondary amines was noted, especially after 4 days of incubation. A 17-fold increase of secondary amines occurred in cornbread contaminated with *F. moniliforme* (Table 4).

In *G. candidum*-inoculated Sabouraud's culture the addition of $NaNO_2$ and diethylamine or methylphenylamine, after a few days of incubation, resulted in the formation of diethylnitrosamine (DENA) or methylphenylnitrosamine (MPNA) as revealed by thin-layer chromatography (TLC) analysis, and negative findings were obtained in uninoculated controls (Table 5). The result gave evidence for the enhancement by fungus of the synthesis of DENA from $NaNO_2$ and diethylamine, and the fungus reduced nitrates and promoted the production of MPNA from its precursors. In fact, the acidity and enzymatic activity in the mouldy food may provide favorable conditions for the synthesis of *N*-nitroso compounds.

Carcinogenic Actions of Pickled Vegetables

Pickled vegetables, commonly consumed in Linxian County, are heavily contaminated by *G. candidum*, containing high amounts of nitrates and nitrites. Dr. Nagao, M., in the laboratory of Dr. T. Sugimura, has shown the presence of a mutagen in these pickles which gave a positive response in the Ames system (*8*). In our previous experiments, among 39 rats fed with extracted or concentrated fluid from pickled vegetables for 330 to 730 days one had adenocarcinoma of the glandular stomach, 4 had fibrosarcoma of the liver and another had angioendothelioma of the thoracic wall, in addition to epithelial dysplasia lesions in the esophagus and the forestomach. No tumor was noted in control rats. Thus, the result indicates the presence of carcinogenic agents in the pickled vegetables (*9*). Recently, we intubated mice with a concentrated fluid of pickles, about 50 mg weekly for each mouse, and observed the development of papillomas in the forestomach after 143 days of treatment (Fig. 8).

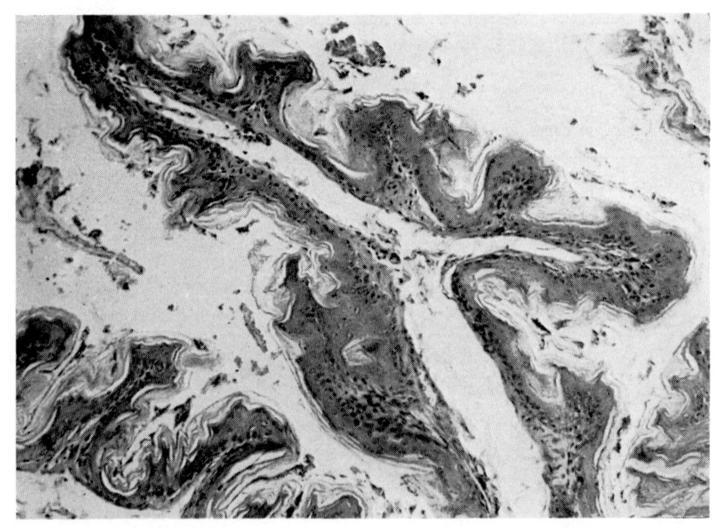

FIG. 8. Papilloma of the forestomach that developed in a mouse (PV-74) treated by gastric intubation of concentrated fluid from pickled vegetables, about 50 mg weekly, for 460 days $\times 100$.

The dichloromethane or ether extract of pickled vegetables was found to contain a nitroso compound as indicated by TLC analysis. Further investigations using GC/MS, mono-ion detection and peak-matching techniques have identified the compound as Roussin red methyl ester (dimethanethiolato-tetranitrosodiiron) which was first synthesized by Roussin in 1858, although it has not been reported to exist in natural products (*10*). The chemical structure of Roussin red is as follows:

NO NO
Fe
CH_3—S S—CH_3
Fe
NO NO

Roussin red had a weak mutagenic action in the Ames test. This compound may provide $[NO]^+$ ions which would react with secondary amines forming nitro-

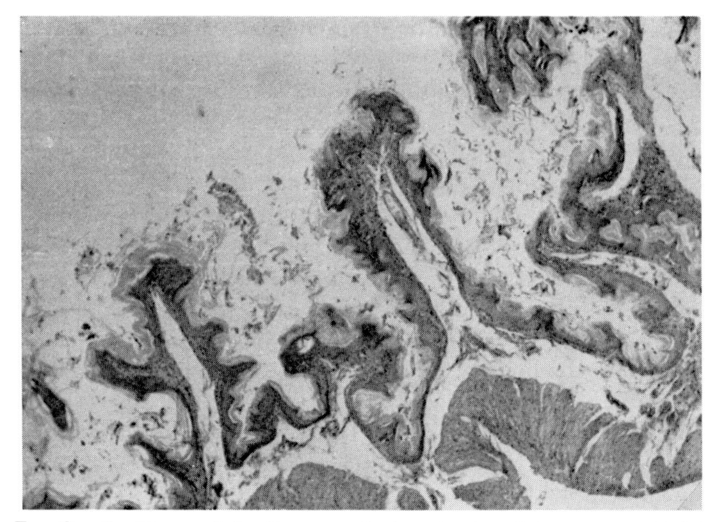

FIG. 9. Papilloma of the forestomach that developed in a mouse (FeS-20) intubated twice weekly with 4 mg Roussin red for 260 days. ×40.

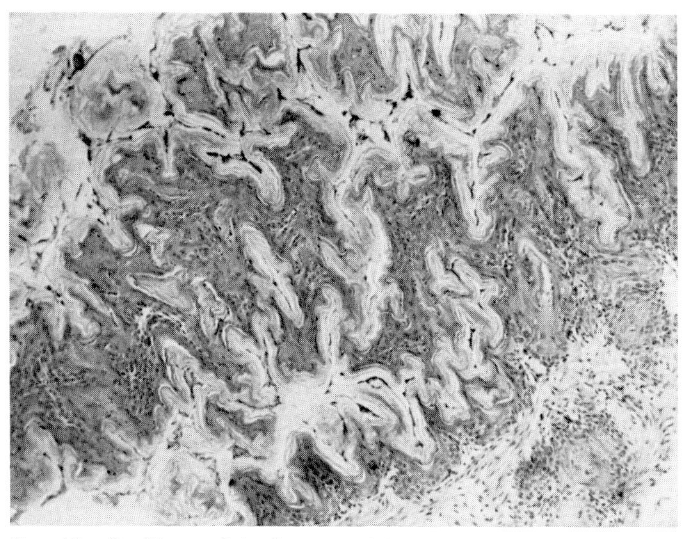

FIG. 10. Papilloma of the forestomach in a mouse (FeS+S-30) intubated twice weekly with 4 mg Roussin red and 50 mg sarcosine for 269 days. ×100.

samines. In animal experiments, the intubation of Roussin red alone or plus sarcosine induced epithelial hyperplasia of the esophagus and forestomach in mice, and papilloma of the forestomach developed in 194–269 days (Figs. 9 and 10). Further studies on the mutagenicity and carcinogenicity of pickled vegetables are in progress.

REFERENCES

1. Atlas of Cancer Maps of the People's Republic of China. China Map Press, 1980.
2. The Co-ordinating Group for Research on the Etiology of Esophageal Cancer of North China. The epidemiology of esophageal cancer in North China and preliminary results in the investigation of its etiological factors. Sci. Sin., *18*: 131–148, 1975.
3. Dept. Epidemiology, Cancer Institute, Dept. Statistics, Inst., of Med. Information, Chinese Acad. Med. Sci., and Institute of Microbiology, Academia Sinica. A preliminary survey on epidemiological factors of esophageal cancer in China. Research on Cancer Treatment and Control, *1977* (2): 1–8, 1977 (in Chinese).
4. Dept. Chemical Etiology and Carcinogenesis, Cancer Inst., Chinese Acad. Med. Sci. Further studies on the etiology of esophageal cancer. *In*; J. M. Birch (ed.), Adv. Med. Oncol., Res. and Educ., Vol. 3, pp. 39–44, Pergamon Press, Oxford and New York, 1979.
5. Li, M. H., Lu, S. H., Ji, C., Wang, M., Cheng, S., and Jin, C. Formation of carcinogenic *N*-nitroso compounds in corn-bread inoculated with fungi. Sci. Sin., *22*: 471–477, 1979.
6. Lu, S. H., Li, M. H., Ji, C., Wang, M., Wang, Y., and Huang, L. A new *N*-nitroso compound, *N*-3-methylbutyl-*N*-1-methyl-acetonylnitrosamine, in corn-bread inoculated with fungi. Sci. Sin., *22*: 601–608, 1979.
7. Mirocha, C. J. and Christensen, C. M. Oestrogenic mycotoxins synthesized by *Fusarium*. *In*; I.F.H. Purchase (ed.), Mycotoxins, pp. 129–148, Elsevier Sci. Publ. Co., Amsterdam, 1974.
8. Miller, R. W. Cancer epidemics in the People's Republic of China. J. Natl. Cancer Inst., *60*: 1195–1203, 1978.
9. Dept. Chemical Etiology and Carcinogenesis, Cancer Inst., Chinese Acad. Med. Sci. Preliminary observation on the carcinogenicity of pickled vegetable extracts from Linxian County. Research on Cancer Treatment and Control, *1977* (2): 46–49, 1977 (in Chinese).
10. Wang, G. H., Zhang, W. X., and Chai, W. G. The identification of natural Roussin red methyl ester. Acta Chim. Sin., *38*: 95–102, 1980.

GENETIC FACTORS

GENETIC AND ENVIRONMENTAL FACTORS IN EXPERIMENTAL AND HUMAN CANCER,
H. V. GELBOIN ET AL. (EDS.), JAPAN SCI. SOC. PRESS, TOKYO, PP. 151–160, 1980

Cancer Mortality and Morbidity among 23,000 Unselected Twin Pairs

Rune CEDERLÖF*[1] and Birgitta FLODERUS-MYRHED*[2]

Department of Environmental Hygiene, the Karolinska Institute, Stockholm, Sweden[1] *and Department of Environmental Hygiene, the National Swedish Environment Protection Board, Stockholm, Sweden*[2]

Abstract: The twin method in cancer epidemiology is presented. Some results concerning concordance rates and behavioral correlates are given with respect to cancer of all sites, cancer of the stomach and the cervix uteri. Data are taken from the Swedish Twin Registry, comprising *ca.* 23,000 unselected twin pairs, and the Swedish Cancer Registry, comprising *ca.* 450,000 records.

A significantly increased concordance rate is found among monozygotic pairs with regard to cervical cancer. Significant associations are shown between this cancer diagnosis and a series of behavioral characteristics.

The implications of the findings are discussed from a methodological point of view.

The Swedish Twin Registry was set up in 1961 with the primary aim of studying the relationship between smoking and health (*1*). The idea was to search for smoking discordant monozygotic and dizygotic twin pairs with the hope that such a matched pair design, similar to the co-twin control method, would enhance the group comparability between exposed and non-exposed individuals, which is a major problem in epidemiological research. The expectation was that twins due to genetics as well as their common childhood environment would share a series of habitual characteristics which otherwise would act as confounding factors in the analysis of cause and effect. This expectation was proved to be largely true as smokers and non-smokers in monozygotic twin pairs were more similar with respect to socioeconomic factors and several habits such as alcohol drinking, than smokers and non-smokers chosen at random. The rationale of the twin method as well as the results from the smoking analyses have previously been published (*2*).

Even though the main interest in the research program has focused on environmental rather than genetic factors, the so-called classical twin method, aimed at elucidating the impact of heritability on disease development, has also been used in some investigations. For example, in a study on cancer (*3*), concordance analysis suggested that cancer of all sites did not reveal any noteworthy heritability, while certain specific diagnoses seemed to speak in favor of a genetic hypothesis.

The present paper will present some preliminary results obtained by the classical, as well as the co-twin control method with respect to cancer of all sites, stomach cancer and cervical cancer.

THE SWEDISH TWIN REGISTRY

The Twin Registry covers about 11,000 same-sexed twin pairs born in 1886–1925 (compiled in 1961, the "older cohorts") and about 13,000 same-sexed twin pairs born in 1925–1958 (compiled in 1971, the "younger cohorts"). The twin pairs were found through scanning of birth records from the total Swedish population and the Registry can be estimated to cover well above 95% of all twin pairs that were alive as unbroken pairs at the times of compilation (*1*, *4*). Baseline epidemiological information was collected mainly through mailed questionnaires, covering a series of socioeconomic, environmental, habitual, psychosocial, and medical items.

The assessment of zygosity was based on a questionnaire item phrased "were you as children like two peas in a pod, or did you have family likeness only?" Pairs who concordantly answered "like two peas in a pod" were classified as monozygotic, while a concordant response of "family likeness only" were classified as dizygotic. More than 90% of the pairs could be classified according to this principle. Although simple, the method proved to be highly reliable in that 99% of the monozygotic and 92% of the dizygotic pairs in a subsample of 200 pairs were correctly classified as judged by serological blood typing (*5*).

The distribution of the pairs in the two cohorts of the registry by sex, year of birth and zygosity is shown in Table 1.

TABLE 1. The Swedish Twin Registry by Zygosity, Sex, and Year of Birth

	Monozygotic pairs			Dizygotic pairs			Number of pairs in registry
	Total	Men	Women	Total	Men	Women	
Older cohorts							
Born in							
1886–1900	702	323	379	1,162	440	722	1,864
1901–1910	947	436	538	1,871	826	1,045	2,845
1911–1925	1,980	890	1,090	3,817	1,717	2,100	5,797
Total 1886–1925	3,656	1,649	2,007	6,850	2,983	3,867	10,506
Younger cohorts							
Born in							
1926–1935	1,109	508	601	1,873	823	1,050	2,982
1936–1945	1,569	709	960	2,391	1,109	1,282	3,960
1946–1958	2,204	1,022	1,182	3,442	1,702	1,740	5,646
Total 1926–1958	4,882	2,239	2,643	7,706	3,634	4,072	12,588

METHODS

Data on cancer in the Twin Registry has been obtained by two sources, *i.e.*, the Swedish Cancer Registry and the Central Death Registry for Sweden.

Information on mortality for the Swedish population is compiled in the Central Death Registry, in accordance with the WHO recommendation. Since 1961 the older twin cohorts have continuously been followed up with respect to cause-specific mortality. The procedure and its validity has previously been described in detail (*6*, *7*).

The Swedish Cancer Registry, maintained by the National Board of Health and Welfare, was established in 1958. This registry is nationwide and is based on reports from physicians working in hospitals or other establishments for medical treatment under public or private administration. Today, the registry covers cancer incidence from 1958 up to 1974, *i.e.*, about 450,000 records. The structure of the registry and reliability of data is given in the 1973 incidence report (*8*). Both cohorts of the Twin Registry were matched against the cancer files and, in total, 2,136 twins were identified as cancer cases.

Available data on cancer to be used in the present analyses appear schematically below:

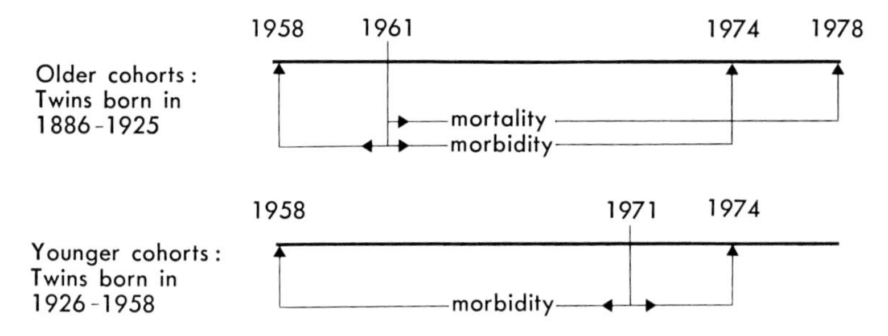

Data on cancer *mortality* from the older cohorts were used to assess how representative the twin registry is with regard to cancer occurrence. The method of preference should be the use of proportional mortality rates, as the absolute number of deaths might be under-represented in the Swedish Twin Registry for years close to the time of compilation, especially for those born before 1900. On the basis of the above-mentioned Central Death Registry the year by year proportional mortality rate for each 5-year population cohort of each sex was calculated with respect to cancer of all sites and certain cause-specific entities. Knowing the year-by-year total mortality in the Swedish Twin Registry it was then possible to calculate the expected number of deaths for these diagnoses. Table 2 shows the expected and ob-

TABLE 2. Cancer Deaths Observed and Expected, during 1961–1978, for Twins Born in 1886–1925

	Men		Women	
	Obs.	Exp.	Obs.	Exp.
Number of death	2,387		2,317	
Deaths due to cancer of all sites (ICD: 140–239)	540	550.85	657	633.71
Cancer of the stomach (ICD: 151)	72	71.81	56	54.95
Cancer of the lung (ICD: 162)	92	92.07	29	30.27
Breast cancer (ICD: 174)			108	108.28

served deaths for cancer of all sites, stomach cancer, lung cancer, and breast cancer. It can be seen that the observed values are very close to their expectation.

Morbidity for the older cohorts can be studied prospectively for the period from 1961 through 1974. Morbidity incurred before 1961 can retrospectively be obtained from 1958 (start of the Cancer Registry). For the younger cohorts, the morbidity incurred after the time of compilation (1971) can be studied prospectively up to 1974. Morbidity incurred before the time of the register compilation has an uncertain frame of reference. However, for cervical cancer this bias is almost insignificant as cases before the age of 32 (the oldest age group in the twin series when the Cancer Registry started its operation in 1958) are very rare in the population. A similar assessment of how representative it is with respect to morbidity, as was done for mortality, has not yet been performed.

The diagnoses for which results will be presented, *i.e.*, stomach and cervical cancer, were chosen because of comparatively frequent occurrence. Cancer of the stomach corresponds to the ICD-code (8th revision) 151, cervical cancer to the code 180 and cancer of all sites covers the codes 140–239.

The two specific diagnoses of cancer defined above were related to some behavioral factors, based on items in different questionnaires.

The factors focused on in connection with cancer of the stomach are smoking and drinking habits, for which data were obtained in 1961 and 1965, respectively. The variables used are defined as follows:

Smoking I——Present and former smokers

Smoking II——Present smokers, with a consumption exceeding 10 cigarettes/day

Drinking I——A "yes" to the question whether or not it sometimes happens that the subject consumes more than a half bottle of liquor on the same occasion

Factors studied in relation to cervical cancer are based on questionnarie data from 1972. One marital status variable, *i.e.*, divorce, and 2 familial variables, *i.e.*, having children and having more than 2 children, are included in the analysis. Smoking and drinking habits were treated in accordance with the definitions given above. Another 7 variables are defined below:

Drinking II——Subjects consuming on an average more than 250 g of 100% alcohol/month

Extraversion——Subjects classified as extroverts on the basis of a short form of the Eysenck Personality Inventory (*9*)

Restoratives——Consumption now and then or regularly during longer or shorter periods

Prescription-free analgesics——Consumption now and then or regularly during longer or shorter periods

Sleeping pills——Consumption now and then or regularly during longer or shorter periods

Tranquillizers——Consumption now and then or regularly during longer or shorter periods

Oral contraceptives——Present or former use

RESULTS

1. Concordance rates

The rationale of the classical twin method is to compare monozygotic and dizygotic twins with respect to the degree of intrapair similarity, the so-called concordance. A higher degree of similarity within monozygotic pairs is generally understood as being caused by genetic factors along with the assumption that monozygotic twins share their environment to the same extent as do dizygotic twins. Concordance rates can be calculated in several ways. We have used the proband concordance rate which is regarded as the best measure to use for quantal data (*10*).

The concordance rates with respect to morbidity of cancer of all sites are shown

TABLE 3. Concordance Rates for Cancer Morbidity of All Sites in the Swedish Twin Registry (Men)

	Monozygotes					Dizygotes				
	Number of pairs			Cum. inc. rate	Conc. rate	Number of pairs			Cum. inc. rate	Conc. rate
	Total	Disc.	Conc.			Total	Disc.	Conc.		
Older cohorts										
1886–1900	323	73	8	13.8	18.0	440	115	11	15.6	16.1
1901–1910	436	66	2	8.0	5.7	826	114	5	7.5	8.1
1911–1925	890	39	1	2.3	4.9	1,717	83	0	2.4	—
Total 1886–1925	1,649	178	11	6.1	11.0	2,983	312	16	15.9	9.3
Younger cohorts										
1926–1935	508	10	0	1.0	0.0	823	8	0	0.5	0.0
1936–1945	709	11	0	0.8	0.0	1,109	5	0	0.2	0.0
1946–1958	1,022	3	0	0.2	0.0	1,702	4	0	0.1	0.0
Total 1926–1958	2,239	24	0	0.5	0.0	3,634	17	0	0.2	0.0
Both cohorts										
1886–1958	3,888	202	11	2.9	9.8	6,617	329	16	2.7	8.9

TABLE 4. Concordance Rates for Cancer Morbidity of All Sites in the Swedish Twin Registry (Women)

	Monozygotes					Dizygotes				
	Number of pairs			Cum. inc. rate	Conc. rate	Number of pairs			Cum. inc. rate	Conc. rate
	Total	Disc.	Conc.			Total	Disc.	Conc.		
Older cohorts										
1886–1900	379	88	9	14.0	17.0	722	139	17	12.0	20.0
1901–1910	538	74	4	7.6	9.8	1,045	180	5	9.1	5.3
1911–1925	1,090	125	9	6.6	12.6	2,100	208	13	5.6	11.1
Total 1886–1925	2,007	287	22	8.2	13.3	3,867	527	35	7.7	11.7
Younger cohorts										
1926–1935	601	36	6	4.0	8.0	1,050	69	3	3.6	8.0
1936–1945	860	41	4	2.8	16.3	1,282	59	3	2.5	9.2
1946–1958	1,182	11	0	0.5	0.0	1,740	25	0	0.7	0.0
Total 1926–1958	2,643	88	10	2.0	18.5	4,072	153	6	2.0	7.3
Both cohorts										
1886–1958	4,650	375	32	4.7	14.5	7,939	680	41	4.8	10.7

in Tables 3 and 4 for men and women, respectively. As the concordance rate is incidence-dependent, this measure is also included in the tables.

The results throughout are negative with respect to a difference in concordance between the two zygosity groups. The incidences are similar enough not to influence the validity of the analysis. The numbers of concordant pairs are small, however, and in the younger cohorts of men no such pairs were found in any subgroup.

The same negative results are seen for stomach cancer morbidity, measured among twins in the older cohorts (Table 5). The accumulated incidence is, however, quite low for men as well as for women.

With respect to cervical cancer (Table 6), the incidence is very low among the youngest twins but reasonably high among the two groups of women born 1926–1935 and 1936–1945, respectively.

If these subgroups are added, the concordance rate among monozygotes is 2.3 times higher than among dizygotes—a figure which is statistically significant ($P<0.05$).

2. *Associations with behavioral items*

The co-twin control method was originally designed to investigate the influence

TABLE 5. Concordance Rates for Stomach Cancer Morbidity (Older Cohorts of the Swedish Twin Registry)

	Monozygotes					Dizygotes				
	Number of pairs			Cum. inc. rate	Conc. rate	Number of pairs			Cum. inc. rate	Conc. rate
	Total	Disc.	Conc.			Total	Disc.	Conc.		
Men										
1886–1900	323	7	1	1.4	22.2	440	22	0	2.5	0.0
1901–1910	436	5	0	0.6	0.0	826	12	0	0.7	0.0
1911–1925	890	3	0	0.2	0.0	1,717	6	0	0.2	0.0
Total 1886–1925	1,649	15	1	0.5	11.8	2,983	40	0	6.7	0.0
Women										
1886–1900	379	9	0	1.2	0.0	722	11	0	0.8	0.0
1901–1910	538	3	0	0.3	0.0	1,045	12	0	0.6	0.0
1911–1925	1,090	4	1	0.3	33.0	2,100	3	0	0.1	0.0
Total 1886–1925	2,007	16	1	0.5	11.1	3,867	26	0	0.3	0.0

TABLE 6. Concordance Rates for Cervical Cancer Morbidity (Younger Cohorts of the Swedish Twin Registry)

	Monozygotes					Dizygotes				
	Number of pairs			Cum. inc. rate	Conc. rate	Number of pairs			Cum. inc. rate	Conc. rate
	Total	Disc.	Conc.			Total	Disc.	Conc.		
Women										
1926–1935	601	25	5	2.9	28.6	1,050	52	3	2.8	10.3
1936–1945	860	30	4	2.2	21.0	1,282	48	3	2.1	11.1
1946–1958	1,182	9	0	0.4	0.0	1,740	17	0	0.5	0.0
Total 1926–1958	2,643	64	9	1.6	2.0	4,072	117	6	1.6	9.3

of different treatments administered to genetically identical individuals, thus avoiding the interindividual dispersion caused by genetic dissimilarities which may confound the results. A similar approach has been adopted in epidemiology with the modification that no treatment is administered; instead, self-inflicted behaviors such as smoking and drinking, or environmental factors, such as air pollution or occupational exposures, constitute the "treatment." Usually the approach is prospective, namely to study the occurrence of disease subsequent to exposure; however, a reverse approach may be employed, namely to study whether behavioral or environmental factors differ between the affected and non-affected individuals in genetically identical pairs, often in comparison with corresponding differences between diseased individuals and their randomly-chosen and genetically-unrelated controls, or appropriate strata from the general population.

The present analysis will employ the last-mentioned approach. Table 7 concerns cancer of the stomach and presents in its first column the "factor frequency" for the studied factors, smoking and drinking, as defined above in the paragraphs on methods. This factor frequency is simply the percent of individuals in the total cohort who smoke or drink. The second column displays the ratio of two factor frequencies, namely the factor frequency in the group of *all* individuals who have stomach cancer divided by the corresponding figure for *all* individuals who are free of stomach cancer.

It can be seen that the ratios indicate a lower frequency of smokers and drinkers among the cancer cases. The next three columns show the corresponding ratios of factor frequencies between the cancer cases and the non-cancer cases among all cancer discordant pairs, the discordant monozygotic and the discordant dizygotic pairs, respectively. It should be mentioned here that the tables contain subjects who cover a wide age range and that due to low numbers no age adjustment has been performed. This will be discussed later, but taken at face value the table does not reveal any association between the cancer and the items studied.

Table 8 concerns cervical cancer and is constructed in exactly the same way

TABLE 7. Factor Frequency Ratios for Affected over Non-affected Subject Groups with Respect to Stomach Cancer (Older Cohorts)[a]

	Twin individuals		Stomach cancer discordant pairs[b]		
	Factor frequency	Ratio	All pairs ratio	MZ ratio	DZ ratio
Men					
Group size	9,264		55	15	40
Smoking I	65.8	0.9	0.9	1.4	0.8
Smoking II	9.6	0.3	0.3	—	0.5
Drinking I	38.2	0.9	0.9	1.3	0.8
Women					
Group size	11,748		42	16	26
Smoking I	19.2	0.6	1.2	1.5	1.0
Smoking II	3.3	0.6	1.0	—	—
Drinking I	5.4	—	—	—	—

[a] Factors are defined in text. [b] MZ, monozygotic; DZ, dizygotic.

TABLE 8. Factor Frequency Ratios for Affected over Non-affected Subject Groups with Respect to Cancer of Cervix Uteri (Younger Cohorts)[a]

	Twin individuals		Cervical cancer discordant pairs		
	Factor frequency	Ratio	All pairs ratio	MZ ratio	DZ ratio
Children	55.1	1.6**	1.0	1.0	1.0
More than 2 children	13.3	2.1**	1.2	1.1	1.2
Divorced	3.5	2.9**	1.1	0.8	1.2
Extraversion	66.0	1.3**	1.3	1.1	1.4**
Smoking I	50.0	1.6**	1.1	1.1	1.1
Smoking II	13.2	2.4**	1.4	1.3	1.3
Drinking I	5.9	2.4*	1.0	1.0	0.5
Drinking II	2.0	1.8**	1.0	—	1.0
Restoratives	47.2	1.1	1.2	1.3	1.1
Prescription-free analgesics	37.7	0.9	0.9	1.0	0.9
Sleeping pills	5.2	1.7**	0.9	1.0	1.0
Tranquillizers	12.2	1.6**	0.8	0.6	1.4
Oral contraceptives	43.6	1.3**	1.0	1.0	1.0

[a] Factors are defined in text. * $P<0.10$; ** $P<0.05$.

as Table 7. Here, however, in the analyses of cancer and non-cancer among all individuals, all items show significant associations with the disease.

Lack of age adjustment emphasizes that the figures should be interpreted with care as to their absolute value. A tabulation (not included in this preliminary analysis) with a breakdown into the three 10-year age-groups has been performed, which reveals that all the significant results from the analysis of individuals also show up in one or two of the subgroups. Smoking, drinking, use of oral contraceptives and extraversion were significantly increased an all three age-groups.

In the series of cancer discordant twin pairs practically no difference is observed between the proband and her co-twin. This was also the case when the cohort was broken down into separate age-groups.

DISCUSSION

It should first of all be pointed out that the presented results are preliminary and that at this stage they rather serve to illustrate some aspects of the twin method in epidemiology than contribute to the scientific understanding of the genetics of cancer. Some shortcomings have already been pointed out, such as the need for age adjustments, when appropriate, or for age breakdowns in the tables. Further analysis will be necessary before the final publication to the extent that numbers allow a closer scrutiny of data. The importance of bias, introduced by the unknown incidence of nonfatal cancers that may have occurred in the cohorts before the start of the cancer registry in 1958, could be estimated on the basis of population statistics. As was pointed out, however, this bias seems to be negligible for cervical cancer, but is at present unknown for stomach cancer and cancer of all sites.

With these reservations in mind, however, some comments can be made with respect to our results. It seems clear that no concordance is present with respect to cancer of all sites. The analysis of stomach cancer from this point of view is not conclusive because of small numbers.

The significantly higher concordance rate among monozygotes than among dizygotes with regard to cervical cancer could indicate a genetic influence, which would be the logistic inference, provided that the monozygotes do not share a more similar environment than do the dizygotes. It is doubtful, however, whether this is true. One possibility is that monozygotic twins compared to dizygotic ones share "environmental" factors to a greater extent.

It is known that monozygotic twins share such "environmental" characteristics as smoking and drinking to a higher extent than dizygotic twins. If such factors are of relevance to the development of disease, a higher cancer concordance would of course be expected even if genetic factors are of no relevance.

All behavioral items used in the analysis proved to be significantly related to cervical cancer when the twin series was regarded as ordinary individuals. The general impression is that women with cervical cancer are characterized by a certain life-style, including smoking, alcohol drinking, use of drugs, a propensity to divorce and use of oral contraceptives, as well as an extraverted behavior.

On the other hand, none of these characteristics were found more often among the cancer-cases in the series of cancer discordant twin-pairs. This may suggest that the proband and her co-twin share these traits, and that the non-diseased partner may be at a higher risk than average to develop cancer of the cervix.

Obviously, this reasoning leads to the conclusion that the higher concordance among monozygotes need not be a proof of a genetic influence on cervical cancer.

Finally, however, might not the lifestyle as such be a genetically-determined trait?

REFERENCES

1. Cederlöf, R. The twin method in epidemiologic studies on chronic disease. Diss. Acad., University of Stockholm, 1966.
2. Cederlöf, R., Friberg, L., and Lundman, T. The interactions of smoking, environment and heredity, and their implications for disease etiology. A report of epidemiological studies on the Swedish Twin Registries. Acta Med. Scand., No. 612 (Suppl.), 1977.
3. Cederlöf, R., Floderus, B., and Friberg, L. Cancer in MZ and DZ twins. Acta Genet. Med. Gemellol., *19*: 69–74, 1970.
4. Medlund, P., Cederlöf, R., Floderus-Myrhed, B., Friberg, L., and Sörensen, S. A new Swedish twin registry. Acta Med. Scand., No. 600 (Suppl.), 1976.
5. Cederlöf, R., Friberg, L., Jonsson, E., and Kaij, L. Studies on similarity diagnosis in twins with the aid of mailed questionnaires. Acta Genet. Med. Gemellol., *11*: 338–362, 1961.
6. Friberg, L., Cederlöf, R., Lorich, U., Lundman, T., and de Faire, U. Mortality in twins in relation to smoking habits and alcohol problems. Arch. Environ. Health, *27*: 294–304, 1973.

7. de Faire, U., Friberg, L., Lorich, U., and Lundman, T. A validation of cause-of-death certification in 1156 deaths. Acta Med. Scand., *200*: 223–228, 1976.
8. Cancer Incidence in Sweden 1973. National Board of Health and Welfare, The Cancer Registry, Stockholm, 1979.
9. Floderus, B. Psycho-social factors in relation to coronary heart disease and associated risk factors. Nord. Hyg. Tidskr., *6* (Suppl.), 1974.
10. Allen, G., Harvald, B., and Sheilds, J. Measures of twin concordance. Acta Genet., *17*: 475–481, 1967.

GENETIC AND ENVIRONMENTAL FACTORS IN EXPERIMENTAL AND HUMAN CANCER,
H. V. GELBOIN ET AL. (EDS.), JAPAN SCI. SOC. PRESS, TOKYO, PP. 161–173, 1980

Inherited Tissue Resistance to the Gene for Retinoblastoma*

Ei MATSUNAGA

Department of Human Genetics, National Institute of Genetics, Mishima, Japan

Abstract: Retinoblastoma is a most suitable cancer for studying the mechanisms of carcinogenesis in man. Studies on hereditary cases have led to the conclusion that inherited tissue resistance plays an important role in the development of retinoblastoma or nonradiogenic osteosarcoma in the gene carriers. The crucial event initiating tumor formation in the gene carriers is not a somatic mutation, but probably an error in differentiation that could be suppressed completely in the resistant group who could remain unaffected, but that could not be suppressed completely in the less resistant group. The tissue resistance can be considered a multifactorial threshold character largely determined by tissue-specific suppressor genes. It is argued that radiogenic osteosarcoma in the gene carriers is ascribable to the accumulation of induced mutations in that suppressor system.

It is generally accepted that the carcinogenic process is usually composed of more than one step, each step probably being controlled or modified by genetic and environmental factors. Then, basic questions are: what is the minimum number of steps needed to transform a normal cell into a cancer cell, what is the nature of each step, and what are the individual factors involved? To answer these questions in human cancers, retinoblastoma is perhaps the most suitable material for study for a number of reasons.

First, a significant proportion of patients with retinoblastoma are diagnosed shortly after birth, indicating initiation of tumor formation during the latter half of the fetal period thus with the minimal number of carcinogenic steps involved. Second, because of its malignant nature almost all children affected with retinoblastoma come to the attention of ophthalmologists, at least in advanced countries, so that a precise incidence figure can be obtained provided the diagnosis is made correctly. Third, the disease, if diagnosed early, is curable and many patients can now survive and procreate, so that a good deal of family data have accumulated

* Contribution No. 1289 from the National Institute of Genetics, Japan.

that are available for genetic analyses. Fourth, the grade of tumor progression can be directly examined by ophthalmoscopy. Fifth, the eyes are a paired organ, hence not only three distinct phenotypes, *i.e.*, unaffected and unilaterally and bilaterally affected, can be distinguished, but also in bilateral cases a correlation in tumor development in the right and left eyes can be measured.

Genetics of Retinoblastoma

Retinoblastoma occurs at a rate of one per *ca.* 20,000 newborn. In more than 95% of the cases the occurrence is sporadic, but occasionally two or more sibs or collateral relatives are affected whose parents are apparently normal; this is the original pattern of appearance of this cancer. With the advent of modern ophthalmology, however, cases inherited from the survivors began to appear. Griffith and Sorsby (*1*) analyzed pedigree data from familial cases and arrived at the conclusion, now accepted generally, that these cases follow a pattern of irregular dominant inheritance: they are due to an autosomal dominant gene with incomplete penetrance and expressivity. The term *penetrance* is a measure for the proportion of gene carriers who manifest the disease, and *expressivity* refers to the variation in clinical signs. Thus, a gene carrier can be either unaffected, or unilaterally or bilaterally affected. In contrast with sporadic bilateral cases, most sporadic unilateral cases are nonhereditary. This was first pointed out by Vogel (*2*) and subsequently confirmed by many investigators (*3–6*). Nonhereditary cases may be due to somatic mutation (*2, 7*), because they cannot be distinguished from hereditary unilateral cases in clinical presentation, natural history, and pathological changes.

Other important observations are: (1) occasional occurrence of two or more affected members in distant collateral relatives, most dramatically presented by Macklin (*8*); (2) a partial deletion of the long arm of No. 13 chromosome (*9*) in a few cases; in the great majority of hereditary cases no such deletion can be detected, even by means of banding techniques (*10*); and (3) an increased risk to a second primary tumor, such as osteosarcoma, of patients who have been treated with radiotherapy because of bilateral disease (*11–13*).

Models for the Genesis of Retinoblastoma

Three different models have been proposed to explain certain facets of the pattern of inheritance and of the age-specific incidence of retinoblastoma: two-mutation model (*7*), delayed mutation model (*14*), and host resistance model (*15*). The two-mutation model assumes that the nonhereditary form is due to two mutations occurring in somatic cells, whereas in the hereditary form with one mutation inherited *via* the germ lines, only one somatic mutation is required to occur at random in either or both eyes. If the mean number of tumors formed per gene carrier is m, the proportion of unaffected, unilaterally and bilaterally affected would be, following a Poisson distribution, e^{-m}, $2e^{-m/2}(1-e^{-m/2})$ and $(1-e^{-m/2})^2$, respectively. The delayed mutation model presupposes a *labile* premutation at the retinoblastoma locus. This concept was originally suggested by Auerbach (*16*) who

TABLE 1. Results of Genetic Analyses of the Data from 231 Sibships with Familial Cases of Retinoblastoma (*15*)

Expressivity of carrier parents	Proportion of children affected	Proportion of bilaterality among affected
Unaffected	0.31 ± 0.03	133/246=0.541
Unilateral	0.42 ± 0.05	96/126=0.762
Bilateral	0.49 ± 0.05	79/88=0.898

TABLE 2. Distribution of Unaffected, Unilaterally and Bilaterally Affected among Children Who Inherited the Gene for Retinoblastoma (*15*)

Expressivity of carrier parents	Distribution of children who inherited the gene			
	Unaffected	Unilateral	Bilateral	Total
Unaffected	0.38	0.28	0.34	1.00
Unilateral	0.16	0.20	0.64	1.00
Bilateral	0.02	0.10	0.88	1.00

found a peculiar pattern of inheritance in *Drosophila* treated with chemical mutagens, and later applied to the pedigrees investigated by Macklin (*8*), who reported that the degree of penetrance in the families may be as low as 20%. The host resistance model is in essence very simple: the three phenotypes of the gene carriers are determined largely by inherited host resistance to the major gene for retinoblastoma.

Table 1 summarizes the results of segregation analyses of familial cases in the literature. Strikingly, both penetrance and expressivity in children who received the gene increased consistently with increasing degree of expressivity in the carrier parents. The possibility of multiple alleles with different penetrance at the retinoblastoma locus could be ruled out because, as shown by Macklin (*8*), penetrance varied considerably from sibship to sibship within the same family. The proportion of affected children born to bilaterally affected parents was close to the expectation from a fully penetrant gene, but it must be slightly lower than 0.50 because in one pedigree (*17*) a bilaterally affected parent did have an unaffected child carrying the gene.

Using the data in Table 1 one can construct the distribution of three phenotypes among children who inherited the retinoblastoma gene (Table 2). It is now possible to test for the hypotheses proposed. According to the delayed mutation model, unaffected carriers and some unilaterally affected cases are gonadic mosaics because they presumably carry a labile premutation (*14*). Then, the proportion of affected among their offspring would vary depending on the proportion of gonadal cells with the completely mutated gene. As a result, there would be no regular pattern to the phenotype distribution among their offspring (*15*).

If tumors are distributed according to a Poisson type, the square of the proportion of unilaterally affected among the gene carriers should be equal to four times the product of the proportion of unaffected and that of those bilaterally affected (*15*). A simple calculation shows a deviation from this expectation by a factor 7 to 10. Thus, the assumption of a Poisson distribution must be rejected.

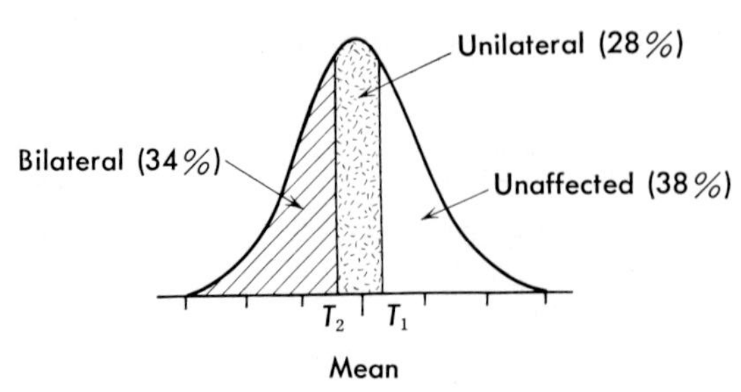

FIG. 1. Distribution of host resistance with two thresholds, T_1 and T_2, among children who inherited the retinoblastoma gene from unaffected carrier parents.

According to the host resistance model, the distribution of the three phenotypes is likely to follow the multifactorial model with two thresholds. Then, assuming a normal distribution of host resistance, the distance between the thresholds is expected to be constant for different groups of children provided the underlying scale is the same. Let T_1 be the threshold beyond which neither eye is affected, and T_2 be the other below which both eyes are affected. For the children with unaffected carrier parents, the deviation of T_1 from the population mean in terms of the standard deviation unit is 0.305 corresponding to the proportion of unaffected carriers (38%), and that of T_2 is -0.412 corresponding to the proportion of bilaterally affected (34%), giving 0.717 for T_1-T_2 (Fig. 1). For the children with unilaterally affected parents T_1-T_2 is 0.636, and for those with bilaterally affected parents it is 0.879. Thus, the three estimates for T_1-T_2 are close to each other, suggesting that the distribution of the three phenotypes is in accordance with the multifactorial model with two thresholds.

While host resistance in general is determined by both genetic and environmental factors, one can imagine, noting a marked shift in the proportion of the three phenotypes among carrier children according to expressivity of their parents (Table 2), that host resistance to the retinoblastoma gene is largely determined by genetic factors. In fact, the heritability, which is a measure for the proportion of variance of the host resistance that is determined by additive genetic variance, was estimated to be about 90% (*15*).

Based on the estimated heritability, it is possible to predict the change in degree of penetrance of the retinoblastoma gene among offspring of unaffected carriers by successive selection. For example, in the fifth generation after such selection, penetrance could decline to as low as 18%. Using the predicted value for penetrance in the offspring of an unaffected carrier, the expected occurrence of retinoblastoma in two collateral relatives was calculated, and this was compared to the observed distribution of pedigrees with collateral relatives affected (Table 3). The latter was found to agree with the expectation if two affected members are only related up to first cousins once removed. There were a small number of pedigrees in excess of the expected number with more distantly related members affected. However, of six such pedigrees, four with only two affected were from the series of Macklin (*8*) who investigated over 4,000 relatives of one particular kindred. Although the probability of the independent occurrence of a second case is as low

TABLE 3. Distribution of Pedigrees Showing Multiple Occurrence of Retinoblastoma among Collateral Relatives Who are Apparently Connected through Unaffected Carriers (*15*)

Affected relatives[a]	No. of pedigrees			
	Total No. of affected		Total	
	2	3 and more	Observed	Expected
Sibs	43	31	74	75.69
Uncle (aunt) and nephew (niece)	—	15	15	17.89
Granduncle and nephew (niece)	1	1	2	6.59
First cousins	4	3	7	4.22
First cousins once removed	2	1	3	1.56
Second cousins	2	1	3	0.58
Second cousins once removed	1	—	1	0.27
Third cousins	1	—	1	0.13
Third cousins once removed	1	—	1	0.07
Total	55	52	107	107.00

[a] Counting from the common ancestor, the first and the second affected belonging to different sibships were taken.

as 1: 20,000, this probability can never be ignored, if one could investigate more and more remote relatives as thoroughly as did Macklin. Therefore, the discrepancy between the observed and expected distributions of pedigrees with distant collateral relatives affected could well be accounted for by coincidence, implying that there is no need to postulate delayed mutation (*15*).

Further Evidence for the Host Resistance Model

The basic assumption for the two-mutation model that in the gene carriers the eyes acquire tumors independently can be examined by a more direct approach (*18*). Table 4 shows the distribution of 215 bilateral cases by interval between diagnosis in the first and second eyes. These cases were ascertained by a national survey carried out in Japan since 1975, supported by the Children's Cancer Association of Japan. In a great majority (92%) of the patients the disease was already bilateral when the diagnosis was first made. This finding itself is by no means

TABLE 4. Distribution of 215 Bilateral Cases by Interval between Diagnosis in First and Second Eyes (*18*)

Interval (months)	No. of cases	Interval (months)	No. of cases
Nil	198	7	1
1	4	8	1
2	1	11	1
3	3	12	1
4	1	20	1
5	1	41	1
6	1	Total	215

new, for the same result has been noted by a number of previous investigators (*19–23*), but it is a striking finding, strongly suggesting a high correlation between age of the patients with bilateral cases at onset in the right and left eyes. If the eyes acquire tumors independently in the probabilistic sense, as assumed by Knudson (*7*), then there should be no correlation, provided each eye is examined independently.

Using the data, a correlation coefficient was calculated to be 0.94, which is obviously biased to an overestimate to some degree because ophthalmoscopy is made usually for both eyes; thus, the chance of detecting a growing tumor in the second eye at the same time is enhanced. To correct errors in the estimate, we must know, on the basis of data for the grade of tumor progression, how much time would elapse before the disease is diagnosed in the second eye if the eyes are examined independently. Fortunately, information about Reese group of tumors was available for 139 cases; this is a classification of tumors proposed by Reese (*24*), taking into account the size of tumors and prognosis.

Table 5 gives the distribution of these cases by Reese group. Only 15 persons (10.8%) were still free of detectable tumors in the second eye, whereas in 20 patients (14.4%), tumors in both eyes were of the same group, which suggests a correlation not only with respect to the time when the tumor is detected in each eye but also

TABLE 5. Distribution of 139 Bilateral Cases by Reese Group When the Diagnosis was First Made in Either Eye (*18*)

Right eye		Left eye						
		No tumor yet	Reese group					Total
			I	II	III	IV	V	
No tumor yet				1		2	4	7
Reese group	I		3	1		4	19	27
	II		4	1	3	2	12	22
	III		4	4	1	3	4	16
	IV	1	2	1	3	6	5	18
	V	7	11	12	9	1	9	49
Total		8	24	20	16	18	53	139

TABLE 6. Reese Group of Tumors in Some of the Bilateral Cases for Which the Diagnosis in the Second Eye was Delayed (*18*)

Case No.	First diagnosed eye		Second eye	
	Age at diagnosis (month)	Reese group	Age at diagnosis (month)	Reese group
50183	18	II	19	I
50269	5	V	6	I
50175	2	V	3	II
50299	19	V	20	II
52008	5	V	7	III
50414	2	V	9	I

with respect to the grade of progression. If we assign zero value to the eye with no tumor yet and one to group I and so on, then the intra-class correlation for Reese group between eyes is 0.445, which is significantly ($P<0.01$) different from zero.

Table 6 presents data from some of the bilateral cases in which the diagnosis in the second eye was delayed. From the interval between ages of the patients at diagnosis for each eye, one can estimate the maximum possible length in months needed for a tumor to progress to the respective group. Thus, in the first two patients, tumors of group I were detected in their second eyes 1 month following the diagnosis of the disease in their first eyes, and in the next two patients tumors of group II were detected likewise after 1 month. In the fifth patient, a tumor of group III was detected after 2 months. Therefore, one can infer that the time required for a growing tumor to progress by one grade in the Reese group is less than 2 months and probably about a month. This inference is also supported by the fact that in the present series, 18 patients had bilateral cases in which a tumor of group V was already present in one eye in the first month after birth. The latter finding in turn suggests that tumor development could start as early as the fifth month of fetal life.

Assuming that a growing tumor in the second eye progresses by one grade in the Reese group per month and that parents take their children to a physician only when the tumor has progressed to the same group as did the tumor in the other eye, one can estimate the age at which the diagnosis would have been made in the second eye if the eyes were examined independently. For example, if the tumor in the second eye was of group I when a tumor of group V was detected in the first eye of a patient at 6 months of age, then the age at diagnosis in the second eye would be 10 months. The same correction was applied to the 15 patients with bilateral cases for which the diagnosis in the second eye was delayed, because the second eye in these patients must have been under ophthalmological surveillance after the disease was diagnosed in the first eye. Thus, the intra-class correlation between ages of the 139 patients at diagnosis of the disease in the right and left eyes was reestimated at 0.819 ($P<0.001$). The correlation was 0.893 if no correction was made, and it was 0.684 if corrected by the assumption that 2 months are needed for a growing tumor to progress by one grade. As the latter is certainly an overcorrection, it is safe to conclude that the true value for intra-class correlation is not smaller than 0.70.

The above finding implies that in bilateral retinoblastoma age of the patient when the tumor is formed in each eye is largely determined by host factors common to both eyes. The same may be true for patients with hereditary unilateral cases that can be regarded as a less susceptible group according to the host resistance model. If genetic factors were involved in the variation in age at onset, we should expect that ages at diagnosis of inherited cases would vary according to the degree of expressivity of the carrier parents. As shown in Table 7, the mean age of the patients at diagnosis of the hereditary unilateral cases was 21.5 months if the parent was an unaffected carrier, and 15.7 months if the parent was unilaterally affected. The same trend is seen for the bilateral cases. The mean age of the patients at diagnosis of the bilateral cases with a bilaterally affected parent was close to that

TABLE 7. Mean Age of Patients at Diagnosis of Inherited Cases of Retinoblastoma according to Expressivity of Carrier Parents (*18*)

Expressivity of carrier parents	Affected children					
	Bilateral			Unilateral		
	No. of children	Mean age at diagnosis (months)	SD	No. of children	Mean age at diagnosis (months)	SD
Unaffected	17	13.9	15.41	16	21.5	14.53
Unilateral	20	9.4	8.46	6	15.7	13.49
Bilateral	12	9.3	11.42	1	1	—
Total	49	10.9	11.92	23	19.1	14.41

of the patients with a unilaterally affected parent. However, several bilaterally affected parents in the present series had spouses who were also blind, so that the time when the parents first noticed something wrong with their child's eye tended to be delayed, resulting in higher ages of the children at diagnosis. Taking this into account, the observed pattern of variation in the mean age of the patients at diagnosis of the inherited cases according to parental phenotype is wholly consistent with the expectation from the host resistance model. Since heritability of the host resistance was estimated from segregation data at approximately 90% (*15*), we may conclude that variation in age at onset in the gene carriers is largely determined by *inherited* host factors (*18*).

Nature of the Second Hit

The high correlation in bilateral cases allows us to infer something of the nature of the presumed second hit initiating tumor formation. We agree with Knudson (*7*) that a second hit must occur because in bilateral cases only a few of the millions of retinal cells with a first mutation develop into a tumor. However, because mutation is a change in the genome occurring randomly not only with respect to locus but also with respect to the structure of DNA, it is highly improbable that a second mutation with the same effect occurred in each eye *almost simultaneously* in most persons with bilateral cases. The possibility that retinoblastoma cells are homozygous for the gene seems to be even more remote. However, if one accepts the prevailing view that cancer is basically a disorder of differentiation of some cells of the body and that differentiation is ultimately a question of determination of gene functions, then the second hit is likely to be an error arising not from a mutational process but from a loosening or changing of the normal determination state of single cells or groups of cells. In the individuals carrying the retinoblastoma gene, every retinal cell can be regarded as liable to such an error. However, because differentiation must be biologically stable, genetic mechanisms could have evolved by natural selection to ensure the stability. Such a safety device may be complete in the resistant group who could remain unaffected, but it may not be complete in the less resistant group; thus, a few retinal cells could escape from it. Then one can explain the high correlation between eyes with respect to the ages of the persons at onset of bilateral cases by assuming that retinal cells in each eye are, as a rule, *equally* liable at a given stage of development, although the liability may decline with increasing age. The high correlation (93%) between total finger ridge counts in the right and left hands (*25*) clearly indicates that differentiation in the development of bilateral organs proceeds generally at an equal rate.

Host Resistance to Nonradiogenic Osteosarcoma in the Gene Carriers

As reviewed earlier, survivors of bilateral disease, but not the majority of survivors of unilateral disease, are prone to develop second primary tumors, mainly osteosarcoma, not only in the area of irradiation but also outside this zone (*11–13*). From this, Kitchin and Ellsworth (*13*) argued that the increased risk to second primary tumors is a pleiotropic effect of the retinoblastoma gene. According to the host resistance model, unaffected carriers are inherently resistant to tumor formation, whereas persons with bilateral cases are the most susceptible, hereditary unilateral cases being the intermediate. Therefore, if the susceptibility to nonradiogenic osteosarcoma is a pleiotropic effect of the retinoblastoma gene, a question is raised as to whether host resistance to the gene effects is tissue-specific or not.

Surveying the literature, data were collected from 26 cases of nonradiogenic osteosarcomas that developed in patients with retinoblastoma as well as in their close relatives (*26*). Table 8 gives the distribution of these cases by phenotypes with respect to retinoblastoma, as compared with the proportion expected from the assumption that the patients are all carriers of the gene and the risk to osteosarcoma does not differ by phenotypes. Of the four patients who were free from retinoblastoma, one could be defined as a *definite* carrier from the family data, and the three were considered *probable* carriers because they had at least one sib affected with retinoblastoma. However, the prevalence of nonradiogenic osteosarcoma in the general population is of the order of 10^{-5} (*27*), which is increased to about 1.2% among survivors of bilateral retinoblastoma (*13*). Therefore, it is reasonable to assume that the three who were considered *probable* carriers were in fact carriers of the retinoblastoma gene. Since the risk to nonradiogenic osteosarcoma does not differ by the phenotypes, host resistance to the gene effects is tissue-specific. In other words, there is no correlation in the liability to malignant transformation between retinal cells and bone cells of the gene carriers. Owing to this finding, it may be better to replace the term "host resistance" by "tissue resistance."

TABLE 8. Distribution of 26 Patients Carrying the Retinoblastoma Gene, by Phenotypes, Who Developed Nonradiogenic Osteosarcoma (*26*)

	Phenotype with respect to retinoblastoma			Total
	Unaffected	Unilateral	Bilateral	
Observed	4[a]	2	20	26
Expected	5.4	3.1	17.4	25.9

[a] Three of the four are probably the gene carriers.

Furthermore, we have found evidence that the liability of the gene carriers to osteosarcoma could be regarded as a multifactorial threshold character, and the heritability was estimated at 71% (*26*). This is comparable to the value of 90% estimated for the heritability of tissue resistance of the gene carriers to retinoblastoma (*15*). The lower heritability of the liability to osteosarcoma may be due to the fact that osteosarcoma develops later than does retinoblastoma, therefore environmental factors could play a greater role in the manifestation of the disease. Although allowance should be made for possible confounding of environmental factors common to the patients and their close relatives, the 71% heritability implies that *inherited* tissue susceptibility plays an important role in the development of osteosarcoma, as it was true for the development of retinoblastoma. This conclusion is supported by the occasional familial occurrence of osteosarcomas (*28*).

Radiogenic Osteosarcoma in the Gene Carriers

In their follow-up investigation of 243 patients with bilateral retinoblastoma who had been treated with radiotherapy, Sagerman *et al.* (*11*) found 9 (3.70%) who developed osteosarcoma in the irradiated area. As osteosarcoma was most common among various histologic types of radiogenic malignant neoplasia in their series, and because carriers of the retinoblastoma gene are inherently susceptible to nonradiogenic osteosarcoma, there is no doubt that, as first suggested by Jensen and Miller (*12*), patients with hereditary retinoblastoma are more susceptible than usual to the induction of osteosarcoma by X-irradiation. Strong (*29*) showed that the latency period for radiogenic osteosarcoma was shorter in patients with bilateral retinoblastoma than in those with other childhood cancers. As the short interval is apparently not a biologic characteristic of radiogenic sarcomas in general, she concluded that the patients with bilateral retinoblastoma represent a unique subgroup who have inherited the first mutational step predisposing to sarcoma, and interpreted this finding, in the light of the two-mutation model (*7*), as evidence compatible with radiation providing the *second* mutation in the carcinogenic pathway. How then could the role of radiation, which is the most potent mutagen, in the induction of osteosarcoma in the gene carriers be accounted for by the host resistance model, according to which the crucial hit initiating tumor formation is not a mutation but a host-dependent error in differentiation?

Although the retinoblastoma gene has a potent carcinogenic effect in the retinal and/or bone tissues, not all of the gene carriers develop cancers; about 13% of the gene carriers in the general population who received a new mutation at this locus do not manifest retinoblastoma (*15*), and 98% of the gene carriers do not develop osteosarcoma unless they received radiation. Consequently, there must be variation in tissue resistance to the gene effects. We have found evidence that the tissue resistance is largely determined by genetic factors. In the gene carriers, every retinoblast or osteoblast can be regarded as liable to malignant transformation. However, as only a few of the millions of differentiating cells at risk develop into a tumor, the genes involved in the tissue resistance must be suppressors of an error in differentiation, which is presumably a crucial event in malignant transformation. There-

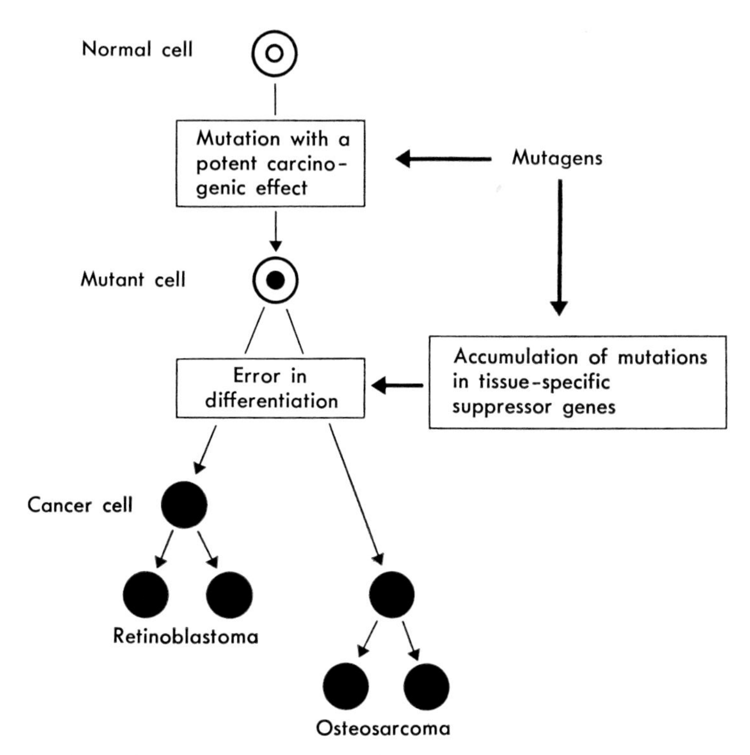

FIG. 2. A diagram relating steps in the genesis of retinoblastoma and osteosarcoma to mutagens. Presumed contents of the black boxes are shown in rectangles. Since prenatal X-irradiation could increase the risk to all types of embryonal cancers in irradiated offspring (*30*), it is assumed that mutagens can also induce a dominant cancer mutation.

fore, it is reasonable to conclude that radiogenic osteosarcoma in the gene carriers is ascribable to the induction of somatic mutations in that tissue-specific suppressor system (Fig. 2). The number of genes in the system is probably not large, and if induced mutations have accumulated in the cell at risk to an extent that exceeds a certain threshold, then an error in differentiation would follow. The number of induced mutations required to exceed the threshold is perhaps few, but may be different from individual to individual according to the number of mutant genes that have been inherited *via* the germinal lines. This interpretation is compatible with the long interval usually observed for radiation-induced cancers, and also with the multistep mutation model proposed for adult cancers (*31*), except in one point: the final step in carcinogenesis is assumed, in our model, to be an error in differentiation resulting from germinal and/or somatic mutations in the suppressor system.

CONCLUSION

The questions raised at the outset of this article can be answered as follows: the minimal number of carcinogenic steps is two; the first step is a mutation having a potent carcinogenic effect, while the second step is probably a host-dependent error in differentiation. The host resistance to the second hit is a multifactorial threshold character largely determined by tissue-specific suppressor genes.

REFERENCES

1. Griffith, A. D. and Sorsby, A. The genetics of retinoblastoma. Br. J. Ophthalmol., *28*: 279–293, 1944.
2. Vogel, F. Neue Untersuchungen zur Genetik des Retinoblastoms (Glioma retinae). Z. Menschl. Vererb. Konstitutionsl., *34*: 205–236, 1957.
3. Schappert-Kimmijser, J., Hemmes, G. D., and Nijland, R. The heredity of retinoblastoma. Ophthalmologica, *151*: 197–213, 1966.
4. Nielsen, M. and Goldschmidt, E. Retinoblastoma among offspring of adult survivors in Denmark. Acta Ophthalmol., *46*: 736–741, 1968.
5. Briard-Guillemot, M. L., Bonaïti-Pellié, C., Feingold, J., and Frézal, J. Étude génétique du retinoblastoma. Hum. Genet., *24*: 271–284, 1974.
6. Matsunaga, E. and Ogyu, H. Retinoblastoma in Japan: Follow-up survey of sporadic cases. Jpn. J. Ophthalmol., *20*: 266–282, 1976.
7. Knudson, A. G., Jr. Mutation and cancer: Statistical study of retinoblastoma. Proc. Natl. Acad. Sci. U.S., *68*: 820–823, 1971.
8. Macklin, M. T. A study of retinoblastoma in Ohio. Am. J. Hum. Genet., *12*: 1–43, 1960.
9. Lele, K. P., Penrose, L. S., and Stallard, H. B. Chromosome deletion in a case of retinoblastoma. Ann. Hum. Genet., *27*: 171–174, 1963.
10. Ladda, R., Atkins, L., Littlefield, J., and Pruett, R. Retinoblastoma: Chromosome banding in patients with heritable tumour. Lancet, *ii*: 506, 1973.
11. Sagerman, R. H., Cassady, J. R., Tretter, P., and Ellsworth, R. M. Radiation induced neoplasia following external beam therapy for children with retinoblastoma. Am. J. Roentgenol., *105*: 529–535, 1969.
12. Jensen, R. D. and Miller, R. W. Retinoblastoma: Epidemiologic characteristics. N. Engl. J. Med., *285*: 307–311, 1971.
13. Kitchin, F. D. and Ellsworth, R. M. Pleiotropic effects of the gene for retinoblastoma. J. Med. Genet., *11*: 244–246, 1974.
14. Herrmann, J. Delayed mutation as a cause of retinoblastoma: Application to genetic counseling. Birth Defects: Orig. Artic. Ser., *12*(1): 79–90, 1976.
15. Matsunaga, E. Hereditary retinoblastoma: Delayed mutation or host resistance? Am. J. Hum. Genet., *30*: 406–424, 1978.
16. Auerbach, C. H. A possible case of delayed mutation in man. Ann. Hum. Genet., *20*: 266–269, 1956.
17. Ellsworth, R. A. The practical management of retinoblastoma. Trans. Am. Ophthalmol. Soc., *67*: 462–534, 1969.
18. Matsunaga, E. Hereditary retinoblastoma: Host resistance and age at onset. J. Natl. Cancer Inst., *63*: 933–939, 1979.
19. Böhringer, H. R. Statistik, Klinik und Genetik der Schweizerischen Retinoblastomfälle (1924–1954). Arch. Klaus Stift. Vererbungsforsch., *31*: 1–16, 1956.
20. Bech, K. and Jensen, O. A. Bilateral retinoblastoma in Denmark. Acta Ophthalmol., *39*: 561–568, 1961.
21. Gordon, H. Family studies in retinoblastoma. Birth Defects: Orig. Artic. Ser., *10*(10): 185–190, 1974.

22. Lennox, E. L., Draper, G. J., and Sanders, B. M. Retinoblastoma: A study of natural history and prognosis of 268 cases. Br. Med. J., *3*: 731–734, 1975.
23. Aherne, G.E.S. and Roberts, D. F. Retinoblastoma—A clinical survey and its genetic implications. Clin. Genet., *8*: 275–290, 1975.
24. Reese, A. B. Tumors of the Eye, 3rd ed., p. 118, Harper & Row, New York, 1976.
25. Holt, S. B. The Genetics of Dermal Ridges, p. 53, C. C. Thomas, Springfield, Ill., 1968.
26. Matsunaga, E. Hereditary retinoblastoma: Host resistance and second primary tumors. J. Natl. Cancer Inst., *65*: 47–51, 1980.
27. Glass, A. G. and Fraumeni, J. F. Epidemiology of bone cancer in children. J. Natl. Cancer Inst., *44*: 187–199, 1970.
28. Epstein, L. I., Bixler, D., and Bennett, J. E. An incident of familial cancer, including 3 cases of osteosarcoma. Cancer, *25*: 889–891, 1970.
29. Strong, L. C. Theories of pathogenesis: Mutation and cancer. *In*; J. J. Mulvihill, R. W. Miller, and J. F. Fraumeni, Jr. (eds.), Genetics of Human Cancer, pp. 401–415, Raven Press, New York, 1977.
30. Bithell, J. F. and Stewart, A. M. Pre-natal irradiation and childhood malignancy: A review of British data from the Oxford survey. Br. J. Cancer, *31*: 271–287, 1971.
31. Cairns, J. The cancer problem. Sci. Am., *233*(5): 64–78, 1975.

GENETIC AND ENVIRONMENTAL FACTORS IN EXPERIMENTAL AND HUMAN CANCER,
H. V. GELBOIN ET AL. (EDS.), JAPAN SCI. SOC. PRESS, TOKYO, PP. 175–186, 1980

Bloom's Syndrome. IX: Review of Cytological and Biochemical Aspects

James GERMAN and Steven SCHONBERG
The New York Blood Center, New York, N.Y. 10021, U.S.A.

Abstract: Bloom's syndrome (BS), perhaps more than any other recessively inherited disease of man, increases the risk that cancer of some type will occur. (More cancers occur in persons with xeroderma pigmentosum, but there only the skin is affected whereas in BS many different tissues are involved.) The explanation for this is unknown, as is the biochemical defect in BS.

Two features of BS have stimulated most of the laboratory experimentation aimed at explaining the high cancer risk. These are clinical hypersensitivity to sun exposure and an abnormally great amount of chromosome instability in cells in culture. Prompting the experimentation has been the tacit assumption that BS cells become cancerous more readily than normal. In various laboratories, DNA-repair mechanisms have been examined, sensitivity to oncogenic agents assayed, and studies of DNA replication made. In view of the large number of studies made it is noteworthy how few have been "positive." These include decreased host-cell reactivation of UV-irradiated virus in one of the several cell lines examined; excessive sensitivity to the mutagen ethyl methanesulfonate; and, abnormally slow replication-fork progression during semiconservative DNA synthesis.

Many of the phenomena which are detectable as abnormalities in BS cells (*e.g.*, sister-chromatid exchange, chromosome aberrations, mutagen sensitivity) are demonstrable in normal cells also, but at a much lower level. BS therefore conceivably serves as a model for oncogenic transformation in general, with only the probability, and not the ultimate cause, of transformation being different.

A third important feature of BS, proneness to bacterial infection, has stimulated relatively few studies. In these few, however, a severe impairment in immune function has been demonstrated, which raises the possibility that the cancer proneness in BS is on the same basis as that in other immunodeficiency states. Unfortunately, that basis also is unknown. It is noteworthy that both BS and another rare genetic disorder, ataxia telangiectasia, share immune deficiency, chromosome instability, and a striking predisposition to cancer. The relative roles, if any, of mutation and immune deficiency in oncogenic transformation and the emergence of clinical cancer should become apparent by continued analysis of such disorders.

The remarkable cancer predisposition in Bloom's syndrome (BS), first reported in 1965 (*1*) and updated periodically (*2*, *3*), has led to a considerable interest in the disease as a possible model for the study of the development of cancer. Recently the clinical and genetic aspects of BS were reviewed (*4*). Here we review the cytological and biochemical aspects, summarizing the numerous experimental studies which have been made of cells in culture. Also, the few studies of the immunological status of BS will be mentioned. The main objective of these studies has been elucidation of the primary genetic defect in BS, which still remains quite obscure.

TABLE 1. Review of Cytological, Biochemical, and Immunological Findings in BS Cells *in Vitro*

Structure or function evaluated Assay system and parameter studied	Increased	Decreased	As in controls
Chromosomes			
Aberrations			
All types, no treatment	*1*, *5*a)		
X-ray induced		*6*b)	*7*b)
Micronuclei, no treatment	*8*		
Micronuclei, UV-induced	*9*c)		*9*
Micronuclei, EMS-induced	*10*		
Homologous exchange (Qrs, TAs—see text), no treatment	*11*, *12*a)		
Qrs, mitomycin C-induced	*13*d)		
Qrs, after lymphocyte proliferation at 40°C	*14*		
Sister-chromatid exchange			
Baseline	*15*, *16*a,e)		
UV-induced	*9*c)		*9*
EMS-induced	*10*		
Mitomycin C-induced			*17*f)
Caffeine effect			g)
Protease inhibitors effect (antipain, leupeptin)			h)
Puromycin effect			i)
Hybridization with diploid human fibroblasts		*18*	
Hybridization with rodent cell lines (A9, CHO)		*19*	
Co-cultivation with diploid human fibroblasts		*20*	*21*, *22*
Co-cultivation with rodent (CHO) cells		*23*	
Medium "conditioned" by diploid human fibroblasts		*20*	
Medium "conditioned" by rodent (CHO) cells			*23*
Cell survival after			
γ-irradiation			*24*
UV-irradiation		*25*	*9*, *24*, *26*
EMS treatment		*24*	
MMS treatment			*24*
DNA Repair			
Excision repair			
UDS after UV-irradiation	*6*	*25*	*7*, *27*, *28*
UDS after AAAF±UV			*28*

Continued...

TABLE 1. Continued.

Structure or function evaluated Assay system and parameter studied	Increased	Decreased	As in controls
Photolysis of BrdU incorporated into parental DNA during repair			*28*
Loss of UV-endonuclease sensitive sites			*28*
Host-cell reactivation of			
UV-irradiated HSV-1		*26*[c]	*26*
UV-irradiated adenovirus-2		*9*[c]	*9*
Others			
Repair of X-ray induced single-strand breaks			*29*
Replicative bypass DNA synthesis (post-replication repair)			*30*
Enhancement of γ-irradiated DNA priming activity for DNA polymerase		*31*	
Apurinic-site specific endonuclease activity			*31, 32*
Others			
DNA polymerase activities (α, β, γ)			*32–34*
DNA fork movement (fiber autoradiography)		*35, 36*	
DNA chain maturation		*25*	
"Hybrid" DNA after one round of replication in BrdU	*37*		*38*j)
Cell transformation by SV40 virus			*39*
Production of a factor which increases SCE in other cells	*21*		*19, 20, 22, 23*
Mutagen production (Ames test on whole cell lysates of BS fibroblasts)			k)
Fibroblast proliferation in response to serum, EGF, and FGF		l)	
Immune function			
Proliferative response of lymphocytes to PWM		*40, 41*[m]	*41*[m]
Proliferative response of lymphocytes to PHA		*41*[m]	*40, 41*[m]
Proliferative response in MLC		*40, 41*	
Stimulatory ability in MLC			*40, 41*
Serum immunoglobulin concentrations		*40–42*	

a) Many additional references not included here. b) The discrepancy in these results may be due to differences in protocols employed. Tice *et al.* (*6*) irradiated cells in G_2; Evans *et al.* (*7*) irradiated cells in G_0. c) The response of one particular BS fibroblast cell line, GM-1492, is different from that of several other BS cell lines. d) Schonberg, S., unpublished results. Mitomycin C also induces more TAs in BS lymphocytes than in control lymphocytes. As is the case with Qrs, the frequency of involvement of the various metaphase chromosomes in TAs is similar with and without the drug. e) Present report, Tables 2 and 3. f) Shiraishi and Sandberg (*17*) did report a reduced yield of mitomycin C-induced SCE relative to baseline SCE in BS cells as compared to control cells. The absolute increase in SCE was similar in BS and control cells with varying concentrations of mitomycin C. g) Present report, Table 3. h) Schonberg, S. and German, J., manuscript in preparation. i) Our unpublished observations. j) Loveday, K. S. and Latt, S. A., personal communication. k) Dubroff, L. M. and Hemphill, E., personal communication. l) Lechner, J. F., Groden, J., and German, J., manuscript in preparation. m) Weemaes *et al.* (*41*) found decreased lymphocyte proliferative responses to PHA and PWM in one family (two sibs with BS) but normal responses in a second family (two sibs with BS). Abbreviations: AAAF, acetoxy acetylaminofluorene; BrdU, bromodeoxyuridine; BS, Bloom's syndrome; EGF, epidermal growth factor; EMS, ethyl methanesulfonate; FGF, fibroblast growth factor; HSV-1, herpes simplex virus-1; MLC, mixed lymphocyte culture system; MMS, methyl methanesulfonate; PHA, phytohemagglutinin; PWM, pokeweed mitogen; Qr, quadriradial configuration; SCE, sister-chromatid exchange; TA, terminal association; UDS, unscheduled DNA synthesis; UV, ultraviolet light, 254 nm.

Table 1 summarizes the findings in the various cytological studies, biochemical studies, and analyses of immune function that have been made of BS cells. Space limitation prevents the discussion of many of these findings in the text of this review, and the reader is referred to the relevant reports for further details. Discrepancies in the findings will be commented upon either in footnotes to Table 1 or in the text. Discussion in the text will concentrate on "positive" findings in experimental reports of BS cells in culture, including some recent results from our laboratory.

Cytogenetic Studies

In 1965, German *et al.* (*1*) reported a high frequency of chromosomal breakage and rearrangement in blood lymphocytes in short-term cultures from persons with BS. Subsequently, this chromosomal instability has been found also in dermal fibroblast lines (*8*), freshly aspirated bone marrow cells (*43*, and our unpublished results), and some Epstein-Barr virus (EBV)-transformed lymphoblastoid cell lines (LCLs) derived from persons with this syndrome. Typical of the instability of BS cells are isochromatid gaps and breaks, sister-chromatid reunions, transverse breakage at centromeres, dicentric chromosomes, and acentric fragments. A highly characteristic finding in BS cells in culture is a symmetrical quadriradial configuration (Qr) (Fig. 1a, d, and e). (The Qrs characteristic of BS have been interpreted as cytological evidence for mitotic crossing-over *in vitro* (*44*).) Analyses of the distribution of Qrs with respect to chromosome length have indicated that Qrs do not

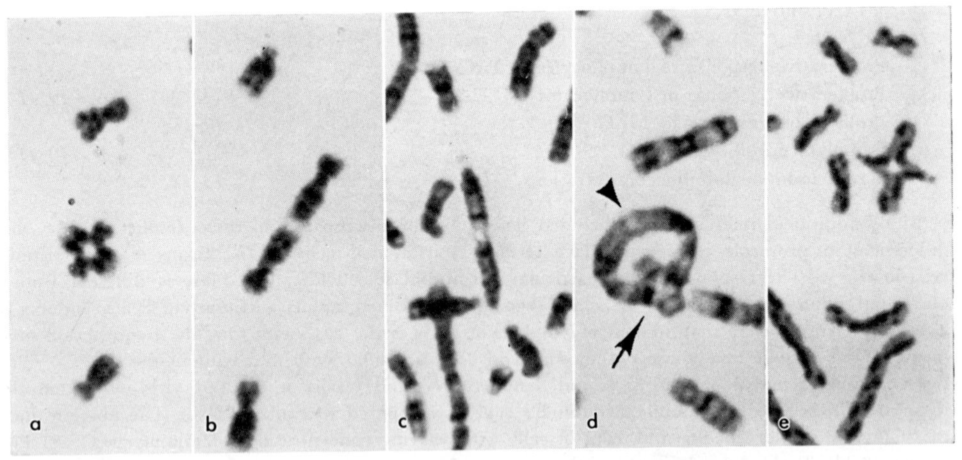

FIG. 1. Five examples of aberrations affecting homologous chromosomes in Bloom's syndrome (BS) lymphocytes. (a)–(d) G banding. (a) Quadriradial configuration (Qr) formed between chromosome Nos. 11. An exchange had occurred between the long arms, at or near the centromeres. (b), (c) Terminal associations (TAs) formed between the long arms of chromosome Nos. 9 (b) and Nos. 13 (c). (d) Homologues, the Nos. 1, displaying both a Qr (arrow) due to exchange between the long arms and a TA (arrowhead) between the short arms. This is the only configuration of this type we have detected in the several hundred Qrs and TAs observed in this laboratory. (e) Qr in cell stained for sister-chromatid differentiation. A full description of our interpretation of the sister-chromatid staining pattern in configurations such as this is to be found in Ref. *15*.

occur at random per unit length (*11*, *45*; Schonberg and German, manuscript in preparation). Recently we reported a new cytogenetic finding in BS cells, terminal association (TA) of homologous chromosomes at metaphase (*12*). Affected homologues assume a reverse tandem position (pq:qp or qp:pq), without the cross formation characteristic of Qrs (Fig. 1b,c, and d). The frequency of involvement of the various metaphase chromosomes (identified by banding) in TAs has been analyzed and found to be nonrandom: an excess of long-arm associations occurs, affecting particularly the smaller chromosomes in the complement; and, when association of short arms does occur, those chromosomes bearing the longest short arms are affected preferentially.

In 1974, BS cells were reported to show a great increase over normal cells in their baseline level of sister-chromatid exchange (SCE) (*15*) (Fig. 2). In a subsequent study of SCE in lymphocytes from 21 individuals with BS (*16*), it was demonstrated that, while each individual did have blood lymphocytes with increased SCE, five of them simultaneously exhibited in a proportion of their lymphocytes an amount of exchange similar to that found in controls. Subsequently, three more individuals with BS have been found to demonstrate this as yet unexplained phenotypic dimorphism (*10*, *46*, and footnote "a" to Table 3). Increased SCE has now been found to be characteristic of several different types of BS cells in culture, including dermal fibroblasts (*22*, *47*, *48*), bone marrow cells (*43*), and some but not all EBV-transformed LCLs (*49*). Samples of data collected in our laboratory for these various cell types are presented in Table 2.

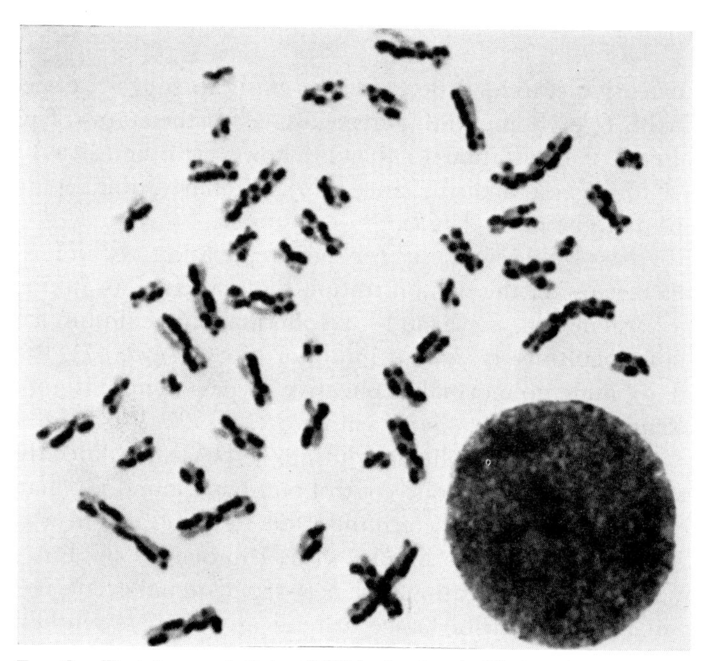

FIG. 2. Phytohemagglutinin (PHA)-stimulated BS lymphocyte in metaphase stained to differentiate sister chromatids. In our laboratory, PHA-stimulated lymphocytes from normal individuals display a mean of 8–10 sister-chromatid exchanges, whereas 152 can be counted here.

TABLE 2. Baseline SCE in Fibroblast Lines (F), Lymphoblastoid Cell Lines (LCLs), and Freshly-aspirated Bone Marrow Cells (BM) from Individuals with BS

Cell type[a]	Source of cells[b]	Designation	SCE	
			Mean±SEM	Range (No. of cells)
F	53(StAs)	HG 916	47.4±2.4	29–66 (21)
F	26(SaTi)	HG 369	61.7±2.4	34–89 (30)
F	71(HaEn)	HG 1290	44.7±1.6	31–61 (25)
F[c]	71(HaEn)	HG 1290	45.3±2.6	20–70 (25)
LCL[d]	15(MaRo)	HG 1036	42.5±1.8	28–72 (34)
LCL[d]	15(MaRo)	HG 1037	41.9±2.2	22–62 (18)
LCL[d]	15(MaRo)	HG 1270	64.6±3.0	35–99 (25)
LCL[e]	11(IaTh)	HG 1100	7.5±0.7	3–16 (24)
BM	32(MiKo)	—	49.0±3.2	25–66 (16)

[a] All cell types cultured in the presence of BrdU (10 μM) for two cell generations and stained by standard procedures for sister-chromatid differentiation. [b] Cell donors identified as in Bloom's Syndrome Registry (*2, 3*). [c] SCE determined on cell line HG 1290 on two occasions. [d] Data from three different LCLs established from one blood sample. [e] LCL derived from an individual displaying the phenotypic dimorphism with respect to SCE reported earlier (*16*) and discussed in the text. Line was apparently derived by Epstein-Barr virus transformation of a circulating B lymphocyte with a low level of SCE.

The fortuitous finding in BS cells of Qrs demonstrating sister-chromatid differentiation (*15*) (Fig. 1e) has lent considerable support to the idea that these configurations actually do represent exchange between homologous chromosomes.

Studies of DNA Metabolism

Despite an extensive search, little evidence exists to suggest a specific defect in DNA repair in BS cells. Some ambiguity exists as to the status of unscheduled DNA synthesis after UV irradiation (Table 1); however, other assays of excision repair have failed to suggest a disturbance (*28*). Similarly, the repair of single-strand breaks appears to be normal (*29*).

Replicative bypass DNA synthesis (or post-replication repair) is of special interest in BS cells because of the demonstration that caffeine, an inhibitor of such synthesis in some systems (*e.g.* "variant" xeroderma pigmentosum cells, rodent cells), under certain conditions is also an inhibitor of SCE (*50, 51*). BS fibroblasts have been found to undergo normal replicative bypass synthesis following UV irradiation; the synthesis is not caffeine sensitive (*25, 30*). We have studied the effect of caffeine on SCE in phytohemagglutinin (PHA)-stimulated lymphocytes from two persons with BS and from one control and have found no change in SCE (Table 3). Caffeine has also been determined not to have an effect on host-cell reactivation of UV-irradiated herpes simplex virus-1 in one BS cell line (*26*). Thus, though SCE is by necessity a replicational or post-replicational event, no association of the high SCE in BS with a disturbance in replicative bypass synthesis has been demonstrable. The relation, if any, of SCE to such synthesis remains obscure.

Though BS cells appear to be proficient at DNA repair, a considerable amount of evidence does exist for a disturbance of some type in DNA metabolism. The

TABLE 3. Effect of Caffeine on SCE in BS and Control (C) Lymphocytes

Source of cells	Caffeine	SCE	
		Mean±SEM	Range (No. of cells)
BS, 50(JeBl)	+	78.3±3.4	55–107 (18)
BS, 50(JeBl)	−	82.6±4.6	54–128 (20)
BS, 59(FrFi)	+	108.9±3.1	83–144 (25)
BS, 59(FrFi)	−	118.4±4.2	54–154 (25)[a]
C, StSch	+	10.6±0.6	5–18 (25)
C, StSch	−	10.6±0.5	4–22 (54)

Caffeine (1 mM) and BrdU were added to cultures 24 hr after initiation. Cells from 50(JeBl) and from StSch were grown for 72 hr in the presence of caffeine and BrdU (100 μM). Cells from 59(FrFi) were grown for 48 hr in the presence of caffeine and BrdU (10 μM), following a newer protocol for sister-chromatid differentiation. [a] One cell from 59(FrFi) had 7 SCEs, indicating that this individual manifests the phenotypic dimorphism with respect to SCE (*16*). This cell was excluded from the calculations.

bulk of this evidence is from cytogenetic studies, reviewed above. In addition to the cytogenetic disturbance, a slower than normal progression of the DNA-replication-fork is demonstrable in BS cells both by DNA fiber autoradiography (*35, 36*) and by pulse-chase experiments with tritiated thymidine (*25*). Replication unit length, incidence of bidirectional replication, and degree of initiation synchrony were all found to be normal by fiber autoradiography (*36*). The slow fork progression is apparently not due to a defect in any of the known DNA polymerases; activities for polymerases α, β, and γ have been found normal (*33, 34*).

Studies of mutagen hypersensitivity have been positive for the DNA alkylating agent ethyl methanesulfonate (EMS), as measured by several different parameters (*10, 24*) (Table 1). UV hypersensitivity has been reported for two diploid lines of BS fibroblasts (*25*), while other lines have been found normal in this respect (*9, 24, 26*). It is not clear whether this discrepancy is due to differences between the cell lines. (The possibility of genetic heterogeneity in BS has not been eliminated; cell hybridization studies addressing this issue are in progress in our laboratory.)

Co-cultivation Experiments

The apparently normal repair of DNA in BS cells has prompted a search for other mechanisms to explain the chromosome aberrations, in particular the elevated SCE. Thus, several groups have co-cultivated BS with non-BS cells to determine whether an (any possible) effect on SCE in either cell type is demonstable. These experiments have resulted in the following conflicting reports.

Tice *et al.* reported data indicating that BS cells produce a factor capable of increasing SCE in other human cells (*21*). BS cell SCE itself did not change upon such co-cultivation with normal cells. However, Rüdiger *et al.* (*20*) found no effect on SCE in diploid human cells upon co-cultivation with BS cells, instead reporting that SCE in BS cells decreases (*20*). We have shown that SCE remains normal in diploid human cells that have been in obligatory metabolic coupling with BS cells over two cell division cycles (*22*). In addition, we have repeated and have been unable to confirm one of the experiments of Tice *et al.* (*21*) in which it was reported that medium "conditioned" by BS cells elevates SCE in other cells. We conclude from our experiments that BS cells do not produce and transfer to other cells a factor capable of increasing SCE in those cells. Unlike the findings of Bartram *et al.* (*20*), our experiments resulted in no diminution of SCE in BS cells. However, the conditions of our co-cultivation system may have precluded such an effect.

Van Buul *et al.* have reported that, upon co-cultivation of BS cells with a permanent cell line of Chinese hamster ovary cells, SCE decreases in the BS cells (*23*). It is unclear whether these results are directly comparable to those discussed above, all of which employed diploid human fibroblasts.

Studies of Immune Function

Repeated and serious bacterial infection during early life in most persons with BS have suggested the possibility of immune defectiveness. A few reports of the immunological status in BS have appeared (*40–42*). Most of the studies made have disclosed a severe immune defect, but the spectrum of defects varies somewhat between individuals. Abnormalities reported include reduced serum immunoglobulins (IgG, IgA, IgM), decreased proliferative response of blood lymphocytes to pokeweed mitogen, and poor response in the mixed lymphocyte culture.

Prospects

The "positive" findings discussed above—increased chromosomal aberrations and SCE, abnormally slow DNA replication-fork progression, and hypersensitivity to the mutagen EMS—are currently the most important clues toward the elucidation of the primary genetic defect in BS. Also of possible significance and of great theoretical interest is the as-yet-unexplained phenotypic dimorphism with respect to SCE in the lymphocytes of some persons with BS. SCE is apparently an S-phase event (*52, 53*); the increased SCE and slow DNA-fork movement in BS cells make semiconservative DNA synthesis a likely site of operation of the primary defect. Though polymerase activities have in some respects been found to be normal, further efforts should be made to study these and other enzymes concerned with semiconservative synthesis.

The growth characteristics of BS cells in culture have not as yet been adequately described. BS fibroblasts are notorious for their slow growth and apparent lack of vigor, which makes experiments with them difficult. This probably is partially responsible for the discrepancies in certain of the studies tabulated above. Data from our laboratory (Lechner, J. F., Groden, J., and German, J., manuscript in preparation) indicate the existence of a severe defect in the proliferative response of BS fibroblasts to increasing serum concentrations and to epidermal and fibroblast growth factors.

With regard to the striking cancer predisposition in BS, important questions which remain to be answered are whether BS cells are hypermutable, hypertransformable, or both. One report of normal rates of transformation by SV40 virus has appeared (*39*).

As noted above, complementation studies to search for genetic heterogeneity in BS are in progress. Experiments reported by Bryant *et al.* (*18*) in which BS fibroblasts were fused with normal human fibroblasts and by Alhadeff *et al.* (*19*) fusing BS with hamster fibroblasts (the result in both cases being a normal level of SCE in the BS chromosomes) demonstrated that such complementation studies are now possible, but formidable because of the poor growth of BS fibroblasts in culture.

Concluding Comment

BS cells show a remarkable and apparently unique pattern of chromosome instability. Chromatid exchange occurs at a rate far exceeding that in cells from other people. The molecular basis for the tendency to chromatid exchange remains unexplained despite a considerable effort by a number of laboratories to understand it. The effort seems justified and worth continuing in view of the magnitude of two major questions remaining unanswered. In relation to BS specifically, the two questions are (i) what is the biochemical defect? and (ii) why do increased numbers of cancers occur? These two questions correspond to important general questions: (i) what is the molecular explanation and significance of chromatid exchange in somatic cells and (ii) what primary event(s) occurs in the conversion of a normal to a neoplastic cell? The answers to the two questions posed specifically for BS might very well contribute to the answering of the two more general questions.

ACKNOWLEDGMENTS

This paper is respectfully dedicated to Dr. David Bloom on the twenty-fifth anniversary of the publication of his description of "congenital telangiectatic erythema resembling lupus erythematosus in dwarfs"—Bloom's syndrome (Am. J. Dis. Child., *88*: 754–758, 1954).

Our research on Bloom's syndrome is supported partially by grants from the American Cancer Society and the National Institutes of Health (HD 04134, HL 09011).

REFERENCES

1. German, J., Archibald, R., and Bloom, D. Chromosomal breakage in a rare and probably genetically determined syndrome of man. Science, *148*: 506–507, 1965.
2. German, J., Bloom, D., and Passarge, E. Bloom's syndrome. V. Surveillance for cancer in affected families. Clin. Genet., *12*: 162–168, 1977.
3. German, J., Bloom, D., and Passarge, E. Bloom's syndrome. VII. Progress report for 1978. Clin. Genet., *15*: 361–367, 1979.
4. German, J. Bloom's syndrome. VIII. Review of clinical and genetic aspects. *In*; R. M. Goodman and A. G. Motulsky (eds.), Genetic Diseases Among Ashkenazi Jews, pp. 121–139, Raven Press, New York, 1979.
5. Schroeder, T. M. and German, J. Bloom's syndrome and Fanconi's anemia. De-

monstration of two distinctive patterns of chromosome disruption and rearrangement. Humangenetik, *25*: 299–306, 1974.

6. Tice, R. R., Rary, J. M., and Bender, M. A. An investigation of DNA repair potential in Bloom's syndrome. *In*; P. C. Hanawalt, E. C. Friedberg, and C. F. Fox (eds.), DNA Repair Mechanisms, pp. 659–662, Academic Press, New York, 1978.
7. Evans, H. J., Adams, A. C., Clarkson, J. M., and German, J. Chromosome aberrations and unscheduled DNA synthesis in X- and UV-irradiated lymphocytes from a boy with Bloom's syndrome and a man with xeroderma pigmentosum. Cytogenet. Cell Genet., *20*: 124–140, 1978.
8. German, J. and Crippa, L. P. Chromosomal breakage in diploid cell lines from Bloom's syndrome and Fanconi's anemia. Ann. Génét., *9*: 143–154, 1966.
9. Krepinsky, A. B., Rainbow, A. J., and Heddle, J. A. Studies on the ultraviolet sight sensitivity of Bloom's syndrome fibroblasts. Mutat. Res., *69*: 357–368, 1980.
10. Krepinsky, A. B., Heddle, J. A., and German, J. Sensitivity of Bloom's syndrome lymphocytes to ethyl methanesulfonate. Hum. Genet., *50*: 151–156, 1979.
11. German, J., Crippa, L. P., and Bloom, D. Bloom's syndrome. III. Analysis of the chromosome aberration characteristic of this disorder. Chromosoma, *48*: 361–366, 1974.
12. Schonberg, S., German, J., and Chaganti, R.S.K. A new cytogenetic finding in Bloom's syndrome: Terminal association of homologous chromosomes at metaphase. Genetics, *88*: s88–s89, 1978.
13. Kuhn, E. M. Mitotic chiasmata and other quadriradials in mitomycin C-treated Bloom's syndrome lymphocytes. Chromosoma, *66*: 287–297, 1978.
14. Schroeder, T. M. and Stahl Mauge, C. Spontaneous chromosome instability, chromosome reparation and recombination in Fanconi's anemia and Bloom's syndrome. *In*; H. Altmann (ed.), DNA Repair and Late Effects, Int. Symp. IGEGM, pp. 35–50, Edition Roetzer, Eisenstadt, 1976.
15. Chaganti, R.S.K., Schonberg, S., and German, J. A manyfold increase in sister chromatid exchanges in Bloom's syndrome lymphocytes. Proc. Natl. Acad. Sci. U.S., *71*: 4508–4512, 1974.
16. German, J., Schonberg, S., Louie, E., and Chaganti, R.S.K. Bloom's syndrome. IV. Sister-chromatid exchanges in lymphocytes. Am. J. Hum. Genet., *29*: 248–255, 1977.
17. Shiraishi, Y. and Sandberg, A. A. Effects of mitomycin C on sister chromatid exchange in normal and Bloom's syndrome cells. Mutat. Res., *49*: 233–238, 1978.
18. Bryant, E. M., Hoehn, H., and Martin, G. M. Normalisation of sister chromatid exchange frequencies in Bloom's syndrome by euploid cell hybridisation. Nature, *279*: 795–796, 1979.
19. Alhadeff, B., Velivasakis, M., Pagan-Charry, I., Wright, W. C., and Siniscalco, M. High rate of sister chromatid exchanges of Bloom's syndrome chromosomes is corrected in rodent human somatic cell hybrids. Cytogenet. Cell Genet., *27*: 8–22, 1980.
20. Rüdiger, H. W., Bartram, C. R., Harder, W., and Passarge, E. Rate of sister chromatid exchanges in Bloom's syndrome fibroblast reduced by co-cultivation with normal fibroblasts. Am. J. Hum. Genet., *32*: 150–157, 1980.
21. Tice, R., Windler, G., and Rary, J. M. Effect of cocultivation on sister chromatid exchange frequencies in Bloom's syndrome and normal fibroblasts. Nature, *273*: 538–540, 1978.
22. Schonberg, S. and German, J. SCE frequency in cells metabolically coupled to Bloom's syndrome cells. Nature, *284*: 72–74 1980.

23. Buul, P.P.W. van, Natarajan, A. T., and Verdegaal-Immerzeel, E.A.M. Suppression of the frequencies of sister chromatid exchanges in Bloom's syndrome fibroblasts by co-cultivation with Chinese hamster cells. Hum. Genet., *44*: 187–189, 1978.
24. Arlett, C. F. and Harcourt, S. A. Cell killing and mutagenesis in repair-defective human cells. *In*; P. C. Hanawalt, E. C. Friedberg, and C. F. Fox (eds.), DNA Repair Mechanisms, pp. 633–636, Academic Press, New York, 1978.
25. Giannelli, F., Benson, P. F., Pawsey, S. A., and Polani, P. E. Ultraviolet light sensitivity and delayed DNA-chain maturation in Bloom's syndrome fibroblasts. Nature, *265*: 466–469, 1977.
26. Selsky, C., Weichselbaum, R., and Little, J. B. Defective host-cell reactivation of UV-irradiated herpes simplex virus by Bloom's syndrome skin fibroblasts. *In*; P. C. Hanawalt, E. C. Friedberg, and C. F. Fox (eds.), DNA Repair Mechanisms, pp. 555–558, Academic Press, New York, 1978.
27. Cleaver, J. E. DNA damage and repair in light-sensitive human skin disease. J. Invest. Dermatol., *54*: 181–195, 1970.
28. Ahmed, F. E. and Setlow, R. B. Excision repair in ataxia telangiectasia, Fanconi's anemia, Cockayne syndrome, and Bloom's syndrome after treatment with ultraviolet radiation and *N*-acetoxy-2-acetyl-aminofluorene. Biochim. Biophys. Acta, *521*: 805–817, 1978.
29. Vincent, R. A., Jr., Hays, M. D., and Johnson, R. C. Single-strand DNA breakage and repair in Bloom's syndrome cells. *In*; P. C. Hanawalt, E. C. Friedberg, and C. F. Fox (eds.), DNA Repair Mechanisms, pp. 663–666, Academic Press, New York, 1978.
30. Lehmann, A. R., Kirk-Bell, S., and Jaspers, N.G.J. Postreplication repair in normal and abnormal human fibroblasts. *In*; W. W. Nichols and D. G. Murphy (eds.), DNA Repair Processes, pp. 203–215, Symposia Specialists, Inc., Miami, 1977.
31. Inoue, T., Hirano, K., Yokoiyama, A., Kada, T., and Kato, H. DNA repair enzymes in ataxia telangiectasia and Bloom's syndrome fibroblasts. Biochim. Biophys. Acta, *479*: 497–500, 1977.
32. Moses, R. E. and Beaudet, A. L. Apurinic DNA endonuclease activities in repair-deficient human cell lines. Nucl. Acids Res., *5*: 463–473, 1978.
33. Parker, V. P. and Lieberman, M. W. Levels of DNA polymerases α, β, and γ in control and repair-deficient human diploid fibroblasts. Nucl. Acids Res., *4*: 2029–2037, 1977.
34. Bertazzoni, U., Scovassi, A. I., Stefanini, M., Giulotto, E., Spadari, S., and Pedrini, A. M. DNA polymerases α, β and γ in inherited diseases affecting DNA repair. Nucl. Acids Res., *5*: 2189–2196, 1978.
35. Hand, R. and German, J. A retarded rate of DNA chain growth in Bloom's syndrome. Proc. Natl. Acad. Sci. U.S., *72*: 758–762, 1975.
36. Hand, R. and German, J. Bloom's syndrome: DNA replication in cultured fibroblasts and lymphocytes. Hum. Genet., *38*: 297–306, 1977.
37. Waters, R., Regan, J. D., and German, J. Increased amounts of hybrid (heavy/heavy) DNA in Bloom's syndrome fibroblasts. Biochem. Biophys. Res. Commun., *83*: 536–541, 1978.
38. Loveday, K. S. and Latt, S. A. Search for DNA interchange corresponding to sister chromatid exchanges in Chinese hamster ovary cells. Nucl. Acids Res., *5*: 4087–4104, 1978.
39. Webb, T. and Harding, M. Chromosome complement and SV40 transformation of cells from patients susceptible to malignant disease. Br. J. Cancer, *36*: 583–591, 1979.

40. Hütteroth, T. H., Litwin, S. D., and German, J. Abnormal immune responses of Bloom's syndrome lymphocytes *in vitro*. J. Clin. Invest., *56*: 1–7, 1975.
41. Weemaes, C.M.R., Bakkeren, J.A.J.M., ter Haar, B.G.A., Hustinx, T.W.J., and van Munster, P.J.J. Immune responses in four patients with Bloom's syndrome. Clin. Immunol. Immunopathol., *12*: 12–19, 1979.
42. Schoen, E. J. and Shearn, M. A. Immunoglobulin deficiency in Bloom's syndrome. Am. J. Dis. Child., *113*: 594–596, 1967.
43. Shiraishi, Y., Freeman, A. I., and Sandberg, A. A. Increased sister chromatid exchange in bone marrow and blood cells from Bloom's syndrome. Cytogenet. Cell Genet., *17*: 162–173, 1976.
44. German, J. Cytological evidence for crossing-over *in vitro* in human lymphoid cells. Science, *144*: 298–301, 1964.
45. Kuhn, E. M. Localization by Q-banding of mitotic chiasmata in cases of Bloom's syndrome. Chromosoma, *57*: 1–11, 1976.
46. Hustinx, T.W.J., ter Haar, B.G.A., Scheres, J.M.J.C., Rutten, F. J., Weemaes, C.M.R., Hoppe, R.L.E., and Janssen, A. H. Bloom's syndrome in two Dutch families. Clin. Genet., *12*: 85–96, 1977.
47. Sperling, K., Goll, U., Lüdtke, E. H., Tolksdolf, M., and Obe, G. Cytogenetic investigation in a new case of Bloom's syndrome. Hum. Genet., *34*: 47–52, 1976.
48. Ved Brat, S. Sister chromatid exchange and cell cycle in fibroblasts of Bloom's syndrome. Hum. Genet., *48*: 73–79, 1979.
49. Henderson, E. and German, J. Development and characterization of lymphoblastoid cell lines (LCLs) from "chromosome breakage syndromes" and related genetic disorders. J. Supramol. Struct., (Suppl. 2): 83, 1978.
50. Kato, H. Induction of sister chromatid exchanges by UV light and its inhibition by caffeine. Exp. Cell Res., *82*: 383–390, 1973.
51. Kato, H. Mechanisms for sister chromatid exchanges and their relation to the production of chromosomal aberrations. Chromosoma, *59*: 179–191, 1977.
52. Kato, H. Possible role of DNA synthesis in formation of sister chromatid exchanges. Nature, *252*: 739–741, 1974.
53. Wolff, S., Bodycote, J., and Painter, R. B. Sister chromatid exchanges induced in Chinese hamster cells by UV irradiation of different stages of the cell cycle: The necessity to pass through S. Mutat. Res., *25*: 73–81, 1974.

DNA REPAIR

GENETIC AND ENVIRONMENTAL FACTORS IN EXPERIMENTAL AND HUMAN CANCER,
H. V. GELBOIN ET AL. (EDS.), JAPAN SCI. SOC. PRESS, TOKYO, PP. 189–199, 1980

DNA Repair in Mammalian Cells Exposed to Combinations of Carcinogenic Agents

R. B. Setlow and F. E. Ahmed*

Brookhaven National Laboratory, Upton, N.Y. 11973, U.S.A.

Abstract: Cells defective in one or more aspects of repair are killed and often mutagenized more readily than normal cells by DNA damaging agents, and humans whose cells are deficient in repair are at an increased carcinogenic risk compared to normal individuals. The excision repair of UV-induced pyrimidine dimers is a well-studied system, but the details of the steps in this repair system are far from being understood in human cells. We know that there are a number of chemicals that mimic UV in that normal human cells repair DNA damage from both these agents and from UV by a long patch excision repair system, and that xeroderma pigmentosum (XP) cells defective in repair of UV are also defective in the repair of damage from these chemicals. The chemicals we have investigated are *N*-acetoxy-2-acetylaminofluorene, 4-nitroquinoline 1-oxide, 7,12-dimethylbenz(a)anthracene 5,6-oxide, and acridine mustard. The repair of UV and these chemicals seems to be controlled coordinately. We describe experiments, using several techniques, in which DNA excision repair is measured after treatment of various human cell strains with combinations of UV and these agents. If two agents have a common rate limiting step then, at doses high enough to saturate the repair system, one would expect the observed repair after a treatment with a combination of agents to equal that from one agent alone. Such is not the case for normal human or excision deficient XP cells. In the former repair is additive and in the latter repair is usually appreciably less than that observed with either agent alone. Models that attempt to explain these surprising results involve complexes of enzymes and cofactors.

There are estimates that 80–90% of human cancers are the result of environmental carcinogens or our life style. There have been many experiments attempting to detect environmental carcinogens by assays such as mutagenic assays using

* Present address: Pharmacopathic Research Laboratories, Inc., 9705 North Washington Blvd. Laurel, Maryland, Laurel, Maryland 20810, U.S.A.

bacteria, carcinogenic assays with laboratory animals, and epidemiological studies. Such assays can indicate which agents are potentially dangerous to humans but they are not satisfactory for assaying the long-term effects of low level exposures. The latter problem is one that has defied a unique solution even for the effects of ionizing radiation. On the other hand, the effects of ambient UV in producing skin cancer have been assessed by epidemiological surveys in the United States (*1*). Part of the confidence in such assessment lies in the fact that a great deal is known about the molecular and cellular changes resulting from UV (*2*, *3*).

In a number of inherited human disorders the affected individuals are cancer prone (*4*, *5* and othe papers in this symposium). The prevalence of cancer among such individuals may be orders of magnitude higher than in the general population. Several of these disorders are associated with defects in the ability of cells to repair certain kinds of physical or chemical damage to their DNA. The identification of such disorders is direct evidence that damage to DNA can be carcinogenic and is the best available evidence for a causal connection between mutagenic and carcinogenic agents. The support for the connection is further strengthed by the observation that mutations occur in xeroderma pigmentosum (XP) cells at lower doses of carcinogens than those affecting normal cells (*6*).

An analysis of the molecular defects in repair-deficient diseases should give strong clues as to the molecular nature of the changes responsible for killing, mutagenesis and carciogenesis and hence give estimates of the probablility that such changes result in biological effects. Since human risks are difficult to assess directly, they must be assessed from data on molecules and animals extrapolated to humans. The extrapolations will involve scanty epidemiological data. Hence, it is important that there be a good theoretical base for the extrapolation and that means an understanding of the molecular mechanisms involved.

Almost all our ideas about the molecular nature of repair come from studies on bacterial systems because of the large number of well-defined repair-deficient mutants and the relative ease analyzing photochemical and molecular changes in them (*2*, *3*). In mammalian cells we have, at the moment, only the naturally occurring human inherited diseases. Hence, mammalian repair studies have taken their clues from the bacterial ones. Even in the bacterial world the problems are complex, although the concepts may be simple, because of the difficulty in purifing the first repair enzyme in the UV excision repair sequence—the so-called UV endonuclease (*7*). The endonuclease in *Escherichia coli* contains several different proteins and recent evidence indicates that the endonuclease may also have a glycosylase activity associated with it (*8*). In mammalian systems a number of endonucleases that attack UV-irradiated DNA have been described but most of them do not act on pyrimidine dimers but on some minor and as-yet-undefined photochemical products. An endonuclease activity from calf-thymus glands has been described that does work on pyrimidine dimers, but the enzyme is very unstable (*9*).

Bacteria that are defective in repairing UV damage are also defective in repairing some chemical damages to their DNA. Hence, it is not surprising that the same thing is true for human cells and that the repair of damaged DNA is more

TABLE 1. Ways in Which Some Chemical Damages Mimic UV Damage in Human Cells

1.	UV-sensitive cells (XP) are more sensitive to the chemical than normal cells.
2.	Chemically treated viruses show a higher survival on normal cells than on XP cells.
3.	XP cells deficient in repair of UV damage are also deficient in excision of chemical damage.
4.	Excision repair of UV and of chemical damage involves long patches (approx. 100 nucleotides).

general than just the special case of repair of pyrimidine dimers (*2–5*). There are a number of chemical agents whose damage in human cells seems to mimic those of UV in the ways indicated in Table 1.

It is apparent that the chemicals that mimic UV radiation are repaired by mechanisms that are coordinately controlled; that is, if cells are defective in UV repair they are defective in repairing these chemical damages. Since there are many experimental lines of evidence suggesting that the rate limiting step in excision repair of UV damage is the initial endonucleolytic step, early proposals to explain the results illustrated in Table 1 hypothesized that the rate limiting step—an endonuclease—was common for all these damages. Since human enzymes involved in excision repair are poorly characterized, it is not possible directly to compare the ability of such enzymes to repair chemical or physically damaged DNA. Hence, we have measured the ability of chemicals to compete for the UV repair system in human cells. Such experiments involve large doses, doses that saturate the individual repair systems. Although such experiments give information about the rate limiting steps and the similarity of rate limiting steps of chemicals to UV, it is important to recognize that they are not directly applicable to the real world where people are also exposed to combinations of carcinogens, since in the latter case the doses and dose rates are much lower than those used experimentally.

Methods

We have used three general methods to measure repair from individual agents or combinations of them. The first applies to UV damage and measures the loss of thymine-containing dimers chromatographically or by the loss of sites sensitive to an endonuclease isolated from *Micrococcus luteus* that is specific for pyrimidine dimers (*10*). Since the endonuclease does not work on other types of damage it is easy to use it as a specific probe for UV damage and its repair in the presence of other types of damage. The second method is unscheduled DNA synthesis—the incorporation of label in treated cells during non-S phases of the cell cycle. Such measurements made radioautographically determine the repair of damage from chemical or UV or from combinations of agents. The third method is the photolysis of bromodeoxyuridine (BrdU) incorporated during repair. The technique gives an estimate of the amount of BrdU incorporated and in this sense it is similar to unscheduled DNA synthesis, but the method has the advantage of giving an estimate of the size of the repaired patch (*11*). The three methods give consistent results for UV-irradiated human cells; that is, the value of unscheduled DNA synthesis calculated from the patch size and the numbers of dimers removed is consistent with that observed (*12*).

Saturation of Repair

Although the amount of UV damage increases with dose, at high doses the number of dimers excised in a given time is a constant. For example, at 20 J/m^2 (a dose that makes approximately one pyrimdine dimer / 4×10^6 daltons) the excision of dimers (measured as the percentage of radioactivity in dimers excised compared to thymine) is 0.050 whereas at 80 J/m^2—a dose that makes four times as many dimers—it is 0.056 (*13*). Similar data are obtained for the loss of endonuclease-

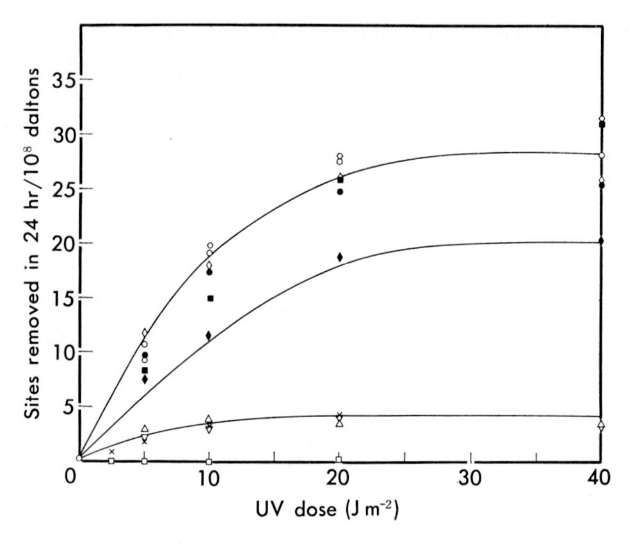

FIG. 1. DNA repair in UV-irradiated cells as a function of initial dose. Repair was measured as the loss in sites sensitive to a UV-endonuclease from *M. luteus* during 24 hr after irradiation. Endonuclease sensitive sites are equivalent to pyrimidine dimers (From Ref. *13*). ⬡ Fanconi's anemia; ■ Cockayne syndrome; ○ normal human; ◇ ataxia telangiectasia; ● XP variant. ◆ XP E; △ XP C; ▽ XP D; × V-79; □ XP A.

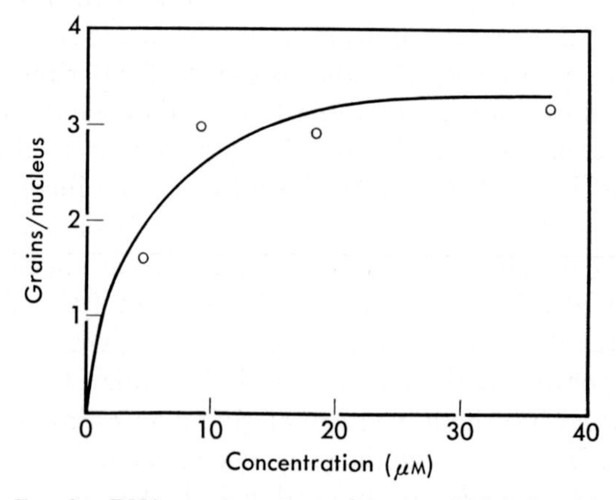

FIG. 2. DNA repair, measured as unscheduled synthesis during 3 hr, in normal human fibroblasts exposed to several concentrations of DMBA-epoxide in serum free medium for 30 min (From Ref. *14*).

sensitive sites (*13*). Figure 1 is an example of such saturation data. At high doses the number of endonuclease-sensitive sites removed becomes constant and independent of dose. The repair system has become saturated. The various repair-deficient XP cells strains also show saturation, but the numbers of sites removed in these strains is appreciably less than in normally excising strains. A second example of the saturation of DNA repair is shown in Fig. 2 where the amount of unscheduled synthesis in normal human cells is shown for cells treated with different concentrations of 7,12-dimethylbenz(a)anthracene 5,6-oxide (DMBA-epoxide) (*14*). Although repair saturates for both UV and epoxide treatment, the saturation level is quite different for the two agents (see below). XP cells are defective in repair of both types of damage (see below) indicating a coordinate control of the repair of two types of damage but obviously the difference in the magnitudes of repair implies some more complicated mechanism than identical pathways for the repair of both types of damage. The smaller amount of unscheduled synthesis for DMBA-epoxide is *not* the result of a smaller patch size (*14*).

Expectation for a Combination of Agents

The two extreme possibilities for the results of treatment of human cells with a combination of agents, such as UV and one of its mimetics, might arise from identical rate limiting steps for the agents or completely different rate-limiting steps for the agents. A schematic diagram of the expectation for each of these possibilities is shown in Fig. 3. Illustrated in the upper portion is the finding that the amount of damage increases proportionately with dose. The lower portions outline the expectations for different rate-limiting steps (one would expect to find twice the amount

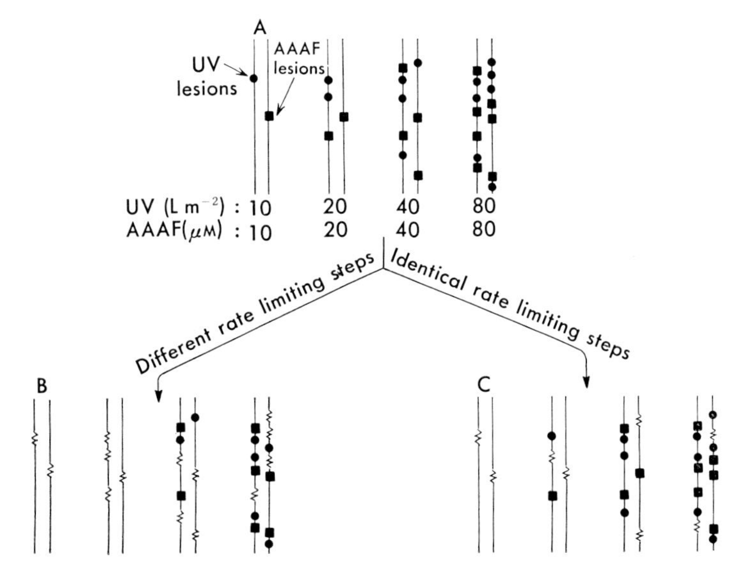

FIG. 3. A schematic diagram illustrating the repair of damage as a function of dose for a combined treatment of cells with UV and AAAF. The lower part shows the expected results for two hypotheses about the rate limiting steps. The repaired regions are represented by ⁓.

of repair from either agent alone), or for identical rate-limiting steps (the maximum amount of repair after treatment with two agents should be no greater than treated with one agent alone). For different rate-limiting steps a UV mimetic should not interfere with the excision of dimers. For identical rate-limiting steps a UV mimetic should inhibit the excision of dimers.

Repair after UV plus N-acetoxy-2-acetylaminofluorene (AAAF)

Figure 4 shows the results of unscheduled synthesis experiments on several human cell strains treated with UV, with AAAF or with a combination of the two (*12*). The concentrations chosen are at saturating levels and it is apparent that in normal human cells repair after the combination is additive; whereas, in the XP strains the total repair after the combination is appreciably less than additive. Normal cells act as if the rate-limiting step for the repair of the two agents is completely different. In XP cells, however, the data indicate that each agent inhibits strongly the repair of the other so that after a combined treatment the total repair is appreciably *less* than the repair of either agent separately. The data can not be explained by general toxicity because, for example, high doses of UV do not inhibit UV repair nor do high doses of AAAF inhibit AAAF repair. We have ob-

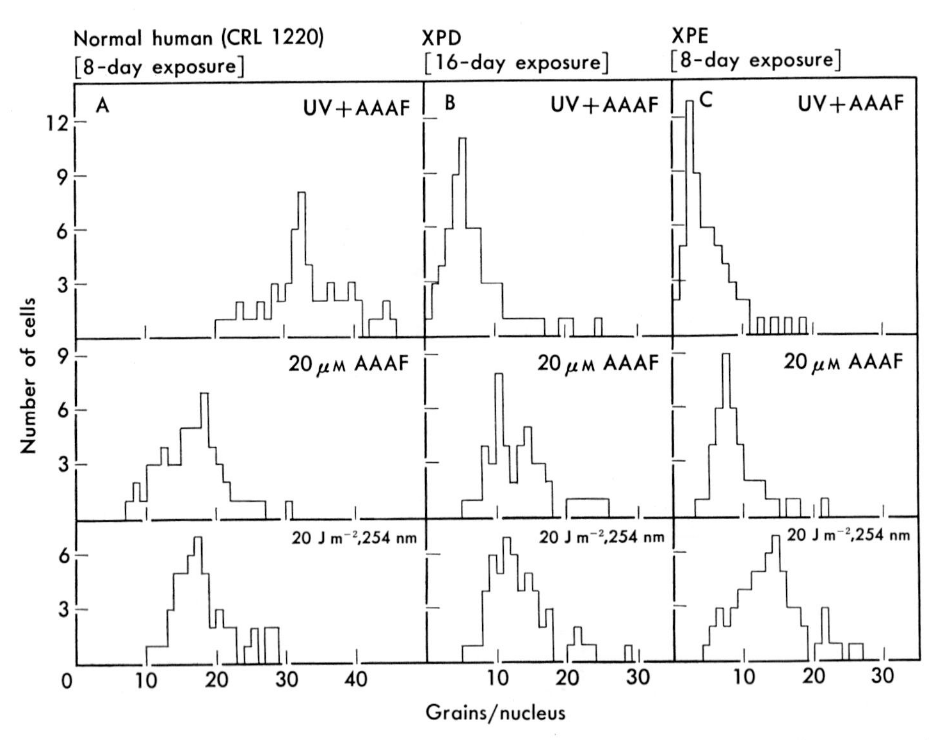

FIG. 4. Unscheduled DNA synthesis (the distribution of grains among cells) in normal human and in XP cells after treatment with saturating doses of UV, AAAF, or a combination of the two (From Ref. *12*).

TABLE 2. Measures of Excision Repair in Human Cells Treated with UV, AAAF, and Combinations

Cell line	Unscheduled synthesis[a]			Endonuclease[b] assay		BrdU photolysis[c]		
	20 J/m²	20 μM	20 J/m²+ 20 μM	20 J/m²	20 J/m²+ 20 μM	20 J/m²	10 μM	20 J/m²+ 10 μM
Normal human								
Par Bel (CRL 1191)	18.6	16.1	33	27.5	27.4	3.5	1.4	4.9
Rid Mor (CRL 1220)	17.4	16.4	32.9	23.1	23.3	3.3	1.4	4.4
Ataxia telangiectasia								
NeNo (CRL 1347)	19.7	16.6	35	24.6	24.2	3.2	2.5	5.4
Se Pan (CRL 1343)	22.3	14	33.2	24.5	24.9	3.0	1.4	3.9
AT 4BI	19.6	14.4	35	26.3	26.4	2.2	1.4	3.4
Fanconi's anemia								
Ce Rel (CRL 1196)	15.4	13.9	28.2	27.5	27.4	3.4	0.9	4.4
Cockayne syndrome								
GM 1098	14.3	19.8	33.3	26.1	26.1	3.2	2.6	6.2
GM 1629				26.9	27.0	4.1	2.1	6.6
Xeroderma pigmentosum								
Variant; Wo Mec (CRL 1162)	23.0	17.1	37.2	24.7	24.0	2.8	1.1	4.3
C; Ge Ar (CRL 1161)	3.7	2.2	1.8	3.7	1.4	1.0	0.1	0.1
D; Be Wen (CRL 1160)	6.4	6.6	3.8	3.9	0.8	1.0	0.2	0.4
E; XP2R0 (CRL 1259)	10.8	6.4	2.8	19.3	9.6	4.8	1.0	3.8

a) Grains/nucleus incorporated in 3 hr (8 days exposure). b) Sites removed in 24 hr/10^8 daltons.
c) $(1/M_w)\times 10^8$ at highest 313 nm dose (12 hr repair).

tained data leading to similar conclusions using the endonuclease-sensitive site assay and the photolysis of BrdU incorporated during repair. A summary of these data (*15*) for a number of human cell strains is shown in Table 2. The generalization indicated above holds for all. Repair is additive in normal cells and strongly inhibitory in cells defective in excision repair.

Repair after UV and Other UV Mimetics

Results very similar to those obtained for UV and AAAF are obtained for UV and other UV mimetics. For example, Fig. 5 shows unscheduled DNA synthesis data for normal human and XP cells treated with UV or DMBA-epoxide or a combination of the two (*14*). The doses used were saturating ones (see Figs. 1 and 2). Nevertheless, the maximum repair after chemical treatment is appreciably less than after UV exposure in both normal and XP cells although repair in XP cells is appreciably less than normal for both agents. For normal cells repair after a combined treatment is approximately additive whereas in XP cells repair is appreciably less than that observed for UV alone. Thus, UV plus DMBA-epoxide falls into the same category as UV and AAAF. A similar conclusion is reached using the other measures of DNA repair.

Similar experiments have been carried out for combinations of UV and 4-nitroquinoline 1-oxide (4-NQO) and UV plus acridine mustard (ICR-170) (*16*). The results of these experiments are summarized briefly in Table 3. In all the cases

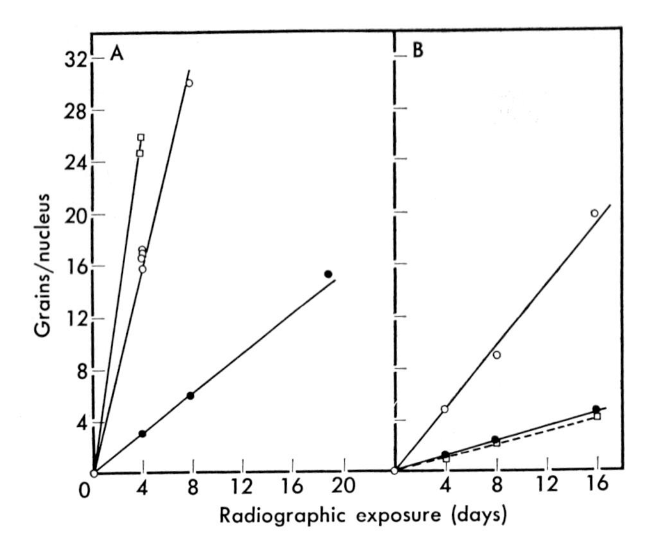

FIG. 5. Unscheduled DNA synthesis (the average number of grains per nucleus) for normal human (A) and XP cells (B) exposed to UV, DMBA-epoxide, or to a combination of the two (From Ref. *14*). ○ 254 nm, 20 J/m² UV; ● 9.2 μM DMBA-epoxide; □ UV+DMBA-epoxide.

TABLE 3. Repair Responses of Human Cells to Combinations of DNA-damaging Agents

	Normal human	XPC
UV+AAAF	Additive	Inhibitory
UV+DMBA-epoxide	Additive	Inhibitory
UV+ICR-170	Additive	Inhibitory
UV+4-NQO	Additive	Additive

we have investigated the so-called UV mimetics do not inhibit repair of UV damage in normal human cells. Except in the case of 4-NQO the chemical agents inhibit UV repair in XP cells. Thus we reach a general conclusion that the low level of repair in XP cells is not the result only of a smaller number of enzymes involved in the rate-limiting step. The fact that combined agents usually act as if they inhibit one another in XP cells, whereas they give an additive result in normal cells implies that the character of the rate-limiting steps are different for normal and XP cells.

Speculations and Conclusions

In the absence of direct knowledge of the enzymes involved in excision repair and a direct knowledge of their biochemical properties, anything we say is really speculative. Nevertheless, we can summarize some of the crucial evidence that leads us to hypothesize the existence of repair complexes in Table 4.

In normal human cells we might think of the repair complex as made up of all the kinds of enzymes that could participate in excision repair (glycosylases, nucleases, polymerases, ligases, proteases, *etc.*). Each chemical damage, within

TABLE 4. Why Hypothesize Enzyme Complexes for Excision Repair in Human Cells?

a) UV mimetic chemicals exist (*11*).
b) XP cells of seven complementation groups (*17*) are defective in excision of UV damage and, where investigated, of damage from these chemicals.
c) Few single stand breaks accumulate during repair (*18*) and there seem to be many fewer single strand breaks in XP cells than in normal ones. Moreover, the addition of exogenous UV endonuclease to XP cells restores, in part, their UV repair activity (*19*).
d) *Hence* a change in any one of seven associated molecules changes the endonuclease function in repair and affects repair of many damages coordinately.
e) A possible explanation is that there is a universal repair endonuclease that is composed of seven or more subunits.
f) If were true, the repair pathways for UV and chemical damages would be the same and chemical treatment would inhibit UV repair at saturating doses. *This is not the case.*

rather broad limits, might have its own endonuclease associated with the other repair enzymes since it seems as if it is the endonuclease that has the specificity for repair. A change in any one of the enzymes in the repair complex, not necessarily the endonuclease, would change the entire activity of the complex and the endonucleolytic activity might fail because of distortion of this multi-enzyme complex. Thus, even a change in an exonuclease might result in an observation of inhibition of endonucleolytic function. Such an explanation easily can explain the observations for normal human cells. It cannot, however, explain the data for XP cells indicating that the various UV mimetic agents often inhibit strongly UV repair so that the resulting repair from combined treatments is much less than that observed from either agent alone. Thus, one must construct more elaborate models to explain the XP results. For example, suppose there were separate endonuclease complexes for the different damages and that in XP cells each complex was present in relatively small numbers. Suppose further that although an endonuclease might bind to damaged sites in DNA, it might not nick them unless two or more cofactors bind to the nuclease itself after it is bound to DNA (*18*). Hence, in UV-irradiated XP cells the small number of endonucleases would bind to DNA and the cofactors would bind to the nuclease and there would be a slow level of excision repair. If the number of cofactors were limited, then after a combined treatment with UV and AAAF the cofactors would distribute themselves among the different endonuclease complexes binding to AAAF and to UV damage. As a result the probability of two cofactors being associated with any one endonuclease complex would be very small. Hence repair would be much lower than expected. These involved speculations are only presented to indicate the complexity of the problem and to illustrate the fact that they will only be solved by understanding more about the basic nature of the repair enzymes involved.

ACKNOWLEDGMENT

This work was supported by the U.S. Department of Energy (Contract DE-AC02-76CH00016).

REFERENCES

1. Scott, E. L. and Straf, M. L. Ultraviolet radiation as a cause of cancer. *In*; H. H. Hiatt, J. D. Watson, and J. A. Winsten (eds.), Origins of Human Cancer, pp. 529–546, Cold Spring Harbor Laboratory, New York, 1977.
2. Setlow, R. B. and Setlow, J. K. Effects of radiation on polynucleotides. Annu. Rev. Biophys. Bioeng., *1*: 293–349, 1972.
3. Hanawalt, P. C., Cooper, P. K., Ganesan, A. K., and Smith, C. A. DNA repair in bacteria and mammalian cells. Annu. Rev. Biochem., *48*: 783–836, 1979.
4. Arlett, C. F. and Lehmann, A. R. Human disorders showing increased sensitivity to the induction of genetic damage. Annu. Rev. Genet., *12*: 95–115, 1978.
5. Setlow, R. B. Repair deficient human disorders and cancer. Nature, *271*: 713–717, 1978.
6. Maher, V. N. and McCormick, J. J. DNA repair and carcinogenesis. *In*; P. L. Grover (ed.), Chemical Carcinogens and DNA, Vol. II, pp. 133–158, CRC Press, Boca Raton, Florida, 1979.
7. Seeberg, E. Reconstitution of an *Escherichia coli* repair endonuclease activity from separated *Uvr* A^+ and *Uvr* B^+/*Uvr* C^+ gene products. Proc. Natl. Acad. Sci. U.S., *75*: 2569–2573, 1978.
8. Grossman, L., Riazuddin, S., Haseltine, W. A., and Lindan, C. Nucleotide excision repair of damaged DNA. Cold Spring Harbor Symp. Quant. Biol., *43*: 947–955, 1979.
9. Waldstein, E. A., Peller, S., and Setlow, R. B. UV-endonuclease from calf thymus with specificity toward pyrimidine dimers in DNA. Proc. Natl. Acad. Sci. U.S., *76*: 3746–3750, 1979.
10. Paterson, M. C. Use of purified lesion-recognizing enzymes to monitor DNA repair *in vivo*. Adv. Radiat. Biol., *7*: 1–53, 1978.
11. Regan, J. D. and Setlow, R. B. Two forms of repair of the DNA of human cells damaged by chemical carcinogens and mutagens. Cancer Res., *34*: 3318–3325, 1974.
12. Ahmed, F. E. and Setlow, R. B. DNA repair in xeroderma pigmentosum cells treated with combinations of ultraviolet radiation and *N*-acetoxy-2-acetylaminofluorene. Cancer Res., *39*: 471–479, 1979.
13. Ahmed, F. E. and Setlow, R. B. Saturation of DNA repair in mammalian cells. Photochem. Photobiol., *29*: 983–999, 1979.
14. Ahmed, F. E., Gentil, A., Rosenstein, B. S., and Setlow, R. B. DNA excision repair in human cells treated with ultraviolet radiation and 7,12-dimethylbenz(a)anthracene 5,6-oxide. Biochim. Biophys. Acta, in press.
15. Ahmed, F. E. and Setlow, R. B. Excision repair in mammalian cells. *In*; P. C. Hanawalt, E. C. Friedberg, and C. F. Fox (eds.), DNA Repair Mechanisms, pp. 333–336, Academic Press, New York, 1978.
16. Ahmed, F. E. and Setlow, R. B. DNA excision in repair proficient and deficient human cells treated with a combination of ultraviolet radiation and acridine mustard (ICR-170) or 4-nitroquinoline 1-oxide. Chem.-Biol. Interact., *29*: 31–42, 1980.
17. Bootsma, D. Xeroderma pigmentosum. *In*; P. C. Hanawalt, E. C. Friedberg, and C. F. Fox (eds.), DNA Repair Mechanisms, pp. 589–601, Academic Press, New York, 1978.
18. Cleaver, J. E. Repair processes for photochemical damage in mammalian cells. Adv. Radiat. Biol., *4*: 1–75, 1974.
19. Tanaka, T., Hayakawa, H., Sekiguchi, M., and Okada, Y. Specific action of T4

endonuclease V on damaged DNA in xeroderma pigmentosum cells. Proc. Natl. Acad. Sci. U.S., *74*: 2598–2962, 1977.

20. Yarosh, D. B. A model for the incision step of excision repair in human cells. Am. Soc. Photobiol., *7*: 157 (Abstr.), 1979.

GENETIC AND ENVIRONMENTAL FACTORS IN EXPERIMENTAL AND HUMAN CANCER,
H. V. GELBOIN ET AL. (EDS.), JAPAN SCI. SOC. PRESS, TOKYO, PP. 201–213, 1980

Quantitative and Qualitative Changes Induced in DNA Polymerases by Carcinogens

Michiko MIYAKI, Noriko AKAMATSU, Kazutoshi SUZUKI,
Mikiko ARAKI, and Tetsuo ONO

Department of Biochemistry, The Tokyo Metropolitan Institute of Medical Science, Tokyo, Japan

Abstract: Effect of direct modification of the pre-existing DNA polymerases by carcinogens on the fidelity of DNA synthesis was examined *in vitro*. Carcinogenic metal cations, which were confirmed to be mutagens in cultured mammalian cells, caused a notable increase in the misincorporation of all four nucleotides by DNA polymerases *in vitro*, whereas noncarcinogenic metals did not change the fidelity. The data seem to suggest that metal mutagenesis results from incorrect nucleotide substitution during DNA replication. Treatment of DNA polymerase I with *N*-methyl-*N'*-nitro-*N*-nitrosoguanidine (MNNG) slightly enhanced the frequency of misincorporation of dAMP and dTMP, while lowering the misincorporation of dCMP and dGMP. Ultraviolet (UV) irradiation of the enzyme did not cause a significant change in the fidelity. Alteration in the fidelity of DNA polymerase II and III from *Escherichia coli* and polymerase α from the spleens of mice were also observed after MNNG treatment of the enzymes.

To test the possibility of inducible enzyme, the changes in DNA polymerase activities caused by exposure of cells to chemical mutagens and UV light was examined. Treatment of *E. coli* polA$^-$ strain with MNNG followed by incubation of cells in growth medium enhanced DNA polymerase II activity by up to 5 times and polymerase III to twice as high as that of control cells. The increased levels of polymerases were maintained for several hours and subsequently dropped to the control levels. This enhancement was inhibited by chloramphenicol addition to the post-treatment incubation medium. The increase in polymerase activities was also observed after treatment of cells with methyl methanesulfonate, hydroxylamine, and UV light. However, no new enzyme other than polymerase II and polymerase III was detected in DEAE cellulose chromatography. The induced enzyme did not exhibit error-proneness when assayed after partial purification. The activities of DNA polymerase α and β were enhanced when HeLa cells were treated with MNNG. These inducible changes in DNA polymerase activities may play a role in some repair pathways.

Involvement of somatic mutation in carcinogenesis is suggested by the fact that inherited defect in repair of DNA damage is a predisposition to human cancer (*1*) and that mutagens for bacteria (*2*) are mostly carcinogenic for animals. Many carcinogens interact with nucleic acid and protein. Modification of the base moiety in DNA is assumed to be the cause of mutation, but the process responsible for mutation fixation is not clear. There are multiple pathways even for UV light-induced mutagenesis (*3*), and inducible error-prone repair systems after UV irradiation of *Escherichia coli* have been postulated (*4–6*). *N*-methyl-*N'*-nitro-*N*-nitrosoguanidine (MNNG), which is one of the most widely used mutagens and carcinogens, induced mutation selectively at the replication point (*7*). Different mutagenic specificities in the different species treated with this chemical suggested participation of inherent cellular function in mutation (*8*), and modification of DNA replication machinery by MNNG has also been suggested (*9*). We have tested the possibility of a change in fidelity of the pre-existing DNA polymerase caused by carcinogens, and another possibility of induction of an enzyme system responsible for error-prone repair after treatment of cells with carcinogens.

The Mechanism of Mutagenesis by Carcinogenic Metal Ions

Metal compounds are known to produce cancer in man (*10*) and animals (*11*). The mutagenicity of metal cations in mammalina cells has not yet been confirmed, although some have been seen to cause mutation in yeast (*12*) and *E. coli* (*13, 14*). First, we tested the mutagenicity of metal cations by measuring the induced mutation frequency at the hypoxanthine guanine phosphoribosyltransferase (HGPRT) locus in Chinese hamster V79 cells (*15*). Beryllium chloride increased the number of 8-azaguanine-resistant colonies to 6 times that of the control as indicated in Table 1. Manganese chloride also increased the mutation frequency. The frequency

TABLE 1. Frequencies of Mutation in V79 Cells Resistant to 8-Azaguanine Induced by Metal Chlorides

Treatment[a]		Survival[b]	Number of AG^r colonies per 10^6 survivals[c]	Induced ratio
Control		100	5.8±0.8	1.0
MNNG	10 μM	69.4	1,451±168	250
MNNG	25 μM	45.6	3,761	648
$BeCl_2$	2.0 mM	56.8	35.0±1.4	6.0
$BeCl_2$	3.0 mM	39.4	36.5±1.7	6.3
$CoCl_2$	0.2 mM	11.3	13.2±0.8	2.3
$MnCl_2$	1.0 mM	6.5	14.1±1.2	2.4
$MnCl_2$	1.5 mM	0.9	5.6±0.2	1.0
$MnCl_2$[d]	1.0 mM	7.2	15.3±1.5	2.6
$MnCl_2$[d]	1.5 mM	2.9	28.2±0.7	4.9
$NiCl_2$[e]	0.4 mM	55.0	7.1±0.2	1.2
$NiCl_2$[e]	0.8 mM	0.4	15.6±2.0	2.7

[a] Expression time was 48 hr. [b] Relative cell number after the expression time, mean value. [c] Mean± SE from three or more (control: from nine) experiments; AG^r: 8-azaguanine-resistant. [d] Expression time was 96 hr. [e] Expression time was 144 hr.

TABLE 2. Nature of 8-Azaguanine-resistant Colonies

Mutant colony		Plating efficiency (PE) (%)[a]	Sensitivity to THAG (survival per 10^6)	HGPRTase activity	
				nmol per mg protein per hr	Relative activity
Wild-type		0	0.93×10^6	152.96	100
Spontaneous	AG1	75	0	0.14	0.09
Spontaneous	AG2	88	0	0.29	0.19
Spontaneous	AG3	104	0	0.08	0.05
MNNG-induced	AG1	106	0	1.01	0.67
MNNG-induced	AG2	117	0	0.18	0.12
$BeCl_2$-induced	AG1	89	0	2.57	1.68
$BeCl_2$-induced	AG2	81	0	4.75	3.11
$BeCl_2$-induced	AG3	48	0	0.04	0.03
$MnCl_2$-induced	AG1	78	1	0.10	0.07
$MnCl_2$-induced	AG2	85	2	0.15	0.10

a) $\frac{\text{PE in AG medium}}{\text{PE in normal medium}} \times 100.$

was slightly increased by cobaltous and nickel chlorides. The extents of induced mutation were low, but the frequencies were reproducible. To confirm the mutation at the HGPRTase locus, about 30 azaguanine resistant colonies were randomly isolated, and tested for sensitivity to amethopterin in thymidine-hypoxanthine-amethopterin-glycine (THAG) medium, and then HGPRTase activity of each mutant colony was measured in cell-free extracts (Table 2). About 75% of the azaguanine-resistant clones were amethopterin-sensitive and 86% of the resistant clones showed less than 3% HGPRTase activity of the original cell line.

The mechanism of metal carcinogenesis or mutagenesis is not clear. A remarkable decrease in fidelity by manganese during DNA synthesis *in vitro*, as reported by Hall and Lehman (*16*), has suggested that misreading during the replication of cellular DNA probably occurred when the concentration of such a mutagenic metal ion was high near the replication point. We examined the effect of metal cations on the DNA polymerase reaction *in vitro*, and observed that carcinogenic or mutagenic metal cations cause a decrease in fidelity of DNA synthesis (*17*).

Two distinct modes of inactivation of *E. coli* polymerase I by divalent cations were observed. The first group including Be, Ca, Cu, Zn, Cd, and Hg ions exclusively inhibited polymerase. These were unable to activate the polymerase (Fig. 1a, b). The second group such as Mn, Co, and Fe ions had two activities. These ions supported the DNA polymerase activity at low concentration but at a higher concentration they inhibited polymerase activity (Fig. 1c, d). The effect of metal ions on the fidelity of DNA synthesis was examined at metal ion concentrations which inhibit DNA polymerase to 1/10 of the activity observed in the presence of 5 mM Mg alone. When 5 mM $MgCl_2$ in the reaction mixture was replaced by 1 mM $MnCl_2$, the increase in the misincorporation of error bases by *E. coli* DNA polymerase I was observed with respect to all four nucleotides, dAMP, dTMP, dGMP, and dCMP (Table 3). As for the misincorporation, competition of Mg

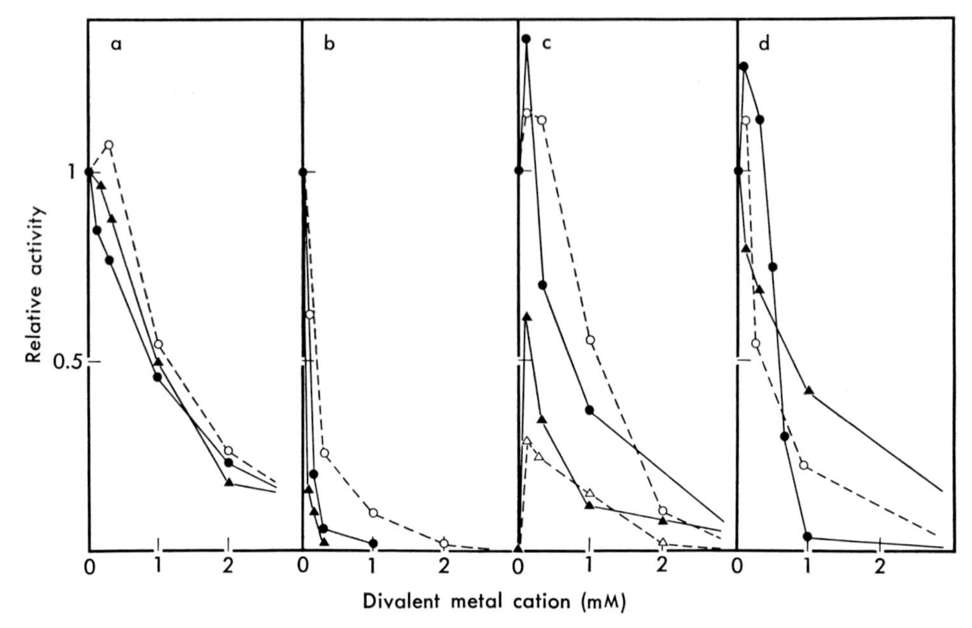

FIG. 1. Effect of metal cations on *E. coli* DNA polymerase I activity. Polymerase activity was measured with different concentrations of divalent metal cations as indicated in figures in the presence or absence of 5 mM $MgCl_2$. Activity at each concentration of metal cation is represented by the value relative to that observed in the presence of 5 mM Mg alone as metal. a: ● Be+Mg; ○ Ca+Mg; ▲ Cu+Mg. b: ● Zn+Mg; ○ Cd+Mg; ▲ Hg+Mg. c: ▲ Mn; ● Mn+Mg; △ Co; ○ Co+Mg. d: ● Fe^{2+}+Mg; ○ Fe^{3+}+Mg; ▲ Ni+Mg.

with Mn was slight, therefore, addition of 5 mM Mg to the reaction mixture containing 1 mM Mn slightly inhibited the misincorporation observed in the presence of 1 mM Mn alone. However, Mn ion did not enhance misincorporation at a concentration of 0.1 mM which is the optimum for activation of the polymerase. This indicated that the binding of the ion at many inhibitory sites of the enzyme besides the active center was necessary to cause a marked change in the fidelity. Be, Fe, Co, and Ni apparently increased misincorporation, and Cd, Hg, and Zn had little effect, but Ca and Cu did not affect the fidelity. Compounds of Be, Fe, Co, Ni, Cd, Hg, and Zn are known to be carcinogens, whereas Ca and Cu have not been reported to be carcinogenic. These results seem to be in agreement with the concept of correlation between carcinogenicity and mutagenicity.

Metal cations have the ability to bind both to the ligands of protein and purine bases. Binding of metal cations at many inhibitory sites of an enzyme seems to cause a decrease in fidelity, and interaction of metal cations with the base moiety in dNTP and in the template (*18*) may also affect base pairing during polymerization. The 3′→5′ exonuclease activity of DNA polymerase is reported to excise noncomplementary nucleotides during polymerization (*19*). Effect of Mn on 3′→5′ exonuclease activity was examined with Klenow's enzyme A (*20*) which has no 5′→3′ nuclease activity. Substitution of 5 mM Mg by 1 mM Mn decreased polymerase activity but increased 3′→5′ exonuclease activity. This indicated that a change in

TABLE 3. Effect of Divalent Metal Cations on Misincorporation by *E. coli* DNA Polymerase I

Metal ion	Incorrect dNMP	Template-primer	Error ($\times 10^{-5}$)[c]
Mg(5)	dAMP	dGdC[a]	6.4
Mg(5)	dTMP	dGdC	2.4
Mg(5)	dGMP	$dAdT_{12-18}$[b]	0.3
Mg(5)	dCMP	$dAdT_{12-18}$	2.7±0.7[d]
Mn(0.1)	dAMP	dGdC	4.5
Mn(0.1)	dTMP	dGdC	0.7
Mn(0.1)	dCMP	$dAdT_{12-18}$	2.3
Mn(1)	dAMP	dGdC	45.1*
Mn(1)	dTMP	dGdC	6.4*
Mn(1)	dGMP	$dAdT_{12-18}$	6.7*
Mn(1)	dCMP	$dAdT_{12-18}$	16.0*
Mn(1) +Mg(5)	dCMP	$dAdT_{12-18}$	9.8*
Be(4) +Mg(5)	dCMP	$dAdT_{12-18}$	8.0*
Ca(3) +Mg(5)	dCMP	$dAdT_{12-18}$	2.6
Cu(3) +Mg(5)	dCMP	$dAdT_{12-18}$	2.6
Zn(0.2)+Mg(5)	dCMP	$dAdT_{12-18}$	3.0
Cd(1) +Mg(5)	dCMP	$dAdT_{12-18}$	4.4*
Hg(0.1)+Mg(5)	dCMP	$dAdT_{12-18}$	4.2*
Fe(0.5) +Mg(5)	dCMP	$dAdT_{12-18}$	13.0*
Co(2)	dCMP	$dAdT_{12-18}$	6.4*
Co(2) +Mg(5)	dCMP	$dAdT_{12-18}$	4.8*
Ni(3) +Mg(5)	dCMP	$dAdT_{12-18}$	7.6*

[a] dGdC=poly(dG) · poly(dC). [b] $dAdT_{12-18}$=poly(dA) · $(dT)_{12-18}$. [c] Error=misincorporation/correct incorporation (pmols), mean of 2 or 3 experiments. [d] Mean of 8 experiments. * Statistically estimated to be significant.

TABLE 4. Effect of Mn on Misincorporation of dCMP into $dAdT_{12-18}$ by *E. coli* DNA Polymerase I, II, or III

Polymerase	Error ($\times 10^{-5}$)	
	Mg (5 mM)	Mn (1 mM)
I	2.5	18.0
II	0.6	1.6
III	0.5	1.4

3′→5′ exonuclease activity was not responsible for the change in misincorporation *in vitro*. The enhanced misincorporation by *E. coli* polymerase II or III in the presence of Mn was also observed (Table 4). However, increases in the misincorporation with these enzymes by 1 mM Mn were less than those with polymerase I because the effect of Mn on polymerase activity differed depending on the enzyme species as indicated in Fig. 2. It is assumed that the number of Mn cations bound to the inhibitory sites of polymerase II or III was less than that bound to polymerase I, although interaction of Mn with dNTP and template was the same in all three polymerization reactions at the same 1 mM concentration.

The misincorporation of error bases by DNA polymerase α from the spleens

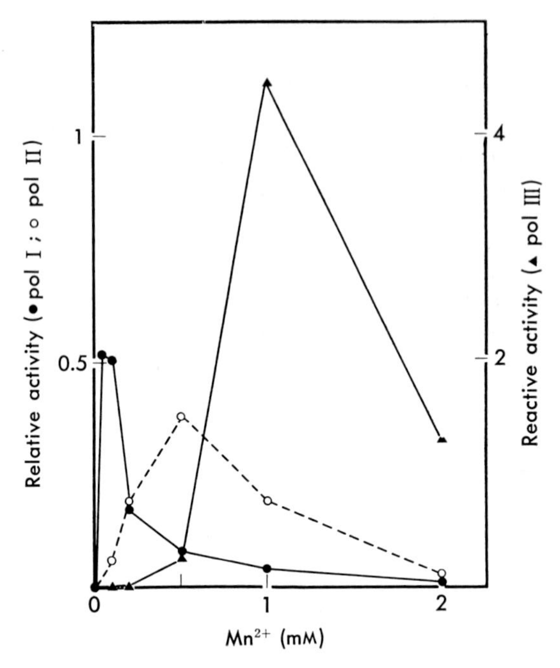

FIG. 2. Effect of Mn on activity of *E. coli* DNA polymerase I, II, or III. Activity of polymerase at each concentration of $MnCl_2$ is represented by the value relative to that observed in the presence of 5 mM $MgCl_2$.

TABLE 5. Effect of Mn on Misincorporation by DNA Polymerase α from Spleens of Mice

Incorrect dNMP	Template-primer	Error ($\times 10^{-5}$)	
		Mg (10 mM)	Mn (1 mM)
dAMP	$dCdG_{12-18}$	19.6	65.8
dTMP	$dCdG_{12-18}$	0.7	34.3
dGMP	$dAdT_{10}$	0.5	36.0
dCMP	$dAdT_{10}$	2.0	7.9

of mice was increased by 1 mM $MnCl_2$ as shown in Table 5. Sirover and Loeb have described similar effects of metals on the fidelity of DNA synthesis with polymerase from avian myeloblastosis virus (*21*).

Change in Fidelity of DNA Polymerase by Treatment with MNNG

Incubation of *E. coli* DNA polymerase I with MNNG *in vitro* caused the binding of nitroamidino group to the enzyme (*22, 23*) and a decrease in the polymerase activity (Fig. 3). DNA polymerase III was more sensitive to MNNG than polymerase I or II. After the enzymes were treated with MNNG to 10% activity of control, the fidelity of DNA synthesis by these enzymes was examined. The frequency of misincorporation of dAMP or dTMP into polydC: dG_{12-18} in place of correct base dGMP by MNNG-treated enzyme was slightly enhanced as indicated

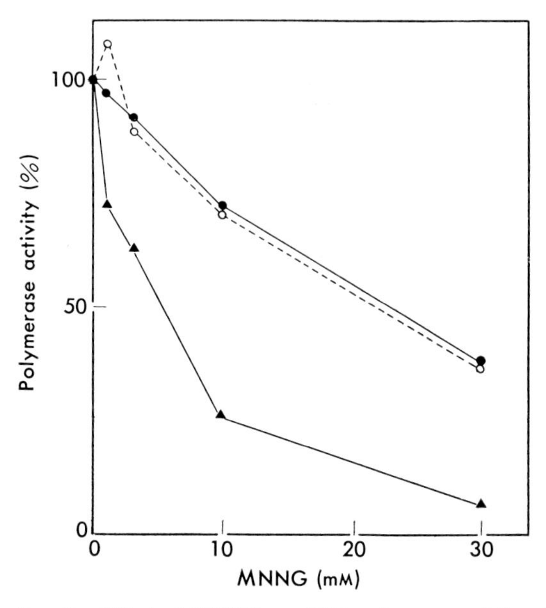

FIG. 3. Effect of MNNG on DNA polymerase of *E. coli*. Enzymes were preincubated with MNNG at 30°C for 60 min and then assayed for DNA polymerase activity. ● polymerase I activity; ○ polymerase II activity; ▲ polymerase III activity.

TABLE 6. Effect of MNNG on Misincorporation by *E. coli* Polymerase I, II, and III

Incorrect dNMP	Template-primer	Polymerase	Error ($\times 10^{-5}$)	
			Control enzyme	MNNG-treated enzyme
dAMP	$dCdG_{12-18}$	I	13.1	19.6
dAMP	$dCdG_{12-18}$	II	16.2	24.0
dAMP	$dCdG_{12-18}$	III	8.4	19.5
dTMP	$dCdG_{12-18}$	I	2.1	3.8
dTMP	$dCdG_{12-18}$	II	0.5	2.0
dGMP	dAT	I	1.9	0.3
dCMP	dAT	I	1.5	0.6
dCMP	dAT	II	1.2	0.7
dCMP	dAT	III	0.5	0.2

Enzymes were pretreated with 10–50 mM MNNG at 37°C for 60 min, dialyzed and assayed for fidelity.

in Table 6. The misincorporation of dCMP or dGMP in poly(dA-dT) was reduced by MNNG treatmant of enzymes.

The change in frequency of misincorporation by polymerase I seemed to correlate with the change in the ratio of 3′→5′ exonuclease to polymerase activity of this enzyme. The effect of MNNG on the 3′→5′ exonuclease activity of DNA polymerase I, measured by the release of dTMP from poly(dA-dT) or polydA: dT_{12-18}, was slightly more intense than on the incorporation of dTMP into template-primer, polydA: dT_{12-18}, by polymerase activity. However, the effect of MNNG on release of dGMP or dCMP from poly(dG-dC) was far weaker than their

TABLE 7. Effect of MNNG on Misincorporation by DNA Polymerase α from Spleens of Mice

Incorrect dNMP	Template-primer	Error ($\times 10^{-5}$)	
		Control enzyme	MNNG-treated enzyme
dAMP	$dCdG_{12-18}$	19.6	11.7
dTMP	$dCdG_{12-18}$	0.7	10.1
dCMP	$dAdT_{10}$	2.0	4.9

The enzyme was pretreated with 1.5 to 5 mM MNNG at 37°C for 60 min, dialyzed and assayed for fidelity.

incorporation into poly(dG-dC). The increased ratio of nuclease to polymerase activity for dGMP or dCMP (above twice that of the control enzyme), caused by MNNG, may result in the decreased misincorporation of dGMP or dCMP as described in the case of the polymerase from a mutant of phage T4 (*24*). It has been described that substitution of specific amino acids at specific sites in the DNA polymerase molecule perturbed the $3' \rightarrow 5'$ exonuclease activity in the case of mutants of phage T4 (*25*). Exposure of DNA polymerase I to UV light in phosphate buffer decreased the polymerase activity, but the fidelity was not changed. The nuclease to polymerase ratio was not altered by UV light for either dTMP or dCMP.

In the case of DNA polymerase α from the spleens of mice, the type of misincorporation enhanced by MNNG treatment was different from those by MNNG-treated *E. coli* polymerase I. Misincorporation of dTMP and dGMP was enhanced (Table 7), and this enzyme is known to have no nuclease activity. Therefore, modification of the replication complex *per se* may result in the decreased fidelity of DNA synthesis.

Induction of DNA Polymerase in the Cells by MNNG Treatment

In the next step, to test the possibility of induction of enzyme responsible for induced repair, we have examined the change in DNA polymerase activities caused by exposure of cells to carcinogens or mutagens (*26*) and UV light. *E. coli* P3478 (pol A^-) were grown in L-broth and then treated with MNNG for 1 hr in phosphate buffer. At this time the colony-forming ability of MNNG-treated cells was 1 to 10% of the control. After 1 hr post-treatment culture in L-broth, the cell density of the control increased to twice that before culture, and the density of MNNG-treated cells increased 1.6 times. The cells were harvested and the extracts from the control and MNNG-treated cells were subjected to DEAE cellulose column chromatography, and the activities of eluted DNA polymerase II and III were measured. Immediately after treatment of bacteria with MNNG, polymerase II and polymerase III activities were the same as those of control cells (Fig. 4b). After 1 hr, the post-MNNG treatment culture resulted in the increase of polymerase II activity up to 5 times as high as that of the control cells (Fig. 4c), the increase in polymerase III activity was not more than 2 times. When chloramphenicol was added to the post-culture medium, no enhancement of polymerases nor multiplication of cells was observed (Fig. 4e). After removal of chloramphenicol, both

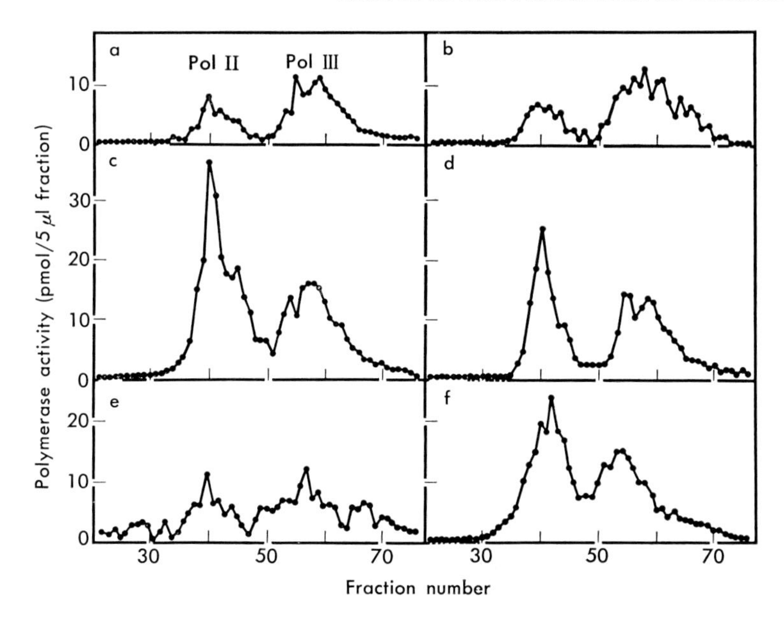

FIG. 4. Chromatography of DNA polymerase II and III from *E. coli* P3478. a, untreated or treated without MNNG; b, immediately after MNNG treatment; c, cultured for 1 hr post-MNNG treatment; d, mixture of protein solution, (a)+(c)=1: 1; e, cultured for 1 hr with chloramphenicol (60 μg/ml) post-MNNG treatment; f, cultured for 2 hr after incubation with chloramphenicol (30°C, 18 hr) post-MNNG treatment.

enhancement of polymerase activity (Fig. 4f) and cell growth were initiated and cell density increased 1.2 times. On the other hand, in the control cells the peak of polymerase II did not increase in spite of 2 times the increase in cell density after removal of chloramphenicol. A striking result was that during the 18 hr incubation of the MNNG-treated cells with chloramphenicol the cellular damage caused by MNNG remained, and an increase in the polymerase was initiated at the time of removal of the antibiotic. Polymerase II and III were also enhanced by treatment of cells with methyl methanesulfonate, hydroxylamine, and UV light.

Then, changes in polymerase activity and mutation to streptomycin-resistance were examined simultaneously in the same post-treatment culture (Fig. 5). The enhanced levels of polymerase II and III were maintained for 1 to 2 hr and subsequently dropped to the control levels. The mutation frequency of streptomycin resistance reached a maximum at 4 hr of the post-culture. Density of MNNG-treated cells increased to more than 2 times that before post-culture indicating that almost all MNNG-damaged cells could divide at least once under the conditions used, although the colony-forming ability was only 1% of the control. These data suggest that enhancement of DNA polymerase activities occurs prior to the fixation of mutation induced by MNNG. The time-course of change in polymerase II and III of *E. coli* WP67 after UV irradiation was similar to that observed after MNNG treatment, but the extent of enhancement of polymerase by UV light was less than that by MNNG. In the case of polymerase II minus strain, H10261, polymerase

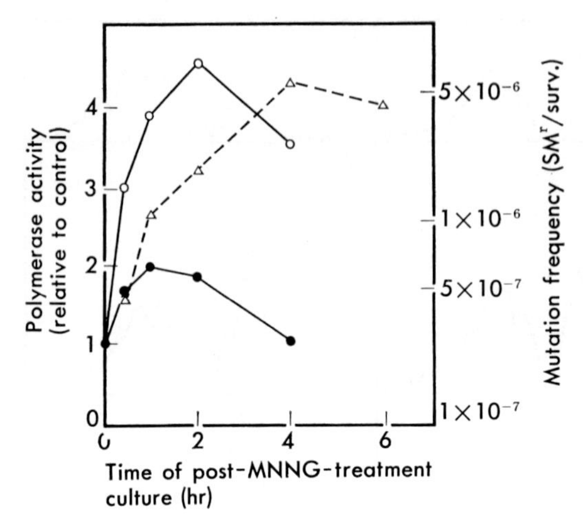

FIG. 5. Time-course of enhancement of polymerase activity and mutation frequency after MNNG treatment of *E. coli* cells. ○ polymerase II activity; ● polymerase III activity; △ mutation frequency of streptomycin resistance.

III was enhanced by MNNG to 2 or 3 times that of the control and mutation to deoxycholate resistance was observed. Therefore, it is possible that the induced polymerase III participates in the fixation of mutation in this strain as described for the UV-induced mutation of *E. coli* (*27*). The extent of misincorporation by the induced enzyme, however, was found to be little different from that by the constitutive enzyme. This result is not in agreement with that described for a *tif* strain

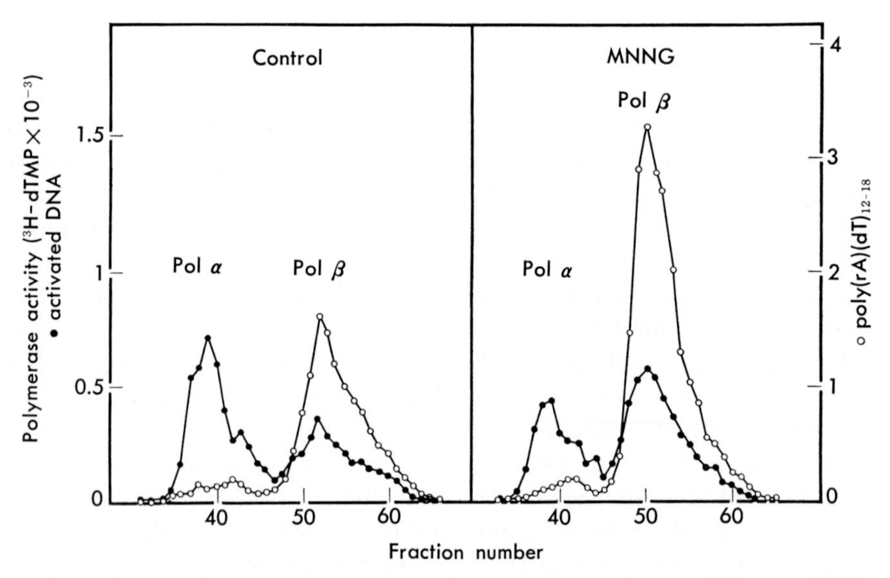

FIG. 6. Phosphocellulose chromatography of DNA polymerases from MNNG-treated HeLa cells. The cells were incubated with MNNG at 37°C for 2 hr in minimal essential medium (MEM), cultured for 16 hr in MEM with 10% fetal calf serum and harvested. The extracts from cells were applied to phosphocellulose column and polymerases were eluted with a linear gradient of 0.05 to 1 M KCl.

of *E. coli*. The crude lysate of this strain grown at 42°C exhibited decreased fidelity of DNA synthesis *in vitro* (*28*).

The effect of MNNG on DNA polymerase activities in HeLa cells was also examined. Treatment of cells with MNNG and subsequent culture for 16 hr enhanced DNA polymerase β to twice of the activity in the control cells (Fig. 6). The curves of response of polymerase α and β to various concentrations of MNNG indicate that a high concentration of MNNG caused an increase in polymerase β and a low concentration of MNNG increased polymerase α.

These inducible changes in DNA polymerase activities may play a role in some repair pathways. Role of polymerase II in repair replication after UV irradiation of *E. coli* (*29*) has been suggested, and polymerase β of lymphocytes has been assumed to be used in repair after X irradiation (*30*). However, further work will be needed to determine whether these polymerases are involved in error-free or error-prone repair systems. It also seems important to see error in DNA synthesis by DNA polymerase treated with mutagen using template modified by mutagen or carcinogen.

ACKNOWLEDGMENT

We are grateful to Dr. S. Kondo for providing *E. coli* strains, and Dr. T. Taguchi for preparing DNA polymerase α from the spleens of mice. We also thank H. Konno and A. Namekai for assistance with a part of the experiments.

REFERENCES

1. Setlow, R. B. Repair deficient human disorders and cancer. Nature, *271*: 713–717, 1978.
2. McCann, J. E., Choi, E., Yamazaki, E., and Ames, B. N. Detection of carcinogens as mutagens in the *Salmonella*/microsome test: Assay for 300 chemicals. Proc. Natl. Acad. Sci. U.S., *72*: 5135–5139, 1975.
3. Smith, K. C. Multiple pathways of DNA repair in bacteria and their roles in mutagenesis. Photochem. Photobiol., *28*: 121–129, 1978.
4. Witkin, E. M. Ultraviolet mutagenesis and inducible DNA repair in *E. coli*. Bacteriol. Rev., *40*: 869–907, 1976.
5. Radman, M. Phenomenology of an inducible mutagenic DNA repair pathway in *E. coli*: SOS repair hypothesis. *In*; L. Prakash *et al.* (eds.), Molecular and Environmental Aspects of Mutagenesis, pp. 128–142, C. C. Thomas, Springfield, Ill., 1974.
6. Ichikawa-Ryo, H. and Kondo, S. Indirect mutagenesis in phage lambda by ultraviolet preirradiation of host bacteria. J. Mol. Biol., *97*: 77–92, 1975.
7. Cerdá-Olmedo, E., Hanawalt, P. C., and Guerola, N. Mutagenesis of the replication point by nitrosoguanidine: Map and pattern of replication of the *Escherichia coli* chromosome. J. Mol. Biol., *33*: 705–719, 1968.
8. Baker, R. and Tessman, I. Different mutagenic specificities in phage S13 and T4: *In vivo* treatment with *N*-methyl-*N*′-nitro-*N*-nitrosoguanidine. J. Mol. Biol., *35*: 439–448, 1968.
9. Yamamoto, K., Kondo, S., and Sugimura, T. Mechanism of potent mutagenic action of *N*-methyl-*N*′-nitro-*N*-nitrosoguanidine on intracellular phage lambda. J. Mol. Biol., *118*: 412–429, 1978.

10. Hewper, W. C. Carcinogenesis in the human environment. Arch. Pathol., *71*: 237–269, 1971.
11. Furst, A. and Haro, R. T. A survey of metal carcinogenesis. Prog. Exp. Tumor Res., *12*: 102–133, 1969.
12. Lindegren, C. C., Nagai, S., and Nagai, H. Induction of respiratory deficiency in yeast by manganese, copper, cobalt and nickel. Nature, *182*: 446–448, 1958.
13. Demerec, M. and Hanson, J. Mutagenic action of manganous chloride. Cold. Spring Harbor Symp. Quant. Biol., *16*: 215–228, 1951.
14. Nishioka, H. Mutagenic activities of metal compounds in bacteria. Mutat. Res., *31*: 185–189, 1975.
15. Miyaki, M., Akamatsu, N., Ono, T., and Koyama, H. Mutagenicity of metal cations in cultured cells from Chinese hamster. Mutat. Res., *68*: 259–263, 1979.
16. Hall, Z. W. and Lehman, I. R. An *in vitro* transversion by a mutationally altered T4-induced DNA polymerase. J. Mol. Biol., *36*: 321–333, 1968.
17. Miyaki, M., Murata, I., Osabe, M., and Ono, T. Effect of metal cations on misincorporation by *E. coli* DNA polymerases. Biochem. Biophys. Res. Commun., *77*: 854–860, 1977.
18. Murray, M. J. and Flessel, C. P. Metal-polynucleotide interactions. A. comparison of carcinogenic and non-carcinogenic metals *in vitro*. Biochim. Biophys. Acta, *425*: 256–261, 1976.
19. Brutlag, D. and Kornberg, A. Enzymatic synthesis of deoxyribonucleic acid. XXXVI. A proof reading function for the 3′→5′ exonuclease activity in deoxyribonucleic acid polymerases. J. Biol. Chem., *247*: 241–248, 1972.
20. Klenow, H. and Henningsen, I. Selective elimination of the exonuclease activity of the deoxyribonucleic acid polymerase from *E. coli* B by limited proteolysis. Proc. Natl. Acad. Sci. U.S., *65*: 168–175, 1970.
21. Sirover, M. A. and Loeb, L. A. Infidelity of DNA synthesis *in vitro*: Screening for potential metal mutagens or carcinogens. Science, *194*: 1434–1436.
22. McCalla, D. R. and Reuvers, A. Reaction of *N*-methyl-*N*′-nitro-*N*-nitrosoguanidine with protein: Formation of nitroguanido derivatives. Can. J. Biochem., *46*: 1411–1415, 1968.
23. Nagao, M., Yokoshima, T., Hosoi, H., and Sugimura, T. Interaction of *N*-methyl-*N*′-nitro-*N*-nitrosoguanidine with ascites cells *in vitro*. Biochim. Biophys. Acta, *192*: 191–199, 1969.
24. Muzyzka, N., Poland, R. I., and Bessman, M. J. Studies on the biochemical basis of spontaneous mutation. I. A comparison of the deoxyribonucleic acid polymerases of mutator, antimutator, and wild-type strains of bacteriophage T4. J. Biol. Chem., *247*: 7116–7122, 1972.
25. Reha-Kranz, L. and Bessman, M. J. Studies on the biochemical basis of mutation. IV. Effect of amino acid substitution on the enzymatic and biological properties of bacteriophage T4 DNA polymerase. J. Mol. Biol., *116*: 99–113, 1977.
26. Miyaki, M., Sai, G., Katagiri, S., Akamatsu, N., and Ono, T. Enhancement of DNA polymerase II activity in *E. coli* after treatment with *N*-methyl-*N*′-nitro-*N*-nitrosoguanidine. Biochem. Biophys. Res. Commun., *76*: 136–141, 1977.
27. Bridges, B. A., Mottershead, R. P., and Sedgwick, S. G. Mutagenic DNA repair in *Escherichia coli*. III. Requirement for a function of DNA polymerase III in ultraviolet light mutagenesis. Mol. Gen. Genet., *144*: 53–58, 1976.
28. Radman, M., Calliet-Fauquet, P. Defais, M., and Villani, G. The molecular mechanism of induced mutations and an *in vitro* assay for mutagenesis. *In*; Screening

Tests in Chemical Carcinogenesis, pp. 537–545, I.A.R.C. Scientific Publications, Lyons, 1976.

29. Masker, W., Hanawalt, P., and Shizuya, H. Role of DNA polymerse II in repair replication in *Escherichia coli*. Nature New Biol., *244*: 242–243, 1973.
30. Bertazzoni, U., Stefanini, M., Noy, G. D., Giulotto, E., Nuzzo, F., Falaschi, A., and Spadari, S. Variations of DNA polymerases-α and -β during prolonged stimulation of human lymphocytes. Proc. Natl. Acad. Sci. U.S., *73*: 785–789, 1976.

GENETIC AND ENVIRONMENTAL FACTORS IN EXPERIMENTAL AND HUMAN CANCER,
H. V. GELBOIN ET AL. (EDS.), JAPAN SCI. SOC. PRESS, TOKYO, PP. 215–229, 1980

Autoradiographic Study of DNA Repair Synthesis *In Vivo* and in Short-term Organ Cultures, with Special Reference to DNA Repair Levels, Aging, and Species Differences

Takatoshi ISHIKAWA, Fumio IDE, and Shozo TAKAYAMA
Department of Experimental Pathology, Cancer Institute, Tokyo, Japan

Abstract: Three methods were devised for measuring DNA repair synthesis at a cellular level *in vivo* or in short-term organ cultures. In these systems unscheduled DNA synthesis could be demonstrated quantitatively by grain counts over the nuclei of cells treated with various carcinogens. Comparative studies were made on DNA repair levels in relation to aging and species differences.

For studies on aging, part of the skull of small aquarium fish (*Oryzias latipes*) of various ages was removed and the brain of the living fish was treated with 4-nitroquinoline 1-oxide (4NQO) and (methyl-^{3}H) thymidine (TdR). Freshly removed retinas and tracheas of rats and golden hamsters of various ages were treated with 4NQO and (methyl-^{3}H) TdR *in vitro*. The level of unscheduled DNA synthesis was studied autoradiographically. These studies showed that there was no age-associated change in the DNA repair levels of the ganglion cells of the central nervous system or the tracheal epithelial cells.

For studies on species differences in DNA repair levels, pieces of retina and trachea from 6 species (mice, rats, golden hamsters, guinea pigs, pigs, and cows; all young adult males) were treated with 3 carcinogens in short-term organ cultures. The results showed that different species differed considerably in their levels of unscheduled DNA synthesis. Moreover, the relative efficiencies of DNA repair in the 6 species depended on the carcinogen tested. For example, mouse cells, which were previously considered to show very low levels of excision repair in cell culture, were found to show active unscheduled DNA synthesis in short-term organ cultures of retinas and tracheas.

There are many indications that damage to DNA in neoplastic transformation. The capacity to repair DNA damage relative to other cellular processes should be an important parameter of the initial steps of carcinogenesis (*1*), but this capacity should be studied in intact animals.

DNA strand breakage and rejoining after treatment with agents that damage DNA have been studied in the liver (*2–6*), brain (*7, 8*), and other organs (*9–15*) of intact animals. However, in this type of study, in which the tissues are treated as a whole without excluding mesenchymal cells, artefactual DNA breakage tends to occur. Moreover, the methods involved seem too complicated for application to a large number of samples. Therefore, it seems important to develop a better method for measuring DNA repair *in vivo*, or conditions mimicking those *in vivo*, to obtain further information on DNA repair *in vivo*. An autoradiographic method should be useful for precise measurement of DNA repair in individual cells of an organ, in which the normal histotypic organization of different types of cells is maintained. There are several reports of autoradiographic demonstration of DNA repair in mammalian cells in organ culture (*16–21*). However, in most of these studies a dose-dependent response was not obtained. Recently, we reported 2 procedures for measuring DNA repair quantitatively in cerebral ganglion cells of an aquarium fish *in vivo* (*22*) and rat retinal ganglion cells in short-term organ culture (*23*). We have also devised a sensitive method for measuring DNA repair in tracheal epithelial cells in organ culture. In these systems unscheduled DNA synthesis could be demonstrated quantitatively by grain counts over the nuclei of cells treated with various carcinogens. This paper reports comparative studies on DNA repair levels in relation to aging of cells and species differences.

Aging and DNA Repair Levels

We approached this problem by using 3 methods to demonstrate unscheduled DNA synthesis. The first 2 methods were reported in detail in previous papers (*22, 23*).

1. Fish cerebral ganglion cells in vivo

Ganglion cells of the central nervous system (CNS) of adult vertebrates are non-dividing cells and they should be useful in studies on changes related to aging. Previously, we demonstrated unscheduled DNA synthesis in ganglion cells of an aquarium fish *in vivo* by autoradiography (*22*). The fish used is a small aquarium fish, whose Japanese name is medaka (*Oryzias latipes*). The medaka, a small egg-laying fish and a member of the family Oryziatidae, is found in some parts of Asia. Adult fish are 3.0–3.5 cm long and weigh 0.5–0.7 g. This fresh-water fish is known to withstand sea water, when gradually adapted to it, and does not seem to be affected by an isotonic solution.

For demonstration of DNA repair, adult fish (5–7 months old) were used. The fish were maintained in the isotonic salt solution prescribed for this fish, composed of 0.13 M NaCl, 27 mM KCl_2, and 0.3 mM $NaHCO_3$, for 3 days before operation. They were anesthetized with 0.01% tricaine methane sulphonate (MS 222, Sandoz). An area (about 3×3 mm) of the upper part of the bony skull was removed with fine knives and scissors under a dissecting microscope. The fish survived in the solution for several days after operation. After operation, fish were transferred to a working solution in small incubation flasks. The working solution consisted of 20 ml of isotonic salt solution, (methyl-^{3}H) thymidine (TdR) (10 μCi/ml) (New England Nuclear, specific activity 57.3 Ci/mmol) and a carcinogen. The following carcinogens were tested at the indicated concentrations: 4-nitroquinoline 1-oxide (4NQO, Wako Pure Chemical Co., Tokyo) (0.25, 1.0, 2.0, 4.0×10^{-5} M), ethylnitrosourea (ENU, a gift from Dr. M. Nakadate, National Institute of Hygienic Science, Tokyo) (0.1, 0.32, 1.0, 3.2×10^{-3} M), and methylmethane sulphonate (MMS, Tokyo Kasei Co., Tokyo) (0.1, 0.32, 1.0, 3.2×10^{-3} M). After incubation for 4 hr at 25°C, the fish were fixed in 10% formaldehyde solution. The brains were embedded in paraffin and frontal sections through the mesencephalon were made at 4 μm thickness. Sections were treated with 3% perchloric acid, dip-covered with NR-M_2 emulsion (Konishiroku Photo Co., Tokyo) and developed after 5 weeks'

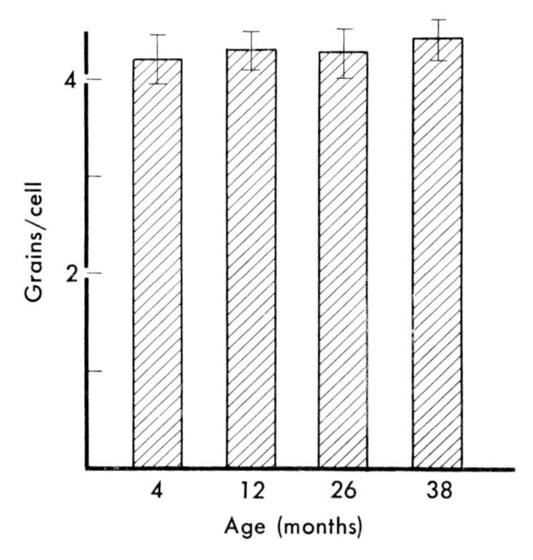

FIG. 1. Age influence on unscheduled DNA synthesis in cerebral ganglion cells of medaka (*Oryzias latipes*). Fish were treated with 2.0×10^{-5} M 4NQO and (methyl-^{3}H)TdR for 4 hr at 25°C.

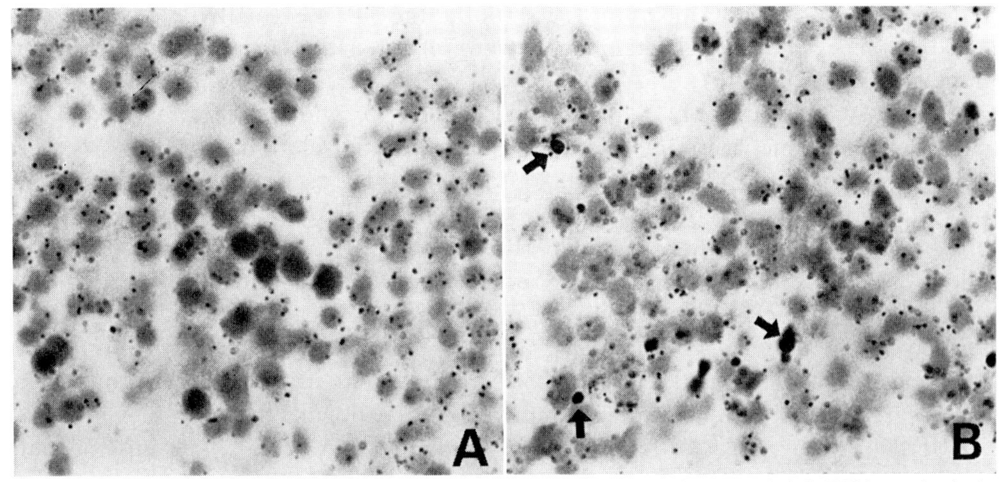

FIG. 2. Examples of autoradiographs. Demonstration of unscheduled DNA synthesis in cerebral ganglion cells of 4-month-old fish (A) and 38-month-old fish (B). Ganglion cells of young and aged fish show similar numbers of grains on the nuclei. Larger brown pigments in B are lipofuscins normally seen in aged fish (arrows). ×1,200.

exposure at 4°C. Preparations were lightly stained with hematoxylin and eosin. Grains were counted under a ×100 objective lens in a Nikon microscope on 100 randomly selected ganglion cell nuclei in each brain. Unscheduled DNA synthesis was clearly demonstrated as silver grains on the nuclei of ganglion cells of groups treated with carcinogens. The highest average grain number in carcinogen-treated groups was about 16 to 17 times that of the control. In order to study the effect of aging on the level of DNA repair, four groups of male fish of 4 to 38 months of age were used. Aged fish (26 and 38 months old), which had been kept in the National Institute of Radiological Science were generously supplied by Dr. H. Matsudaira. Groups of 5–10 fish were treated with a single dose of 4NQO (2.0×10^{-5} M) and (methyl-^{3}H) TdR (20 μCi/ml) for 3 hr at 25°C. Autoradiography was carried out in the same way and the numbers of grains on 300 randomly selected ganglion cells in each brain were counted. The oldest medaka used in this study was considered to be at about the limit of its maximum lifespan (*24*). As shown in Figs. 1 and 2, there was no age-associated change in the ability of repair synthesis in ganglion cells of the medaka.

2. *Rat retinal ganglion cells in organ culture*

The retina is a part of the CNS and it survives in organ culture for several weeks (*25, 26*). DNA repair synthesis in individual cells of different types in the cultures can be determined by autoradiography (*23*). Male Wistar rats (6 months old) were anesthetized with sodium pentobarbital, and their eyes were excised. A circular cut was made around the limbus and the corneal area, and the lens and vitreous humor were removed. The residual tissue containing the retina was immersed in medium L-15 at 37°C. The retina was then gently detached in the medium under a dissecting microscope. These procedures were performed in a darkened room. Pieces of retina (about 4×4 mm) were incubated for 2 hr at 37°C in a small test tube with 1 ml of medium L-15 containing 10 μCi (methyl-^{3}H) TdR and a carcinogen. The following carcinogens were tested at the indicated concentrations: 4NQO (0.032, 0.1, 0.32, 1.0, and 3.2×10^{-5} M); ENU (0.032, 0.1, 0.32, 1.0, and 3.2×10^{-3} M), and MMS (0.032, 0.1, 0.32, 1.0, and 3.2×10^{-3} M). After incubation, the tissues were fixed in 10% formaldehyde solution. Fixed tissues were embedded in paraffin and cut into 4- to 5-μm thick sections. Autoradiographic procedures were as described for studies on fish. In the carcinogen-treated group, unscheduled DNA synthesis was clearly demonstrated as silver grains over the nuclei of the outer nuclear layer of the retina. Grains were counted on 100 randomly selected ganglion cells from the outer nuclear layer of each section. The cells in this layer had uniform numbers of grains. The highest average grain number in carcinogen-treated groups was about 50–70 times that in the control group. The three carcinogens showed similar roughly dose-dependent effects.

To study the effect of aging on the levels of DNA repair, 4 groups of male random-bred Wistar rats of 1 to 24 months old were used. Pieces of retina from 5 rats in each group were treated with a single dose of 4NQO (3.2×10^{-5} M) and (methyl-^{3}H) TdR for 2 hr at 37°C. Grains were counted on 200 randomly selected ganglion cells. Figure 3 shows the levels of unscheduled DNA synthesis in the

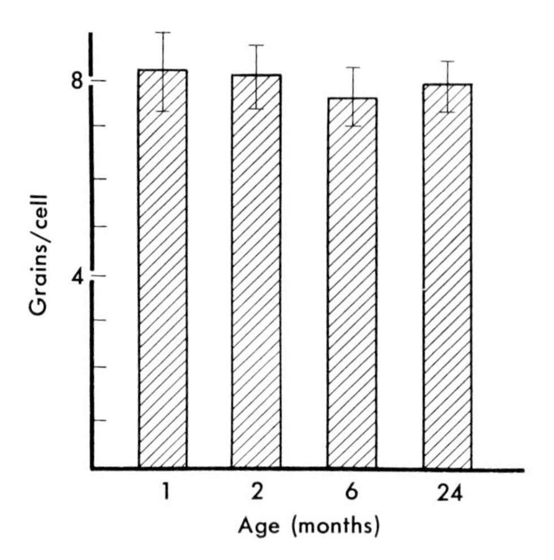

FIG. 3. Influence of age on unscheduled DNA synthesis in rat retinal ganglion cells treated with 3.2×10^{-5} M 4NQO.

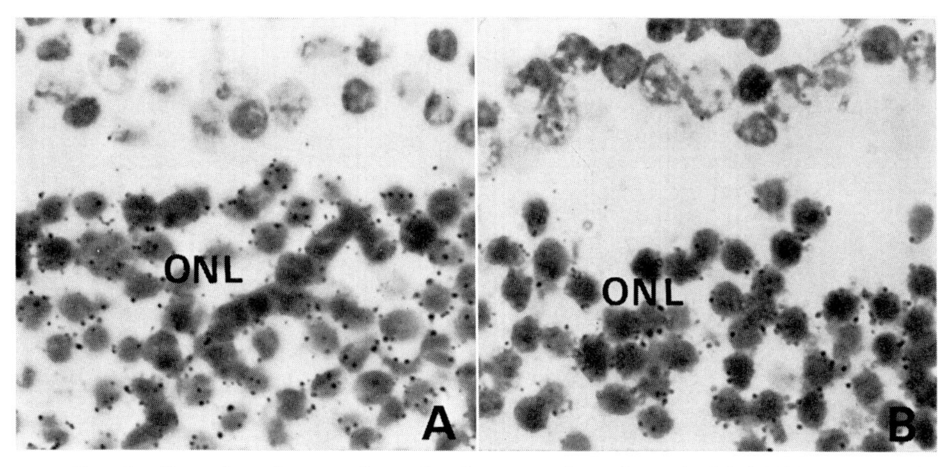

FIG. 4. Examples of autoradiographs. Demonstration of unscheduled DNA synthesis in retinal ganglion cells of the outer nuclear layer of young (1-month-old) (A) and aged (24-month-old) (B) rats, showing similar numbers of grains on the nuclei of the two. ONL: outer nuclear layer. ×1,500.

retina of rats 1, 2, 6, and 24 weeks old after treatment with a single dose of 4NQO. No age-associated change was found in the level of unscheduled DNA synthesis in the retinal ganglion cells (Figs. 3 and 4).

3. Tracheal epithelial cells in organ culture

A system in which a tracheal organ culture of Fischer rats was used in combination with autoradiography was developed for quantitative measurement of DNA repair synthesis in tracheal epithelial cells. Male Fischer rats (8 weeks old), were obtained from Charles River Japan, Inc. (Atsugi-shi). Animals were killed by a blow on the head. After exsanguination, their tracheas were separated from the

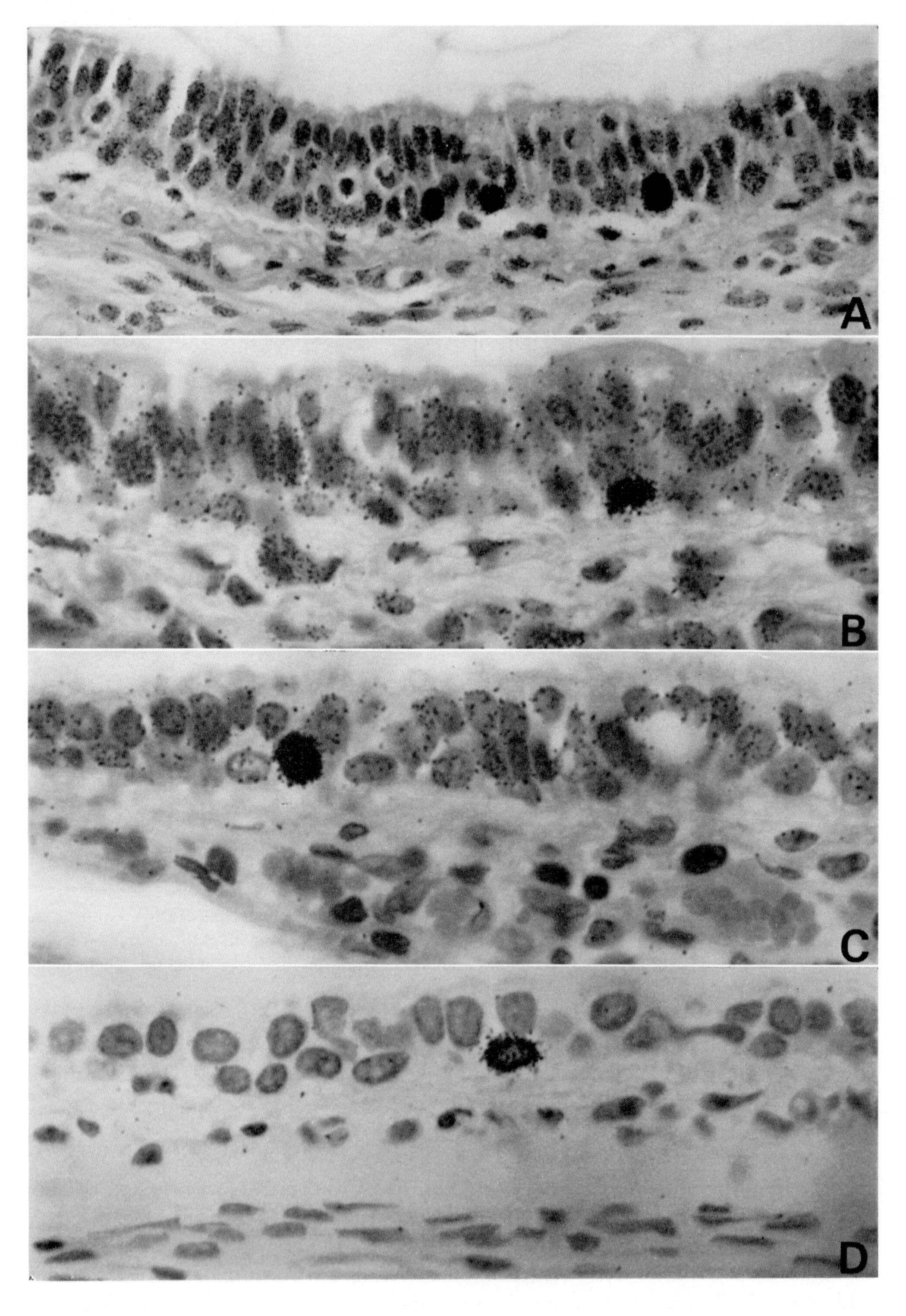
A
B
C
D

lung. Specimens were immediately immersed in L-15 medium at 4°C. The external connective tissue was carefully removed and the trachea was cut into 10–12 thin transverse sections. Five to 6 sections from different animals were incubated together for 2 hr at 37°C in a test tube in 2 ml of L-15 medium containing 20 μCi (methyl-^{3}H) TdR and a carcinogen. Tubes containing the sections were maintained in a chamber in an atmosphere of 50% O_2 and 50% N_2. The following three carcinogens were tested at the indicated concentrations: 4NQO (0.32, 1.0, 3.2, 10, 32, and 100×10^{-6} M); ENU (0.1, 0.32, 1.0, 3.2, and 10×10^{-3} M), and MMS (0.032, 0.1, 0.32, 1.0, 3.2, and 10×10^{-3} M). After incubation, the tissues were fixed in 10% neutral formaldehyde solution. The 5 to 6 fixed tracheal rings in each tube were embedded in the same transverse plane and cut into sections of 4–5 μm thickness. Autoradiographic procedures were as described previously. Grains were counted on 100 randomly selected epithelial cells of each tracheal ring. In carcinogen-treated groups, DNA repair synthesis was clearly demonstrated as silver grains on the nuclei of epithelial cells of the trachea (Fig. 5). The epithelial cells of the entire ring had uniform numbers of grains. Cells with heavily labeled nuclei, occasionally seen in the basal layer of the epithelium, were presumably in the S-phase during incubation. The labeling index was approximately 0.02 and this did not interfere significantly with grain counting. The three carcinogens had similar dose-dependent effects. The highest average grain number in the carcinogen-treated groups was about 70 times that in the control group (0.2–0.3 grains/cell).

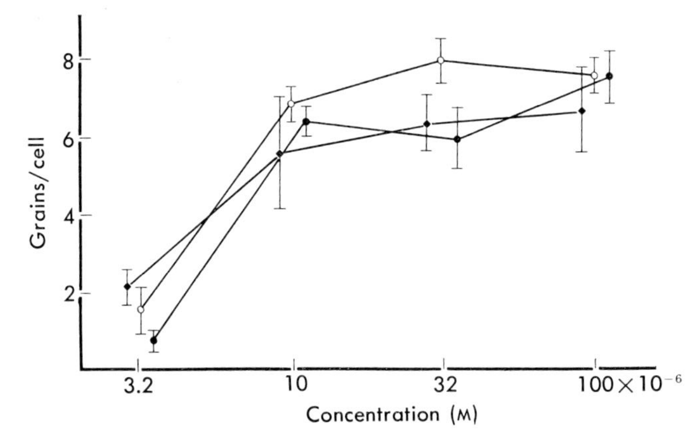

FIG. 6. Influence of age on unscheduled DNA synthesis in hamster tracheal epithelium treated with 4NQO and (methyl-^{3}H)TdR for 2 hr at 37°C. Numbers of grains per cell are averages for 4 tracheas from different animals. Age (months): ◆ 2; ○ 12; ● 18. Bars, SD.

←FIG. 5. Autoradiographic demonstration of unscheduled DNA sysnthesis in rat tracheal epithelium treated with chemical carcinogens. A. Autoradiograph showing silver grains on the nuclei of tracheal epithelium indicative of DNA repair synthesis after treatment with 3.2×10^{-5} M 4NQO and (methyl-^{3}H)TdR. ×250. B. Autoradiograph of trachea after treatment with 3.2×10^{-3} M ENU and (methyl-^{3}H)TdR. ×1,000. C. Autoradiograph of trachea exposed to 3.2×10^{-3} M MMS. ×1,000. D. Control autoradiography showing few background grains. The trachea was treated with (methyl-^{3}H)TdR. Cells with heavily labeled nuclei were in the S-phase during incubation. ×1,000.

We found that unscheduled DNA sysnthesis could be demonstrated in the tracheal epithelial cells of golden hamsters with these carcinogens. A preliminary experiment on aging was conducted with hamsters. The Syrian golden hamsters 12 and 18 months old (Tokyo Experimental Animal Co., Tokyo), used in this work were control animals for other experiments. Tracheal rings from 4 different animals in each group were treated with 4 doses of 4NQO and (methyl-^{3}H) TdR. The results in Fig. 6 show the levels of unscheduled DNA synthesis in animals 2, 12, and 18 months old in response to 4 doses of 4NQO. No age-associated change occurred in the levels of unscheduled DNA synthesis in the tracheal epithelium of hamsters.

Species Differences and DNA Repair

Interspecific differences in DNA repair levels were examined by measuring carcinogen-induced DNA repair synthesis in organ cultures of retinas and tracheas

TABLE 1. Dose-response Relationship for Unscheduled DNA Synthesis in Retinal Ganglion Cells from 6 Species Treated with 3 Carcinogens (4NQO, MMS, and ENU)

Species	Conc. of 4NQO ($\times 10^{-5}$ M)				
	0.1	0.32	1.0	3.2	10
Mouse	0.31±0.09	0.95±0.16	2.49±1.22	8.01±1.73	2.01±0.81
Rat	0.34±0.20	0.83±0.87	2.90±0.65	2.32±0.61	2.30±0.50
Golden hamster	0.27±0.11	0.50±0.26	1.10±0.78	1.55±0.33	0.22±1.44
Guinea pig	0.36±0.10	1.31±0.77	1.53±1.22	5.86±2.15	2.67±0.37
Pig	2.23	6.87	4.17	5.61	2.80
Cow	1.19	1.60	3.45	6.28	5.81
Species	**Conc. of MMS ($\times 10^{-3}$ M)**				
	0.1	0.32	1.0	3.2	10
Mouse	1.05±0.53	1.85±0.24	2.31±0.60	7.97±1.10	1.83±1.27
Rat	1.05±0.47	1.47±0.76	2.27±1.42	8.78±0.73	3.55±1.16
Golden hamster	2.06±0.34	2.94±0.83	3.83±3.17	4.88±1.95	2.67±0.55
Guinea pig	2.28±0.61	2.68±0.45	3.18±0.53	3.81±1.46	2.76±0.37
Pig	1.32	1.91	7.13	5.43	2.26
Cow	2.19	2.90	5.91	2.21	1.80
Species	**Conc. of ENU ($\times 10^{-3}$ M)**				
	0.1	0.32	1.0	3.2	10
Mouse	0.37±0.09	0.89±0.09	2.03±0.28	0.35±0.08	0.21±0.06
Rat	0.20±0.09	0.65±0.10	3.31±1.07	5.27±1.39	0.78±0.22
Golden hamster	0.34±0.09	0.28±0.17	1.02±0.69	2.54±0.65	0.60±0.17
Guinea pig	0.23±0.06	0.77±0.12	2.48±0.41	1.88±0.20	0.55±0.11
Pig	0.72	1.10	1.96	0.89	
Cow	0.55	2.07	1.98	0.71	

Numbers of grains per cell (mean±SD) were conuted against the concentration of carcinogen on a logarithmic scale. Numbers of grains per cell are averages for 4 retinas from different individuals. Averages for 2 animals are shown for the cow and pig.

from 6 different species of animals. The animals used in this study were as follows: 8-week-old male ICR mice (Charles River Japan, Inc., Astugi-shi); 10-week-old male Fischer rats (Charles River Japan, Inc., Atsugi-shi); 10-week-old male Syrian golden hamsters (Tokyo Experimental Animal Co., Tokyo); 3- to 4-month-old male guinea pigs (Saitama Experimental Animal Co., Saitama); 6-month-old castrated pigs (by courtesy of Mr. M. Ogata) and 2-year-old castrated cows (by courtesy of Mr. M. Ogata). These animals were all young adult male animals which had completed approximately 1/15 to 1/30 of their lifespan. The techniques used were essentially as described above, but as we considered that comparison of widely different species should be done under as nearly similar conditions as possible we cut freshly removed retinas and tracheas from laboratory animals into small sections and kept them in L-15 medium at 4°C before use to obtain similar conditions to those used with tissues from cows and pigs. Retinas and tracheas were removed from 2 different cows or pigs immediately after sacrifice at a slaughter-house (Tachikawa-shi) and pieces of tissues were transferred to the laboratory in L-15 medium at 4°C in about 2 hr. Care was also taken to use the same vials of carci-

TABLE 2. Dose-response Relationship for Uuscheduled DNA Synthesis in Epithelial Cells of Trachea from 6 Species Treated with 3 Carcinogens (4NQO, MMS, and ENU)

Species	Conc. of 4NQO ($\times 10^{-5}$ M)				
	0.1	0.32	1.0	3.2	10
Mouse	1.69±0.12	2.20±0.11	4.05±0.42	8.12±0.92	14.42±1.18
Rat	0.87±0.53	2.30±0.49	4.13±0.40	9.09±1.19	1.08±0.24
Golden hamster	1.73±0.21	2.69±0.10	3.89±0.10	7.43±1.00	9.16±0.27
Guinea pig	0.38±0.19	1.12±0.30	3.77±0.34	6.68±1.25	3.34±0.39
Pig	1.81	2.62	7.87	4.71	2.18
Cow	0.19	0.20	0.99	8.62	2.19
Species	**Conc. of MMS ($\times 10^{-3}$ M)**				
	0.1	0.32	1.0	3.2	10
Mouse	1.82±0.17	2.79±0.31	7.62±0.70	12.59±0.48	2.77±0.49
Rat	0.60±0.20	1.23±0.23	3.83±0.31	10.25±0.36	2.63±0.29
Golden hamster	0.21±0.04	2.04±0.16	4.03±0.64	9.26±0.58	8.82±0.51
Guinea pig	0.25±0.15	0.47±0.17	1.49±0.71	5.85±0.82	7.85±1.23
Pig	0.24	0.46	0.53	0.59	4.22
Cow	0.32	0.20	1.18	8.44	0.56
Species	**Conc. of ENU ($\times 10^{-3}$ M)**				
	0.1	0.32	1.0	3.2	10
Mouse	0.89±0.20	1.78±0.15	4.16±0.44	8.35±0.73	3.99±1.10
Rat	1.57±0.32	2.59±0.29	4.94±0.53	10.40±2.05	2.60±0.56
Golden hamster		2.79±0.16	5.60±0.36	9.65±0.42	6.78±0.69
Guinea pig	0.17±0.02	0.34±0.15	0.40±0.13	0.62±0.11	5.22±0.49
Pig	0.21	0.25	0.46	6.27	0.87
Cow	0.30	0.37	0.59	4.37	1.46

Numbers of grains per cell (mean±SD) are averages for 5 tracheas from different individuals. Averages for 2 animals are shown for the cow and pig.

nogens and of (methyl-^{3}H) TdR throughout, and series of experiments were completed within a relatively short time (3 weeks). Three carcinogens (4NQO, MMS, and ENU) at various concentrations were used to induce unscheduled DNA synthesis as shown in Tables 1 and 2. For retinal cultures, pieces of retina from 4 different individuals (except those of cows and pigs) were treated with 5 doses of each carcinogen for 2 hr at 37°C. Tracheal sections from 5 different animals were also treated in the same way. The volume of working solution (1–2 ml) was

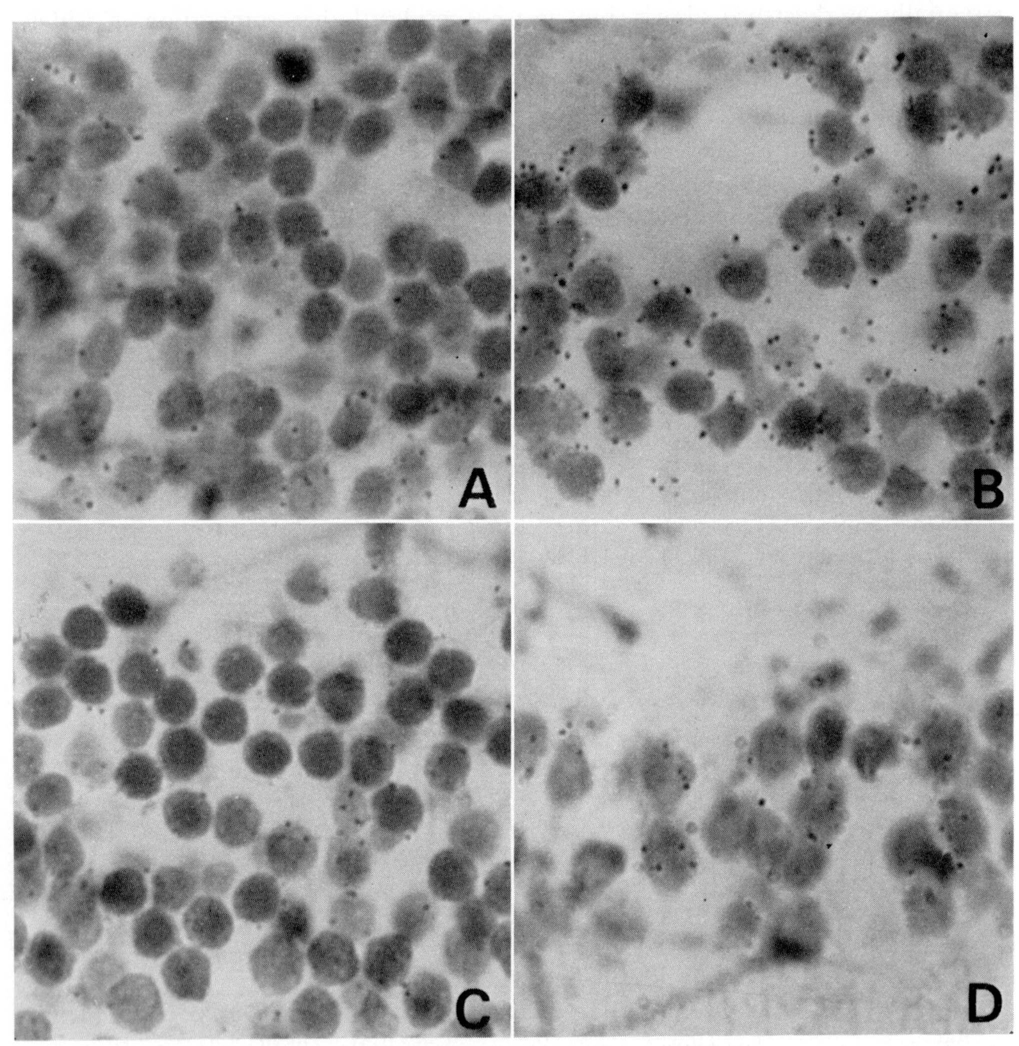

FIG. 7. Examples of autoradiographs of retinal ganglion cells from various animals. A. Autoradiograph showing silver grains on the nuclei of mouse retinal ganglion cells after treatment with 10^{-3} M ENU and (methyl-^{3}H)TdR. B. Autoradiograph of retinal ganglion cells of a rat after treatment with 3.2×10^{-3} M ENU and (methyl-^{3}H)TdR. C. Autoradiograph of retinal ganglion cells of a golden hamster after treatment with 10^{-3} M ENU and (methyl-^{3}H)TdR. D. Autoradiograph of retinal ganglion cells of a pig after treatment with 10^{-3} M ENU and (methyl-^{3}H)TdR. $\times 1,500$.

varied according to the weight of tissue (20–40 mg). All autoradiographic procedures were carried out simultaneously. Grains were counted on 200 randomly selected cells. The mean grain number for controls was 0.2–0.3.

1. Retinal ganglion cells in organ culture

Table 1 shows the grain counts in the 6 species. The 3 carcinogens (4NQO, MMS, and ENU) showed dose-dependent effects on unscheduled DNA synthesis

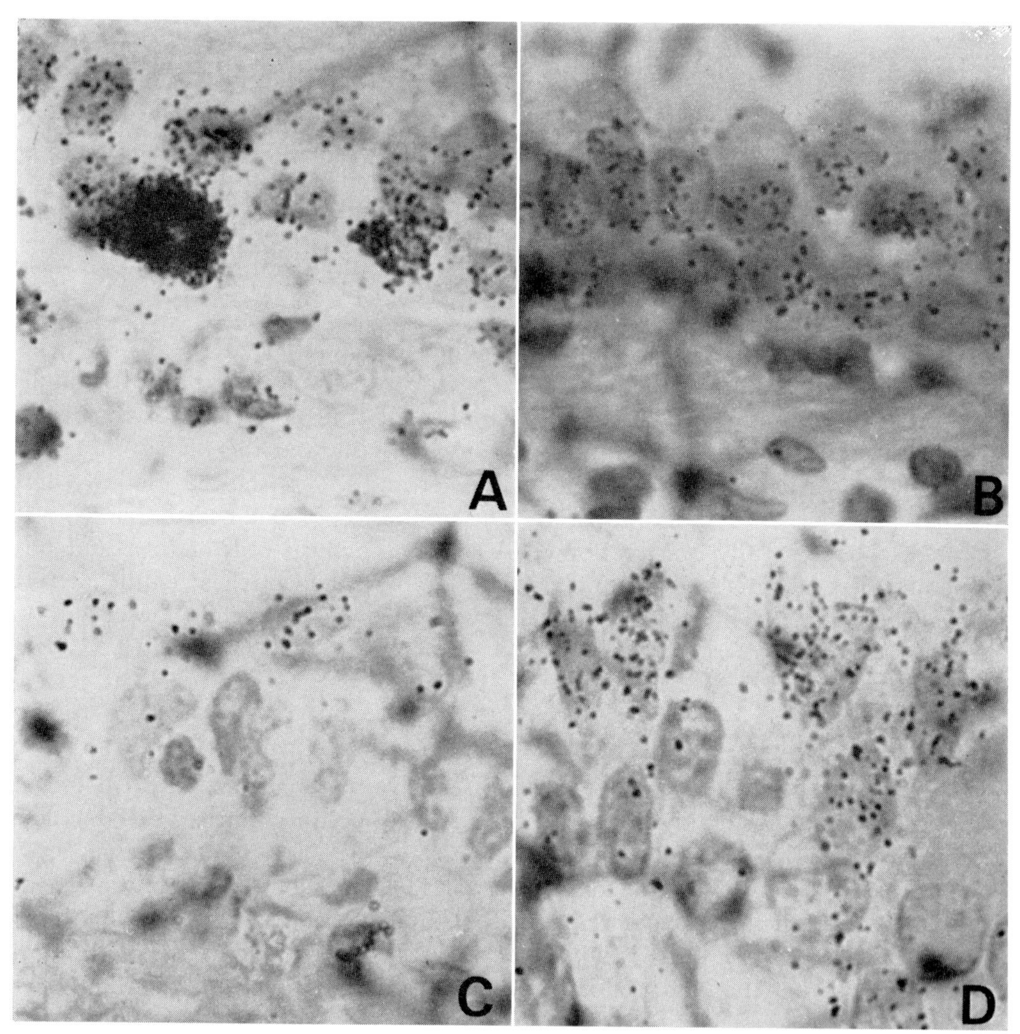

FIG. 8. Examples of autoradiographs of tracheal epithelial cells from various animals. A. Autoradiograph showing silver grains on the nuclei of tracheal epithelial cells of a mouse after treatment with 3.2×10^{-3} M MMS and (methyl-^{3}H)TdR. B. Autoradiograph of tracheal epithelial cells of a rat after treatment with 3.2×10^{-3} M MMS and (methyl-^{3}H) TdR. C. Autoradiograph of tracheal epithelial cells of a pig after treatment with 10×10^{-3} M MMS and (methyl-^{3}H)TdR. D. Autoradiograph of tracheal epithelial cells of a cow after treatment with 3.2×10^{-3} M MMS and (methyl-^{3}H)TdR. $\times$1,500.

in retinal ganglion cells. Examples of autoradiographs are also shown in Fig. 7. We estimated the highest level of each response curve (not illustrated here) as the highest efficiency of DNA repair in each species, since we could not exclude the possibility that the dose-response curve might be shifted by differences in permeability of carcinogen into the cells of the different species.

We found that the DNA repair levels in response to each carcinogen differed from animal to animal as shown in Table 1. It should also be noted that the relative efficiency of DNA repair in the 6 species differed from one carcinogen to another: for instance, mouse cells showed the high responses to 4NQO and MMS, but a low response to ENU; rat cells also showed high responses to MMS and ENU, but little response to 4NQO. Since 4NQO is not a direct carcinogen, DNA repair levels may be modified by the levels of carcinogen metabolism. In most animals the numbers of silver grains decreased at higher doses of carcinogens, possibly due to the toxic effect of the carcinogens. Cows and pigs seemed to be rather sensitive to the toxic effects of carcinogenes. The retinal ganglion cells of golden hamsters and guinea pigs generally showed low DNA repair.

2. *Tracheal epithelial cells in organ culture*

The dose-response relationship for unscheduled DNA synthesis in epithelial cells of tracheas treated with the 3 carcinogens are shown in Table 2. Examples of autoradiographs are shown in Fig. 8. Numbers of grains per cell are averages for 5 tracheas from the different animals. Mouse cells showed the highest responses to 4NQO and MMS, while rat, golden hamster and mouse showed similarly high DNA repair levels in response to ENU. The tracheal epithelial cells of guinea pigs seemed to be generally inactive for DNA repair. Tracheal cells of cows and pigs generally showed intermediate levels of DNA repair. The tracheal epithelial cells of hamsters showed high responses, although their retinal cells showed low or intermediate levels of response.

DISCUSSION

One aim of this study was to develop sensitive methods for analytical studies at cellular levels, preferably *in vivo* or in short-term organ culture. Three methods were developed for measuring DNA repair.

It seemed convenient to consider 2 sorts of questions in studying genetic resistance, or proneness, to cancer in relation to DNA repair. First, we were interested to know whether there is any decrease in DNA repair efficiency with aging and, if so, whether the increased risk of cancer in old age might in part be related to decrease in the capacity for DNA repair. Impairment of the DNA repair system with aging has been suggested from the following observations: 1. Increase in structural change of DNA in various types of mammalian cells with age (*8*, *27*). 2. Decrease in the levels of DNA repair in several types of somatic cells (*28–32*), and 3. Defect in the ability to join DNA strand breaks in progeroid cells (*33*). In the present experiments, we found no age-associated decrease in the DNA repair levels in fish ganglion cells, in rat retinal ganglion cells or in hamster tracheal cells. Our present results, therefore, do not support the idea that a decreased level of DNA repair might trigger the aging of cells or increase proneness to cancer in aged animals.

Secondly, we would like to know if there are species differences in repair capacity. The first comprative studies on this problem were reported in 1974 by Hart and Setlow (*28*). They subcultured early passage numbers of cultures of fibroblastic cells from various species of animals, which differed in their maximum lifespans. They measured unscheduled DNA synthesis autoradiographically after exposing the cells to various intensities of ultraviolet light. They obtained a striking correlation between the extent of excision repair and the normal lifespan of the species. Assuming that the excision repair system is important in restitution of modalities of injuries caused by agents other than ultraviolet light, the results can be extrapolated to chemical carcinogens. Thus, we hoped to find that these observations would also hold true in chemical carcinogen-induced DNA repair in our systems for retinas and tracheas in organ cultures. Although our systems were quite different from those of Hart and Setlow for fibroblastic cells treated with ultraviolet light, we also found that carcinogen-induced DNA repair levels differed considerably from species to species. However, the relative efficiencies of DNA repair of the 6 species tested differed from one carcinogen to another. Moreover, with respect to the correlation between the maximum lifespan and the DNA repair level, our results showed that shorter-lived animals (*e.g.* mice and rats) showed the highest levels of DNA repair in response to some carcinogens. Recently, Peleg *et al.* (*34*) showed that dimer excision in mouse embryo fibroblasts declined abruptly during progressive passages of the cells in culture. Setlow *et al.* (*35*) and Namba *et al.* (*36*) also reported that excision repair does in fact occur in rodents cells, although at a significantly lower rate than in human cells. In this sense mouse cells are "human xeroderma pigmentosum cells." However, as shown in our experiments, mouse cells were very active in excision-repair as measured by unscheduled DNA synthesis. Our observations suggest that the excision-repair efficiency may depend both on the cell type and the type of DNA damaging agent. It seem probable that the rate of repair of certain DNA damage is determined by the interrelation between the cell type and the type of damaging agents. Thus, the various rates of DNA repair efficiency in different species may be an important indication of interspecific differences in carcinogenesis.

ACKNOWLEDGMENT

This work was supported by a Grant-in-Aid for Cancer Research from the ministry of Education, Science, and Culture, and Grants from the Society for Promotion of Cancer Research and the Japan Tabacco and Salt Public Corporation of Japan. We thank Miss K. Kawana for excellent technical assistance.

REFERENCES

1. Setlow, R. B. Repair deficient human disorders and cancer. Nature, *271*: 713–717, 1978.

2. Cox, R., Damjanov, I., Abanoli, S. E., and Sarma, D.S.R. A method for measuring damage and repair in the liver *in vivo*. Cancer Res., *33*: 2114–2121, 1973.
3. Damjanov, I., Cox, R., Sarma, D.S.R., and Farber, E. Pattern of damage and repair of liver DNA induced by carcinogenic methylating agents *in vivo*. Cancer Res., *33*: 2122–2128, 1973.
4. Goodman, J. I. and Potter, V. R. Evidence for DNA repair synthesis and turnover in rat liver following ingestion of 3′-methyl-4-dimethylaminoazobenzene. Cancer Res., *32*: 766–775, 1972.
5. Rajaiakshmi, S. and Sarma, D.S.R. Rapid repair of hepatic DNA damage induced by comptothecin in the intact rat. Biochem. Biophys. Res. Commun., *53*: 1268–1272, 1973.
6. Stewart, B. W. and Farber, E. Strand breakage in rat liver DNA and its repair following administration of cyclic nitrosamines. Cancer Res., *33*: 3209–3215, 1973.
7. Hadjiolov, D. and Venkov, L. Strand breakage in rat brain DNA and its repair induced by ethylnitrosourea *in vivo*. Z. Krebsforsch., *84*: 223–225, 1975.
8. Wheeler K. T. and Lett, J. T. On the possibility that DNA repair is related to age in non-dividing cells. Proc. Natl. Acad. Sci. U.S., *71*: 1862–1865, 1974.
9. Cox, R. and Irving, C. C. Damage and repair of DNA in various tissues of the rat induced by 4-nitroquinoline-1-oxide. Cancer Res., *35*: 1858–1860, 1975.
10. Cox, R. and Irving, C. C. Effect of *N*-methyl-*N*-nitrosourea on the DNA of rat bladder epithelium. Cancer Res., *36*: 4114–4118, 1976.
11. Kanagahngan, K. and Balis, M. E. *In vivo* repair of rat intestinal DNA damage by alkylating agents. Cancer Res., *36*: 2364–2372, 1976.
12. Koropatnick, D. J. and Stich, H. F. DNA fragmentation in mouse gastric epithelial cells by precarcinogens, ultimate carcinogens and nitrosation products. An indicator for the determination of organotrophy and metabolic activation. Int. J. Cancer, *17*: 765–772, 1976.
13. Petzold, G. L. and Swenberg, J. A. Detection of DNA damage induced *in vivo* following exposure of rats to carcinogens. Cancer Res., *38*: 1589–1594, 1978.
14. Wheeler, K. T. and Lett, J. T. Formation of DNA strand breaks in irradiated neurones: *In vivo*. Radiat. Res., *52*: 59–67, 1972.
15. Zubroff, J. and Sarma, D.S.R. A nonradioactive method for measuring DNA damage and its repair in nonproliferating tissues. Anal. Biochem., *70*: 387–396, 1976.
16. Bodell, W. J. and Banerjee, M. R. DNA repair in normal and preneoplastic mammary tissues. Cancer Res., *38*: 736–740, 1978.
17. Iqbal, Z. M., Majdan, M., and Epstein, S. S. Evidence of repair of DNA damage induced by 4-hydroxyaminoquinoline-1-oxide in guinea pig pancreatic slices *in vitro*. Cancer Res., *36*: 1108–1113, 1976.
18. Lieberman, M. W. and Forbes, P. D. Demonstration of DNA repair in normal and neoplastic tissues after treatment with proximate chemical carcinogens and ultraviolet radiation. Nature New Biol., *341*: 199–201, 1973.
19. Pedersen, R. A. and Cleaver, J. E. Repair of UV damage to DNA of implantation-stage mouse embryos *in vitro*. Exp. Cell Res., *95*: 247–253, 1975.
20. Sega, G. A. Unscheduled DNA synthesis in the germ cells of male mice exposed *in vitro* to the chemical mutagen ethyl methanesulfonate. Proc. Natl. Acad. Sci. U.S., *71*: 4955–4959, 1974.
21. Stich, H. F. and Koropatnick, D. J. The adaptation of short-term assays for carcinogens to the gastrointestinal system. *In*; E. Farber *et al.* (eds.), Pathophysiology of

Carcinogenesis in Digestive Organs, pp. 121–134, Japan Scientific Society Press, Tokyo/University Park Press, Baltimore, 1977.

22. Ishikawa, T., Takayama, S., and Kitagawa, T. Autoradiographic demonstration of DNA repair synthesis in ganglion cells of aquarium fish at various ages *in vivo*. Virchows Arch. B. Cell Pathol., *28*: 235–242, 1978.
23. Ishikawa, T., Takayama, S., and Kitagawa, T. DNA repair synthesis in rat retinal ganglion cells treated with chemical carcinogens or ultraviolet light *in vitro*, with special reference to aging and repair level. J. Natl. Cancer Inst., *61*: 1101–1105, 1978.
24. Egami, N. Life span data for the small fish, *Oryzias latipes*. Exp. Gerontol., *6*: 379–382, 1971.
25. Lucas, D. R. and Trowell, O. A. *In vitro* culture of the eye and the retina of the mouse and rat. J. Embryol. Exp. Morphol., *6*: 178–182, 1958.
26. Hansson, H. A. and Sourander, P. Studies on cultures of mammalian retina. Z. Zellforsch., *62*: 26–47, 1964.
27. Modak, S. P. and Price, G. B. Exogenous DNA polymerase-catalysed incorporation of deoxyribonucleotide monophosphate in nuclei of fixed mouse-brain cells. Changes associated with age and X-irradiation. Exp. Cell Res., *65*: 289–296, 1971.
28. Hart, R. W. and Setlow, R. B. Correlation between deoxyribonucleic acid excision repair and life-span in a number of mammalian species. Proc. Natl. Acad. Sci. U.S., *71*: 2169–2173, 1974.
29. Hahn, G. M., King, D., and Yang, S.-J. Quantitative changes in unscheduled DNA synthesis in rat muscle cells after differentiation. Nature New Biol., *230*: 242–244, 1971.
30. Kofman-Alfaro, S. and Chandley, A. C. Radiation-initiated DNA synthesis in spermatogenic cells of the mouse. Exp. Cell Res., *69*: 33–44, 1971.
31. Gledhill, B. L. and Darzynkiewicz, Z. Unscheduled DNA synthesis during spermatogenesis. J. Exp. Zool., *183*: 375–382, 1973.
32. Karran, P. and Ormerod, M. G. Is the ability to repair damage to DNA related to the proliferative capacity of a cell? The rejoining of X-ray-induced breaks. Biochim. Biophys. Acta, *299*: 54–64, 1973.
33. Epstein, J., Williams, J. R., and Little, J. B. Defective DNA repair in human progeroid cells. Proc. Natl. Acad. Sci. U.S., *70*: 977–981, 1973.
34. Peleg, L., Raz, E., and Ben-Ishai, R. Changing capacity for DNA excision repair in mouse embryonic cells *in vitro*. Exp. Cell Res., *104*: 301–307, 1976.
35. Setlow, R. B., Regan, J. D., and Carrier, W. L. Different levels of excision repair in mammalian cell lines. Biophys. Soc. Abstr., *13*: 19a, 1972.
36. Namba, M., Nishitani, K., and Kimoto, T. Comparison of various effects of a carcinogen, 4-nitroquinoline 1-oxide, on normal human cells and on normal mouse cells in culture. Jpn. J. Exp. Med., *47*: 263–269, 1977.

GENETIC AND ENVIRONMENTAL FACTORS IN EXPERIMENTAL AND HUMAN CANCER,
H. V. GELBOIN ET AL. (EDS.), JAPAN SCI. SOC. PRESS, TOKYO, PP. 231–246, 1980

Studies on Cells from Patients Who Are Cancer Prone and Who May Be Radiosensitive

D. G. HARNDEN, M. EDWARDS, T. FEATHERSTONE, J. MORTEN, R. MORGAN, and A.M.R. TAYLOR

Department of Cancer Studies, University of Birmingham, Birmingham, England, U.K.

Abstract: Although ionising radiation increases the incidence of cancer and leukaemia in exposed populations, most of those exposed do not develop a neoplasm. There are several possible explanations, some of which suggest that the population may vary genetically in either sensitivity to radiation or in ability to eliminate a potentially malignant focus. Few studies of radiosensitivity on normal populations have been carried out, but attention has now been focussed on several rare genetically determined conditions where there is some clinical evidence of radiosensitititivity. In ataxia telangiectasia (AT) the clinical radiosensitivity is shown by tissue damage; and susceptibility to cell killing and to induction of chromosome damage by ionising radiation can be demonstrated *in vitro*. There is some evidence to suggest that cells from AT patients may be deficient in one or more DNA repair functions. Patients with basal cell naevus syndrome show an apparent increased susceptibility to the induction of skin tumours by ionising radiation. So far no *in vitro* test has demonstrated radiosensitivity. In the case of retinoblastoma there is clinical evidence for the induction of sarcomas by radiation. We have, however, not been able to confirm reports of radiosensitivity *in vitro* and suggest a need for careful definition of the range of variation in radiosensitivity amongst normal individuals.

The induction of leukaemia and cancer in man by ionising radiation is well established and data from a variety of different sources are in good agreement (reviewed by Jablon (*1*)). Table 1 shows incidence figures for leukaemia in normal and irradiated populations in U.K. and Japan and it can be seen that the relative risks range from 3.7 to 9.5 which represents a very large increase in risk. However, this means that in the population of patients irradiated for ankylosing spondylitis 52 patients developed leukaemia out of a total of 14,554 patients exposed to a mean dose of 372 rads (range 250–2,750 rads). In other words 14,502 patients exposed to a mean dose of 372 rads did not get leukaemia. The same argument can be applied to other exposed groups and to other types of malignancy. The outcome is always

TABLE 1. Leukaemia Incidence in Irradiated and Normal Populations in U.K. and Japan

(a)	A-bomb survivors <10 years	A-bomb survivors >10 years	Spondylitis patients
Total of exposed subjects			
Whole body	4,507	19,472	—
Partial body	—	—	14,554
Mean dose	69 rads	86 rads	372 rads[a]
Observed cases of leukaemia	19	62	52
Expected cases of leukaemia	2.93	16.8	5.48
Relative risk	6.5	3.7	9.5

(b)	Annual incidences/100,000 (all ages, both sexes)	
	U.K. (Birmingham)	Japan (Okayama)
Cancers excluding leukaemia	316.0	210.0
All leukaemias	6.5	4.1
Leukaemias——spondylitis	61.75	—
Leukaemias——A-bomb survivors		
<10 years	—	26.65
>10 years	—	15.17

[a] Partial body. From Refs. *1* and *50.*

the same, namely that while there may be a very large increase in the number of cases occurring in the exposed population they represent only a tiny fraction of the total number of individuals at risk.

The difference between individuals in their response to ionising radiation may reside at any one of several levels. (1) It may be that the carcinogenic effect of radiation is the consequence of altering a cell in a highly specific manner and that such an event is extremely rare among the many millions of potentially damaging events that occur when ionising radiation interacts with the cells of an individual. If the progression of such a specifically altered cell towards full malignancy is then inevitable this, on its own, could explain the relatively rare occurrence of leukaemias and cancers in exposed populations. (2) It may be that interactions between radiation and the cell which have potentially malignant consequences are more likely to occur in some individuals than in others. (3) Even if the frequency of occurrence of a significant event is the same in all exposed individuals progression to malignancy may not be inevitable. The probability of a malignancy occurring would then differ from one individual to another because of variation in either (i) the efficiency of surveillance or (ii) the exposure to promoting agencies.

If the proposition (1) holds good then there is no room for and no need for a genetic hypothesis to explain individual variation to exposure to ionising radiation. Similarly for proposition (3-ii) there is no need for any genetic explanation. If, however, either of the other mechanisms obtain then it is necessary to postulate an inherent, probably genetic, element in individual variation either at the level of increased probability of the occurrence of a significant event or at the level of increased probability of progression towards full malignancy. Since we do not yet understand the mechanisms underlying radiation carcinogenesis, we are not able to choose between these possibilities.

Normal Populations

Ultimately studies on a normal population will be necessary, but any differences may be slight. Very few attempts have been made to study variation in radiosensitivity between different normal individuals, but many clinicians refer anecdotally to such variation in patients exposed to therapeutic radiation. In an attempt to quantitate this variation Weichselbaum *et al.* (*2*) tried to relate variation in the production of erythema by radiation to sensitivity to cell killing by radiation *in vitro*. A fairly wide range of variation in both parameters was observed, but there was no correlation between the two. A cytogenetic study is under way in our laboratory on a group of normal individuals, but results are not yet available.

Susceptibility Syndromes

We have concentrated until recently on patients where there is a recognised genetic susceptibility to develop cancer and where there is either firm or suggestive evidence of sensitivity to ionising radiation.

1. Ataxia telangiectasia (AT)

AT is a recessively inherited syndrome characterised by a progressive cerebellar ataxia and an oculocutaneous telangiectasia. Other features include unusual susceptibility to infection, defects of cell mediated and sometimes humoral immunity and susceptibility to cancer. So far we know of 90 cancers in AT patients; (most of these are reported in Ref. *3*). While the majority of these malignancies are leukaemias or lymphomas the number of cases of other cancers reported in patients with this very rare syndrome suggests that the incidence of these two is elevated in AT patients. There are three reports of radiosensitivity at the clinical level (*4–6*). Several groups (*7–9*) have confirmed the observation of Taylor *et al.* (*10*) that fibroblastic cells from these patients are unusually sensitive to the cell killing effects of both X-rays and γ-rays. Spontaneous chromosome aberrations occur in the lymphocytes of these patients (*11*, *12*). Cells of these patients are also more sensitive than are normal cells to the chromosome damaging effects of ionising radiation (*13*). All types of chromosome aberrations are not increased in proportion especially at higher doses (400 rads), there being an excess of fragments, chromatid breaks and chromatid exchanges following irradiation in the G_0 or early G_1 phases of the cell cycle and this has led Taylor (*14*) to suggest that specific, rare types of DNA lesion may be responsible for the chromosome damage that is observed. Using conventional gradient centrifugation techniques there appears to be no difference between normal cells and AT cells in their capacity to rejoin single and double strand DNA breaks (*10*, *15*). It has also been demonstrated that under oxic conditions extracts from AT skin fibroblasts remove γ-radiation-induced 5′-6, dihydroxydihydrothymine type damage normally from bacteriophage DNA and from chromatin isolated from normal and AT skin fibroblasts (*16*). A decreased ability of some AT cell strains to undergo repair replication following exposure to γ-radiation under anoxic conditions has been reported (*17*). The same workers also observed that under anoxic conditions normal human fibroblasts are capable of removing γ-endonuclease-sensitive sites from DNA after γ-irradiation, but in fibroblasts from the AT patients with defective repair replication this repair process was present at a much lower level under the same anoxic conditions.

Inoue *et al.* (*18*) have demonstrated the existence of a DNA repair enzyme activity in cell-free extracts of normal human fibroblasts. This enzyme when incubated in the presence of γ-irradiated colicin E1 DNA enhances the rate of de-

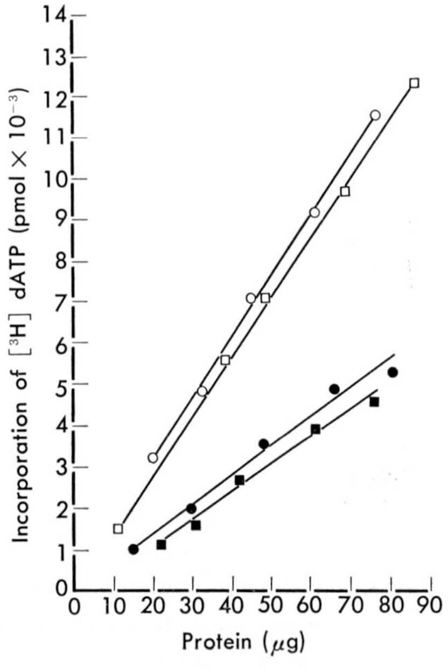

FIG. 1. The increase in DNA polymerase activity after preincubation with γ-irradiated DNA, with increasing concentrations of cell-free extracts from normal (○ Con100BI; □ Con101BI) and two ataxia (● AT5BI; ■ AT7BI) fibroblast cell lines. A cell-free extract of AT and normal cells was prepared by sonication in 50 mM Tris-HCl buffer, pH 7.4 containing 10 mM $MgCl_2$, 5 mM 2-mercaptoethanol. After analysis against 20 mM Tris-HCl, pH 7.4 the non-diffusable material was centrifuged and the supernatant used in the assays. Calf thymus DNA (500 μg/ml) in 20 mM NaCl-10 mM Tris-HCl buffer, pH 8.0, was irradiated with a ^{60}CO γ-ray source (dose rate 20 krads/min) to a total dose of 50 krads. DNA polymerase I from *M. luteus* (Sigma) was repurified and tested for activity. Primer activity of the cell extracts was measured by incubating an appropriate dilution with the irradiated and unirradiated DNA at 37°C for 30 min. This material was then incubated with DNA polymerase I and ^{3}H-labelled dATP for 5 min at 37°C. After terminating the reaction acid-insoluble material was collected on filters and counted in a scintillation counter. Enhancement of activity of DNA polymerase was calculated as $(a-b)-(c-d)$ where a=activity of enzyme-treated, γ-irradiated DNA, b= activity of non-enzyme-treated, γ-irradiated DNA, c=activity of enzyme-treated, non-irradiated DNA, d=activity of non-enzyme-treated, non-irradiated DNA. Reprinted with permission from Biochem. J.

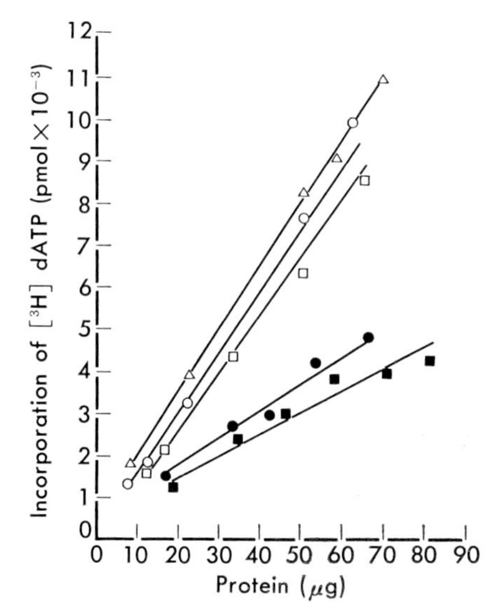

FIG. 2. The increase in DNA polymerase I activity after pre-incubation of γ-irradiated DNA (50krads) with increasing concentration of cell-free extracts from normal (○ LIZ; △ DEN) ataxia telangiectasia (● GM1525; ■ GM1526) and ataxia telangiectasia heterozygote (□ GM736) lymphoid cell lines. The methods are as described for Fig. 1. Reprinted with permission of Biochem. J.

oxyribonucleotide triphosphate incorporation into DNA. Using a single homozygous AT cell strain Inoue *et al.* demonstrated a deficiency in this activity. Using a slightly modified technique we have shown (*19*) that two AT fibroblast cell lines (AT5BI and AT7BI) and three lymphoid cell lines (AT2BI, AT8BI, and GM 717*) are all deficient in this enzyme activity thus confirming the result of Inoue *et al.* (Figs. 1 and 2). The two fibroblast strains studied by us were also studied by Paterson (*9*) who found them to be repair proficient under anoxic conditions. It seems possible therefore, when we consider that the proportion of strand breaks to γ-endonuclease susceptible sites varies under oxic and anoxic conditions, that several different repair pathways may be involved in determining the radiosensitivity of AT cell strains.

These biochemical studies have all been carried out using very high doses of radiation (of the order of 50,000 rads) and techniques are now available for observing the effects of much lower doses of radiation. For example, using the nucleoid sedimentation technique of Cook and Brazell (*20*) the effect of relatively low doses was investigated in this laboratory (*21*). The technique is based on the observation that sedimentation in neutral sucrose gradients of nucleoids released from cells lysed on the top of the gradient is dependent on the degree of supercoiling of the DNA. γ-Irradiation causes strand breaks allowing the supercoiling to relax and the nucleoids sediment to a position higher in the gradient than do nucleoids from comparable unirradiated cells. It has been shown that two strains of AT cells (AT5BI and AT8BI) respond to irradiation in the same way as normal

* A lymphoid cell line obtained from Dr. A. Green, Institute of Medical Research, Camden, New Jersey.

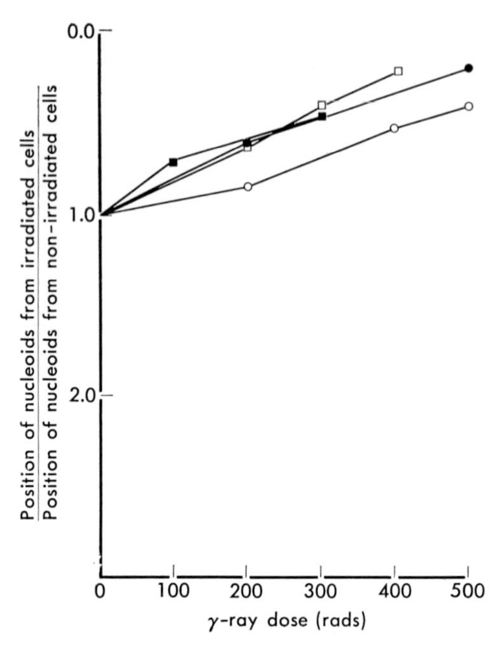

FIG. 3. The effect of γ-irradiation on the sedimentation rate of nucleoids from normal and AT fibroblasts. Normal and AT fibroblasts were irradiated in suspension under oxic conditions with various doses of γ-rays from a ^{60}Co source (dose rate 38.4 krads/hr). Nucleoids were obtained by lysing $4–8 \times 10^5$ cells in a mixture containing Triton X-100 (0.5%, v/v), 1 M NaCl and 0.01 M EDTA, pH 8.0 on top of 15–30% sucrose gradients that contained 1 M NaCl, 0.01 M Tris-HCl and 0.001 M EDTA, pH 8.0. The gradients were centrifuged in a Beckman SW 50.1 rotor for 15 min at 7,000 rpm and were fractionated by upward displacement using an ISCO model 185 gradient fractionater. Nucleoids were located by their absorbance at 254 nm. Cells: ● N1; ○ N2; ■ AT5BI; □ AT8BI.

cells (Fig. 3). Moreover, when time for repair is allowed to elapse the AT cells recover their position in the gradient relative to unirradiated cells to the same extent and at the same rate as do normal cells (Fig. 4). It is concluded that by this method of examining DNA repair AT cells do not differ from normal.

Another method of measuring the affect of relatively low doses of ionising radiation is the filter elution technique of Kohn and Grimek-Ewig (*22*). The technique depends on the fact that DNA of cells lysed on a stack of filters is retained on these filters. An alteration to alkaline pH selectively removes single stranded regions. Irradiation of the cells prior to lysis results in an increased removal of prelabelled single stranded regions from the filter upon alkaline elution (Fig. 5). When time is allowed to elapse for repair after irradiation and before lysis there is an increase in the activity retained on the filter since repair of strand breaks has increased the amount of double strandedness. When AT cells are compared with controls following irradiation with 2.5 krads it is found that there is a considerable variation between individuals, but no consistent difference between AT cells and control cells (Fig. 6). We conclude that by this measure also no difference can be detected in the capacity of AT cells to repair strand breaks.

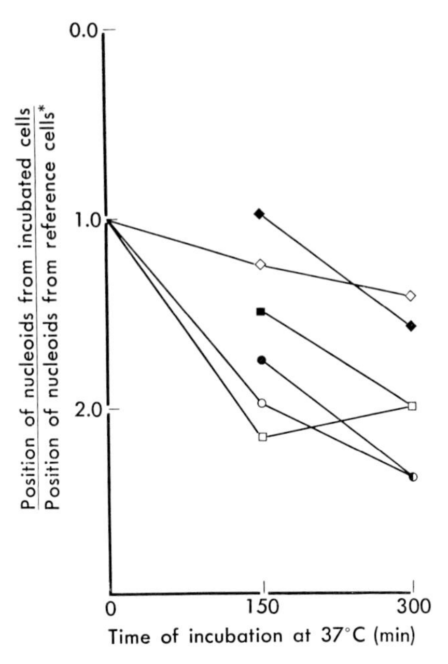

Fig. 4. The effect of long incubation periods on nucleoid sedimentation rate. Normal and AT fibroblasts were irradiated in suspension under oxic conditions at room temperature with 500 rads γ-rays (dose rate 38.4 krads/hr). Control cultures were prepared in parallel, but were not irradiated. Irradiated and non-irradiated fibroblasts were incubated at 37°C in DMEM that contained 0.5% foetal calf serum. Samples were taken after 0, 150, and 300 min of incubation, washed and resuspended in ice-cold phosphate-buffered saline (PBS). Nucleoids were obtained from the fibroblasts as described in the legend to Fig. 3. The position of nucleoids from non-irradiated, non-incubated cells was taken as a control and the position of nucleoids in other gradients expressed relative to it. Cells: ◆ irradiated N1; ● irradiated N2; ■ irradiated AT8BI; ◇ non-irradiated N1; ○ non-irradiated N2; □ non-irradiated AT8BI.

Thus by several different techniques for measuring DNA repair, namely single and double strand rejoining, nucleoid sedimentation, filter elution, excision of dihydroxydihydrothymine type damage and aerobic removal of γ-nuclease-sensitive sites AT cells do not differ from normal. Some strains are deficient in removing γ-nuclease-sensitive sites under anoxic conditions while all those tested so far lack an enzymic activity which primes γ-damaged DNA for the action of DNA polymerase. These findings are consistent with the idea that many different pathways are involved in the repair of γ-ray-induced damage in human cells and that AT cells are proficient in most of these, but may be deficient in one or possibly two of them.

It is also worth noting that AT cells are also sensitive to the cell killing and chromosome damaging effects of bleomycin, a drug with a known chromosomally radiomimetic action (*23*). They are, however, apparently quite proficient at repairing damage by other so called radiomimetic drugs (*24*). Again this is likely to be a reflection of the existence of several different pathways for the repair of the different types of damage induced by ionising radiation.

The frequency of individuals in the general population heterozygous for the

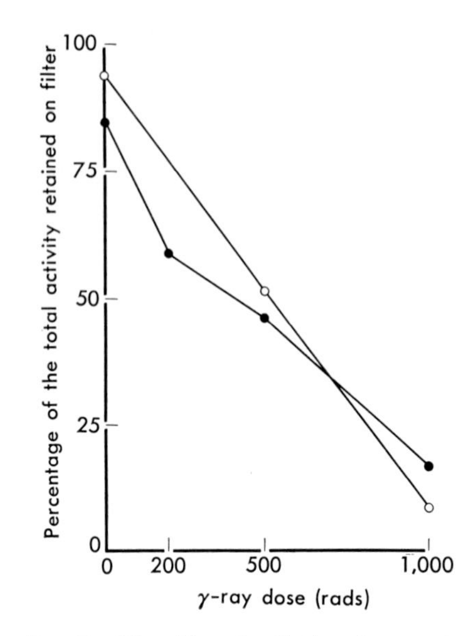

FIG. 5. The effect of radiation dose on the percentage actively remaining on the filter after alkaline elution. Normal and AT fibroblasts were irradiated in suspension under oxic conditions with various doses of γ-rays (dose rate 44 krads/hr), at room temperature. $1.2–1.8 \times 10^6$ cells were layered onto prewashed PVC filters (Millipore 0.22 μM pore size) and washed with PBS, pH 7.3. A lysis mix containing 0.02 M Na_3 EDTA, 2 M NaCl, and 0.2% w/v sodium lauryl sarkosine, pH 8.2 added to isolate the DNA. The DNA retained by the filter was washed with 0.02 M Na_3 EDTA, pH 7.8 and made single stranded by layering alkaline elution buffer, 0.02 M Na_3 EDTA and tetrapropylammonium hydroxide (10% in H_2O), pH 12.2, onto the filter. Alkaline elution was carried out at 0.5 ml/min. One ml fractions were collected during the elution and when elution was complete the filter was removed and 5 ml ice-cold PBS washed through the filter lines. The total activity recovered in an experiment was the sum of the activity of each fraction+activity retained on the filter+the activity recovered in the wash. Cells: ● N1; ○ AT5BI.

AT gene, has been estimated at about 1% (*25*). Swift *et al.* have also reported that these individuals show an increased predisposition to malignancies. In particular blood relatives of AT homozygotes under the age of 45 showed a 5-fold higher risk of dying from any neoplasm. This study has led to others investigating possible radiosensitivity in AT heterozygotes as an unambiguous method for their detection. We, however, have been unable in our laboratory to find any difference in radiosensitivity between eight AT heterozygotes and our range of normal controls, by measuring survival of fibroblasts following γ-irradiation in air (*26*). It has been reported, however, that lymphoblastoid cell lines derived from AT heterozygotes show a radiosensitivity between that of homozygotes and that of normal controls (*27*). This observation would obviously have important implication for the population if it meant that the individuals from which the cells were derived were markedly more radiosensitive. Paterson *et al.* (*28*) have shown that some, but not all, AT heterozygotes tested were more sensitive to anoxic γ-irradiation than strains from normal individuals, and also are partially deficient in γ-ray-induced repair replication. The same heterozygote lines appear to have normal sensitivity following

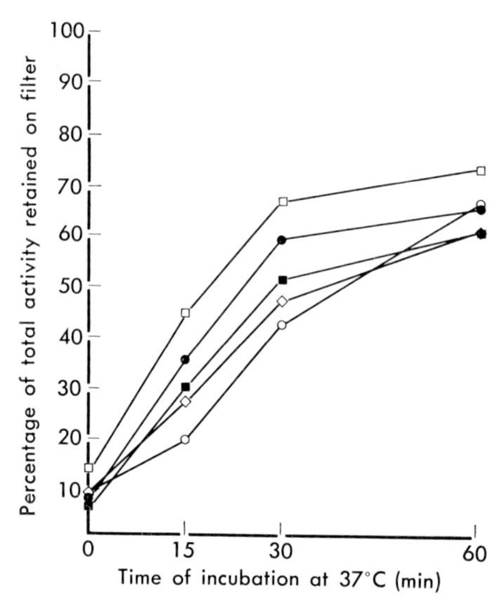

FIG. 6. The effect of post irradiation incubation on the percentage of activity retained on the filter after alkaline elution. Normal, AT, and retinoblastoma fibroblasts were irradiated with 2.5 krads ^{60}Co γ-rays under the usual conditions. Immediately before and immediately after irradiation a sample of the cell-suspension was removed and placed on ice. The remaining irradiated cells were incubated at 37°C. Samples were removed after 15, 30, or 60 min and placed on ice. The samples were resuspended in 2 ml of ice-cold PBS at a concentration of $6–9 \times 10^5$ cells per ml and layered onto a prewashed membrane filter for alkaline elution. The activity retained by the filter when alkaline elution was complete is expressed as a percentage of the total activity recovered in the experiment. Cells: ● N1; ■ N2; ○ AT3BI; □ AT5BI.

irradiation in air. The reasons for these disparate results using different cell types and different conditions is not understood.

2. *Basal cell naevus syndrome (BCNS)*

BCNS is a rare autosomal dominant condition in which penetrance of the gene is high, but its expression is variable. A variety of abnormalities may be present, of which the most common are basal cell naevi, odontogenic keratocysts, various skeletal abnormalities and palmar and plantar pits (*29, 30*). The patients are liable to develop multiple basal cell carcinomas on both sun-exposed and non-exposed cutaneous areas at a distinctly earlier age and in greater numbers than would normally be expected (*31, 32*). They are also thought to have an unusual liability to develop other tumours particularly the childhood tumor medulloblastoma (*33, 34*) and ovarian fibromas (*35*). A number of other neoplasms are thought to be associated with this rare syndrome and it seems likely that the naevoid basal cell carcinomas of this syndrome should be considered as only the main manifestation of an unusual tendency to form neoplasia (*36*).

Clinically there is evidence that these patients may be sensitive in some way to the effects of ionising radiation. In two instances (*37, 38*) treatment of the basal cell carcinomas of BCNS patients using radiotherapy resulted in a disasterous clinical course, the tumours not responding to therapy, but becoming worse with extensive invasion of surrounding tissue. There are avaiable thirteen reports of patients with BCNS who have received radiotherapy after developing medulloblastoma. Of the surviving patients all have developed multiple basal cell carcinomas over the irradiated area within 6 months to 3 years following therapy, which is a much shorter latent period than that seen in radiogenic basal carcinomas appearing in non-BCNS survivors of childhood malignancy (including medulloblastoma). The number of carcinomas that are seen in the BCNS children is also far more than that seen in the non-BCNS children. The basal cell carcinomas develop at a distinctly earlier age and with a distribution unlike that of other family members suffering from BCNS (*39*, *40*).

One of the most unusual cases is that reported by Scharnagel and Pack (*41*) who describe a 5-year-old boy who received X-ray therapy shortly after birth for an enlarged thymus and who developed a great number (about 1,000 separate foci) of basal cell epitheliomas within the irradiated field. While there is no clear diag-

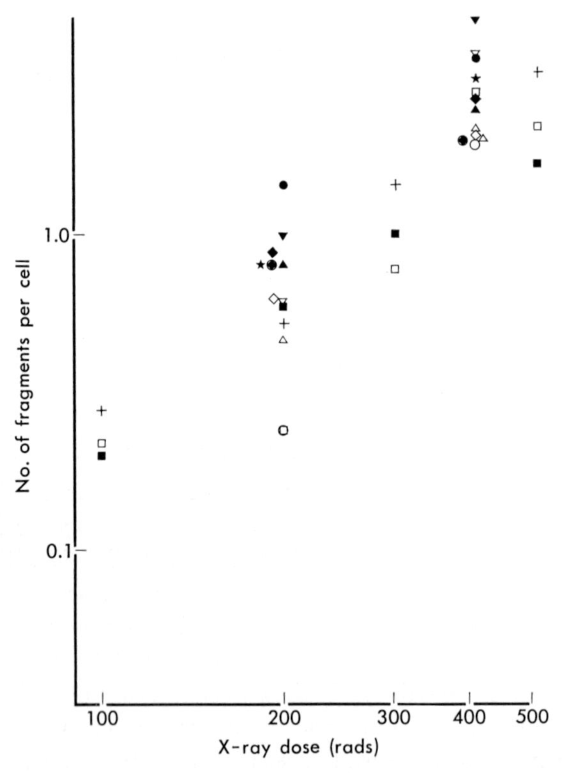

FIG. 7. Increase in chromosome aberrations following irradiation of BCNS and control cells. Human lymphocytes were irradiated in G_0 and cultured with phytohaemagglutinin for 48 hr. Cells were harvested and stained with acetic orcein. All the cells were scored by the same observer for fragments and other rearrangements. Cells: □ Con 636; + Con 653; △ Con 712; ▽ Con 788; ◇ Con 813; ○ Con 884; ■ BCNS 635; ▼ BCNS 787; ▲ BCNS 711; ◆ BCNS 814; ◈ BCNS815; ★ BCNS 816; ● BCNS 883.

nosis of BCNS the description of the child would suggest that he was a case of this syndrome.

There are several reports of " spontaneous " chromosome aberrations in BCNS (*42*), but there is, in some cases, doubt as to whether this is truly spontaneous or the consequence of the treatment received. While Hecht and McCaw (*43*) classify BCNS as a chromosome breakage syndrome the effect certainly does not occur in all patients. There are very few experimental studies on BCNS cells. Taylor *et al.* (*10*) concluded that three BCNS cell lines did not differ from normal controls in their response to the cell killing effects of ionising radiation. Similarly, Lehman *et al.* (*44*) were unable to detect any unusual sensitivitiy to cell killing of BCNS cells by ultraviolet light. They also showed that the levels of excision repair after irradiation with ultraviolet light in BCNS fibroblasts were indistinguishable from normal fibroblasts. Unscheduled DNA synthesis after irradiation with ultraviolet light was also shown by Morgan (*21*) to be normal. In an extensive study of BCNS cell strains carried out in this laboratory Featherstone (*45*) has shown that lymphocytes from seven BCNS patients did not differ from seven normal controls in their sensitivity to the induction of chromosome damage by X-rays following G_0 irradiation (Fig. 7). Similarly using cell survival techniques BCNS fibroblast cell lines were shown to have the same sensitivity as normal cells to γ-rays (5 lines), X-ray (3 lines), ultrasoft carbon X-ray (2 lines), and α-particles (3 lines) (Fig. 8). Further it was shown that following post X-irradiation holding of fibroblasts in stationary phase for 6 hr the repair of potential lethal damage was as good in BCNS cells (3 lines) as in normal cells (2 lines). It must be concluded therefore that

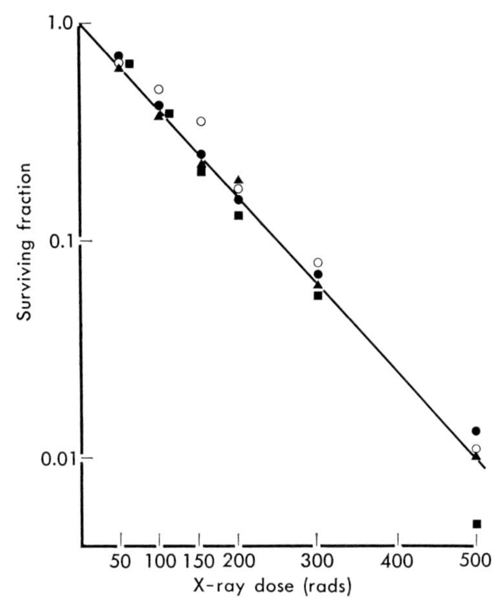

FIG. 8. The fraction of colony-forming cells from control and BCNS patients which survive irradiation with X-rays. Each point is the mean of 4 dishes. Cells: ○ Con 6; ■ BCNS 1BI; ▲ BCNS 3BI; ● BCNS 787.

by the measurement of chromosome damage and cell survival BCNS cells in culture do not differ from normal in their response to ionising radiation.

Studies on mutability and cellular transformation are under way, but are not complete.

3. *Retinoblastoma*

Retinoblastoma is a highly malignant tumour of the retina which occurs almost exclusively in children under 10 years of age. Many cases are sporadic, but roughly half show some kind of familial incidence. In some cases the inheritance quite clearly follows an autosomal dominant pattern, but in other cases penetrance is incomplete, but this is still consistent with the concept that retinoblastoma is an autosomal dominant trait. In the hereditary type of retinoblastoma the tumours are more likely to be multifocal; they are often bilateral and have an earlier age of onset (*46*). For most patients the tumour is the only manifestation of the gene. There is, however, some evidence to suggest that retinoblastoma patients have an increased incidence of osteosarcoma of the long bones. They also develop sarcomas within the irradiation fields where the primary retinoblastomas have been treated by irradiation (*47*). It is likely, however, that the number of radiation induced sarcomas of the orbit is greater than would be expected for a comparable radiation dose in

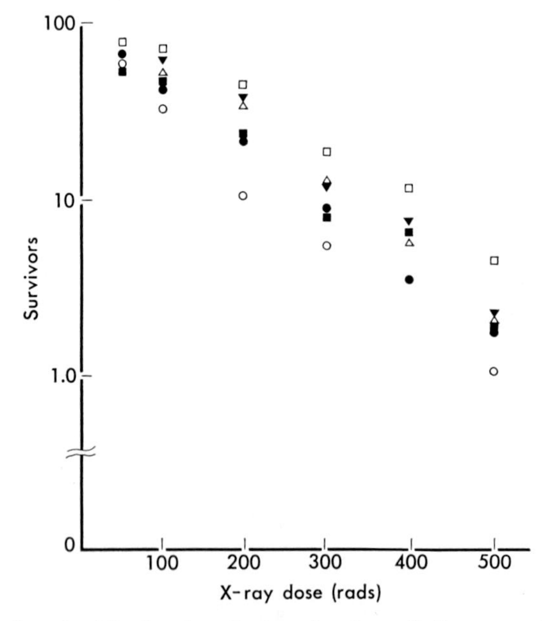

FIG. 9. The fraction of colony-forming cells from control and retinoblastoma patients which survive irradiation with X-rays. Each point is the mean of 4 dishes and usually the mean of several experiments. Cells; □ control (5); ○ control (5); △ control (3); ● 13q-(3); ■ familial bilateral (7); ▼ non-familial (1). The number given after the line designation refers to the number of repeat experiments with that line. The 13q-retinoblastoma line is the same as that reported by Weichselbaum *et al.* (*7*).

normal individuals (*39*, *40*). There is a reasonable supposition therefore that these patients are in some way radiosensitive.

Although most patients with retinoblastoma have a normal chromosome complement, a small number of patients have been described with constitutional chromosome abnormalities many of which involve a deletion of the long arm of chromosome 13. Many of these patients have other congenital abnormalities and it seems probable that the deletion of chromosome 13 leads to a malformation syndrome of which susceptibility to retinoblastoma is a part. In the cases with retinoblastoma part of the band 13q14 is deleted or else there is a translocation with one of the break points in this region. Weichselbaum, Nove and Little (*7*) reported that fibroblasts from a single patient with retinoblastoma with a 13q-deletion were unusually sensitive to the cell killing effects of ionising radiation whereas cells from hereditary retinoblastoma patients without a 13q-deletion showed a range of response between this level and normal. In more recent papers the same group (*48*, *49*) have confirmed and extended these observations and they now suggest, on the basis of a series of slightly different deletions, that there are two loci very close together on the long arm of chromosome 13, one of which is concerned with radiosensitivity and the other with retinoblastoma so that depending on the extent of the deletion the two characteristics may be expressed independently or simultaneously.

This hypothesis is very interesting, but its validity rests on how well defined are the limits of the normal cell killing response to ionising radiation. In an earlier paper Weichselbaum *et al.* (*5*) found a wide range of radiosensitivity in normal subjects and this has been our experience also. Indeed in our hands the cell survival curve of one 13q-deletion retinoblastoma (the same one as that studied by Weichselbaum *et al.*) falls within the range of normal for this laboratory (Morten, J. and Taylor, A.M.R., unpublished). Further we can find no difference between any hereditary retinoblastomas and normal subjects in their sensitivity to the cell killing effects of ionising radiation. We suggest that to resolve this discrepancy a careful study is required on the range of sensitivity expected in the normal populations.

CONCLUSIONS

There is clear evidence from the studies on ataxia telangiectasia that some aspects of radiosensitivity in man may be under genetic control. It seems likely that genetic control over a DNA repair function may be involved though this may not be the basic lesion in ataxia telangiectasia. In the case of basal cell naevus syndrome the *in vivo* radiosensitivity is manifest in tumour induction rather than tissue damage. It may not be too surprising therefore that there is no evidence of cell killing or chromosome damage *in vitro*. Nevertheless, the clinical evidence is good and we will seek other measures *in vitro* of demonstrating in what function these cells are radiosensitive. The clinical evidence of tumour induction by radiation in retinoblastomas is perhaps less firm, but does suggest a need for further studies. However, our studies have failed to demonstrate unusual radiosensitivity to cell killing. The range of normal variation may be wider than has hitherto been suggested and it is of course possible, even probable, that this normal variation is under genetic control. Whether or not this variation has any bearing on tumour induction could form the subject of further investigations.

ACKNOWLEDGMENTS

We would like to thank the Cancer Research Campaign who have supported this research and also many clinical colleagues who have supplied us with material and provided helpful discussions.

REFERENCES

1. Jablon, S. Radiation. *In*; J. F. Fraumeni (ed.), Persons at High Risk of Cancer, pp. 151–165, Academic Press, New York, 1975.
2. Weichselbaum, R. R., Epstein, J., and Little, J. B. *In vitro* radiosensitivity of human diploid fibroblasts derived from patients with unusual clinical responses to radiation. Radiology, *121*: 479–482, 1976.
3. Spector, B. D., Perry, G. S., and Kersey, J. H. Genetically determined immunodeficiency disease (GDID) and malignancy; report from the Immunodeficiency Cancer Registry. Clin. Immunol. Immunopathol., *11*: 12–29, 1978.
4. Gotoff, S. P., Amirmokri, E., and Liebner, E. J. Ataxia telangiectasia: Neoplasia, untoward response to X-irradiation and tuberous sclerosis. Am. J. Dis. Child., *114*: 617–625, 1967.
5. Morgan, J. L., Holcomb, T. M., and Morrissey, R. W. Radiation reaction in ataxia telangiectasia. Am. J. Dis. Child., *116*; 557–558, 1968.
6. Cunliffe, P. N., Mann, J. R., Cameron, A. H., Roberts, K. D., and Ward, H.W.C. Radiosensitivity in ataxia telangiectasia. Br. J. Radiol., *48*: 374–376, 1974.
7. Weichselbaum, R. R., Nove, J., and Little, J. B. Skin fibroblasts from a D-deletion type retinoblastoma patient are abnormally X-ray sensitive. Nature, *266*; 726–727, 1977.
8. Cox, R., Hosking, G. P., and Wilson, J. Ataxia telangiectasia. Evaluation of radiosensitivity in cultured skin fibroblasts as a diagnostic test. Arch. Dis. Child., *53*: 386–390, 1978.
9. Paterson, M. C. Environmental carcinogenesis and imperfect repair of damaged DNA in Homo Sapiens: Causal relation revealed by rare hereditary disorders. *In*; A. C. Griffin and C. R. Shaw (eds.), Carcinogens: Identification and Mechanisms of Action, pp. 251–276, Raven Press, New York, 1978.
10. Taylor, A.M.R., Harnden, D. G., Arlett, C. F., Harcourt, S. A., Lehman, A. R., Stevens, S., and Bridges, B. A. Ataxia telangiectasia: A human mutation with abnormal radiation sensitivity. Nature, *258*: 427, 1975.
11. Hecht, F., Koler, R. D., Rigas, D. A., Dahnke, G. S., Case, M. P., Tisdale, V., and Miller, R. W. Leukaemia and lymphocytes in ataxia telangiectasia. Lancet *ii*: 1193, 1966.
12. Harnden, D. G. Ataxia telangiectasia syndrome: Cytogenetic and cancer aspects. *In*; J. German (ed.), Chromosomes and Cancer, pp. 619–636, J. Wiley & Sons, New York, 1974.
13. Taylor, A.M.R., Metcalfe, J. A., Oxford, J. M., and Harnden, D. G. Is chromatid type damage in ataxia telangiectasia after irradiation at G_0 a consequence of defective repair? Nature, *260*; 441, 1976.

14. Taylor, A.M.R. Unrepaired DNA strand breaks in irradiated ataxia telangiectasia lymphocytes suggested from cytogenetic observation. Mutat. Res., *50*: 407–418, 1978.
15. Lehmann, A. R. and Stevens, S. The production and repair of double strand breaks in cells from normal humans and from patients with ataxia telangiectasia. Biochim. Biophys. Acta, *474*: 49–60, 1977.
16. Cerutti, P. A. and Remsen, J. F. Formation and repair of DNA damage induced by oxygen radical species in human cells. *In*; W. W. Nichols and D. G. Murphy (eds.), Cellular Senescence and Somatic Cells Genetics: DNA Repair Processes, pp. 147–166, Symposium Specialists, Miami, 1977.
17. Paterson, M. C., Smith, B. P., Lohman, P.H.M., Anderson, A. K., and Fishman, L. Defective excision repair of γ-ray damaged DNA in human (ataxia telangiectasia) fibroblasts. Nature, *260*: 444–447, 1976.
18. Inoue, T., Hirano, K., Yokoiyama, A., Kada, T., and Kato, H. DNA repair enzymes in ataxia telangiectasia and Bloom's syndrome fibroblasts. Biochim. Biophys. Acta, *479*: 497, 1977.
19. Edwards, M. J., Taylor, A.M.R., and Duckworth, G. An enzyme in normal and ataxia telangiectasia cell lines which is involved in the repair of γ-irradiation induced DNA damage. Biochem, J., *188*: 677–682, 1980.
20. Cook, P. R. and Brazell, I. A. Supercoils in human DNA. J. Cell Sci., *19*: 261–279, 1975.
21. Morgan, G. R. DNA repair in cultured normal and mutant human fibroblasts. Ph. D. Thesis, University of Birmingham, U.K., 1979.
22. Kohn, K. W. and Grimek-Ewig, R. A. Alkaline elution analysis: A new approach to study of DNA single strand interruptions in cells. Cancer Res., *33*: 1849–1853, 1973.
23. Taylor, A.M.R., Rosney, C. M., and Campbell, J. B. Unusual sensitivity of ataxia telangiectasia cells to bleomycin. Cancer Res., *39*: 1046, 1979.
24. Arlett, C. F. and Lehmann, A. R. Human disorders showing increased sensitivity to genetic damage. Annu. Rev. Genet., *12*: 95–115, 1978.
25. Swift, M., Stolman, L., Perry, M., and Chase, M. Malignant neoplasms in the families of patients with ataxia telangiectasia. Cancer Res., *36*: 209–215, 1976.
26. Taylor, A.M.R. and Rosney, C. M. Is there a difference in the range of radiosensitivity between normal individuals and heterozygotes for ataxia telangiectasia? Br. J. Cancer, *40*: 305, 1979.
27. Chen, P. C., Lavin, M. F., Kidson, C., and Moss, D. Identification of ataxia telangiectasia heterozygotes, a cancer prone population. Nature, *274*: 484–486, 1978.
28. Paterson, M. C., Anderson, A. K., Smith, B. P., and Smith, P. J. Enhanced radiosensitivity of cultured fibroblasts from ataxia telangiectasia heterozygotes manifested by defective colony forming ability and reduced DNA repair replication after hypoxic γ-irradiation. Cancer Res., *39*: 3725–3734, 1979.
29. Rater, C. J., Selke, A. C., and van Epps, E. F. Basal cell naevus syndrome. Am. J. Roentgenol, *103*: 589–594, 1968.
30. Gorlin, R. J. and Sedano, H. O. The multiple naevoid basal cell carcinoma syndrome revisited. Birth Defects: Original Article, Ser. VII, No. 8, 150–148, 1971.
31. Howell, J. B. and Caro, M. R. Basal cell naevus: Its relationship to multiple cutaneous cancers and associated anomalies of development. Arch. Dermatol., *79*: 67–80, 1959.
32. Gilhuus-Moe, O., Haugen, L. K., and Dee, P. M. The syndrome of multiple cysts

of the jaws, basal cell carcinomas and skeletal anomalies. Br. J. Oral Surg., *6*: 211–222, 1968.

33. Cook. W. A. Family pedigree—Cancer, cysts and oligodontia. Dent. Radiol. Photogr., *37*: 27, 1964.
34. Heimler, A., Friedman, E., and Rosenthal, A. D. Naevoid basal carcinoma and Charcot-Marie-Tooth disease. J. Med. Genet., *15*: 288–291, 1978.
35. Clendenning, W. E., Herdt, J. R., and Block, J. B. Ovarian fibromas and mesenteric cysts; their association with hereditary basal cell cancer of skin. Am. J. Obstet. Gynecol., *87*: 1008–1012, 1963.
36. Berlin, N. I., Scott, E. J., Clendenning, W. E., Archard, H. O., Block, J. B., Witkop, C. J., and Haines, H. A. Basal cell naevus syndrome. Ann. Intern. Med., *64*: 403–421, 1966.
37. Happle, R. Naevobasaliom und Ameloblastom. Hautarzt, *24*: 290–294, 1973.
38. Berendes, U. Die klinische Bedeutung der onkotischen Phase des basalzell naevus Syndroms. Hautarzt, *22*: 261–263, 1971.
39. Strong, L. C. Theories of pathogenesis: mutation and cancer. *In*; J. J. Mulvihill, R. W. Miller, and J. F. Fraumeni (eds.), Genetics of Human Cancer, pp. 401–416, Raven Press, New York, 1977.
40. Strong, L. C. Genetic and environmental interaction in the devolpment of multiple primary tumours. Cancer, *40*: 1861–1866, 1977.
41. Scharnagel, I. M. and Pack, G. I. Multiple basal cell epitheliomas in a 5 year old child. Am. J. Dis Child., *77*: 647–651, 1949.
42. Happle, R. and Kupferschmidt, A. A further case of basal cell naevus syndrome and structural chromosome abnormalities. Humangenetik, *15*: 287–288, 1972.
43. Hecht, F. and Kaiser McCaw, B. Chromosome instability syndromes. *In*; J. J. Mulvihill, R. W. Miller, and J. F. Fraumeni (eds.), Genetics of Human Cancer, pp. 105–123, Raven Press, New York, 1977.
44. Lehmann, A. R., Kirk-Bell, S., Arlett, C. F., Harcourt, S. A., de Weerd Kastelein, E. A., Keijzer, W., and Hall-Smith, P. Repair of ultraviolet light damage in a variety of human fibroblast cell strains. Cancer Res., *37*: 904–910, 1977.
45. Featherstone, T. A study of cultured cells from basal cell naevus syndrome patients. M. Sc. Thesis, University of Birmingham, U. K., 1979.
46. Knudson, A. G., Hethcote, H. W., and Brown, B. V. Mutation and childhood cancer: a probabilistic model for the incidence of retinoblastoma. Proc. Natl. Acad. Sci. U.S., *72*: 5116–5120, 1975.
47. Francois, J. Retinoblastoma and osteogenic sarcoma. Ophthalmologica, *175*: 185–191, 1977.
48. Weichselbaum, R. R., Nove, J., and Little, J. B. X-ray sensitivity of diploid fibroblasts from patients with hereditary or sporadic retinoblastoma. Proc. Natl. Acad. Sci. U.S., *75*: 3962–3964, 1978.
49. Nove, J., Little, J. B., Weichselbaum, R. R., Nichols, W. W., and Hoffman, E. Retinoblastoma, chromosome 13 and *in vitro* cellular radiosensitivity. Cytogenet. Cell Genet., *24*: 176–184, 1979.
50. Waterhouse, J.A.H., Muir, C., Correa, P., and Powell, J. Cancer Incidence in Five Contients, vol. III, IARC Scientific Publication No. 15, International Agency for Research on Cancer, Lyons, 1976.

GENETIC AND ENVIRONMENTAL FACTORS IN EXPERIMENTAL AND HUMAN CANCER,
H. V. GELBOIN ET AL. (EDS.), JAPAN SCI. SOC. PRESS, TOKYO, PP. 247–257, 1980

Human Tumor Strains with Abnormal Repair of Alkylation Damage

Rufus S. DAY, III, Chuck H. J. ZIOLKOWSKI, Dominic A. SCUDIERO, Sharon A. MEYER, and Michael R. MATTERN

Nucleic Acids Section, Laboratory of Molecular Carcinogenesis, CIP, DCCP, NCI, NIH, Bethesda, Md. 20205, U.S.A.

Abstract: Nine of 39 human tumor cell strains are deficient in host-cell reactivation of *N*-methyl-*N'*-nitro-*N*-nitrosoguanidine (MNNG)-treated adenovirus 5. We propose to call the phenotype so identified the Mer^- phenotype. None of 22 strains of normal human skin fibroblasts tested showed such deficient repair, while two strains of SV40 transformed human fibroblasts did. The Mer^- phenotype is accompanied by increased sensitivity to inactivation by MNNG as measured by post-treatment colony-forming ability, but normal host-cell reactivation of UV-irradiated adenovirus and normal post-UV colony-forming ability.

Human beings most probably are protected from carcinogenesis by cellular mechanisms that repair damaged DNA. This is evident from the work initiated by Cleaver on xeroderma pigmentosum (XP) (*1*), by Sasaki and coworkers on Fanconi's anaemia (*2*), and by Taylor and coworkers (*3*) and Patterson (*4*) on ataxia telangiectasia. These syndromes are genetically transmitted and predispose the people afflicted by them to cancer.

We considered the possibility that tumors composed of repair-deficient cells might arise in people not known to be genetically predisposed to cancer: if DNA repair were important in reducing the probability that humans undergo *in vivo* carcinogenesis, then tumor formation might sometimes be due to carcinogenesis occurring under conditions that exclude the action of the normal repair mechanisms. One such case would arise if somatically produced repair-deficient cells were the targets for tumorigenesis. The resultant tumor would likely be composed of repair-deficient cells.

This report describes some of the properties of nine human tumor cell strains likely to be deficient in the repair of MNNG-damaged DNA. However, we have no data differentiating between the possibility that the defect leads to tumorigenesis and the possibility that tumor cells become repair-deficient during the carcinogenic process.

RESULTS

1. Identification of the Mer⁻ phenotype

Four of 13 strains prepared from human astrocytomas and gliomas were found to be defective in their ability to support the growth of *N*-methyl-*N′*-nitro-*N*-nitrosoguanidine (MNNG)-treated human adenovirus 5 (*5*) (Fig. 1). To inactivate the virus, 0.9 ml samples of adenovirus 5 (in 0.3 M Tris, pH 9.0) were treated with 0.1 ml aliquots of ethanol solutions of MNNG (at ten-fold the desired final concentrations). Samples were incubated at 37°C for 30 min. The remaining MNNG was reduced by the addition of 0.1 ml 0.5 M *N*-acetyl-L-cysteine, and further incubation for 10–15 min. In this study, strains that showed normal ability to support the growth of MNNG-damaged adenovirus included the remaining 9 astrocytoma and glioma strains, 5 strains prepared from other tumors, 14 strains of human skin fibroblasts, and one strain of human embryonic kidney cells. Because such cell-dependent differential survival of an experimentally treated set of virus samples has been interpreted in the past as reflective of differences in cellular ability to repair damaged DNA (*6*), we believed that the strains that showed such decreased host-cell reactivation were deficient in the ability to repair MNNG-damaged DNA. For ease of reference, we propose to term the phenotype of decreased host-cell reactivation of MNNG-treated adenovirus 5 the *Mer⁻* phenotype (for *N*-methyl-NNG repair minus).

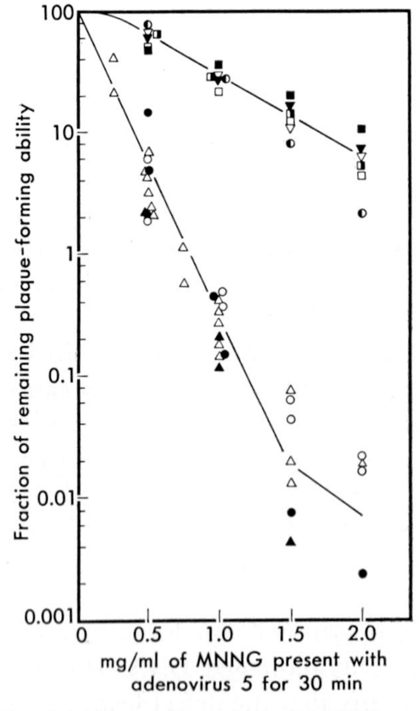

FIG. 1. Decreased survival of MNNG-treated adenovirus 5 using brain tumor cell strains. ▲ A382; ● U-87MG; ○ U-105MG; △ A172; ◐ CRL1187; ▽ H4; ■ U-373MG; ▽ 118MG; □ U-178MG; ◨ human embryonic kidney cells. See Ref. *5* for sources of strains, Table 1 for notes on strains. Reprinted from Ref. *5* with permission from Nature.

2. Mer⁻ strains arise from a variety of tumors

Our suspicion that *Mer*$^-$ cell strains arise from tumors other than brain tumors was correct (Table 1, Fig. 2). In a study of thirty more strains, 4 strains, one prepared from a melanoma (A101D), one from a neck carcinoma (A253), one from a colon carcinoma (BE), and one from a lung carcinoma (A427) were identified as *Mer*$^-$. Seventeen other tumor strains, 8 human fibroblast strains, and one strain of MNNG-transformed human fibroblasts showed the normal *Mer*$^+$ level of host-cell reactivation of MNNG-treated adenovirus 5.

3. Strain A498 is Mer⁻ after MNNG treatment

One human kidney carcinoma cell strain, A498, showed a slightly depressed level of host-cell reactivation of MNNG-treated adenovirus 5: when 2 mg/ml MNNG were used to treat the virus, its survival (compared to untreated virus) was 0.1% in contrast to the 1% survival noted in *Mer*$^+$ strains (*7*). A498 cells were pretreated for 1 hr with MNNG at selected concentrations 24 hr prior to being infected with MNNG-treated adenovirus 5 (see Fig. 3). MNNG pretreatment of these cells caused them to show the *Mer*$^-$ phenotype. The 32 μM pretreatment blocked little more recovery than did the 16 μM pretreatment, indicating that the

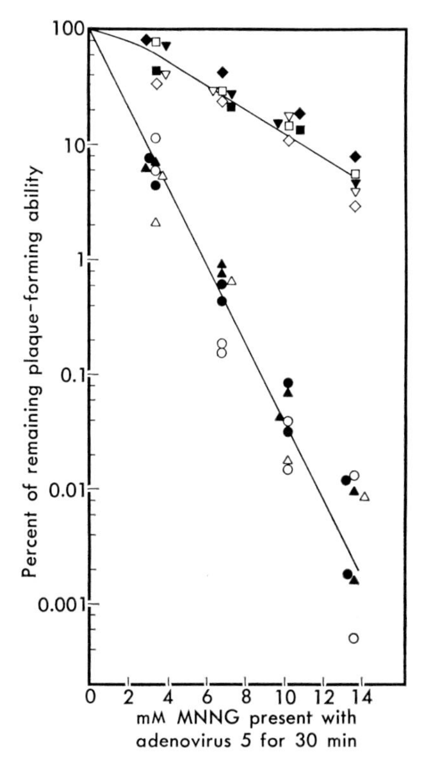

FIG. 2. Decreased survival of MNNG-treated adenovirus 5 by melanoma, lung, colon, and neck tumor cell strains. ○ A253; ▲ A427; ● A101D; △ BE; □ Hs768; ■ A673; ▽ Hs255; ▼ GM449; ◇ HT1080; ◆ A388. See Table 1 for notes on strains.

TABLE 1. Strains Whose Ability to Perform Host-cell Reactivation of MNNG-treated Adenovirus 5 Was Studied

	Notes on origin	Mer^+/Mer^-
Tumor strains		
A172	Astrocytoma, gr. IV	−
A382	Astrocytoma	−
U-105MG	Astrocytoma, gr. III	−
U-87MG	Astrocytoma, gr. II	−
118MG	Astrocytoma	+
U-251MG	Astrocytoma, gr. I	+
U-373MG	Astrocytoma, gr. III	+
U-178MG	Astrocytoma, gr. II	+
U-138MG	Astrocytoma, gr. IV	+
T-98	Astrocytoma, gr. IV	+
Hs683	Glioma	+
H4	Neuroglioma	+
Hs783T	Neuroblastoma	+
BE	Colon carcinoma	−
Hs255	Colon carcinoma	+
HT-29	Colon carcinoma	+
Hs675	Colon carcinoma	+
Hs237	Colon carcinoma	+
A101D	Melanoma	+
Hs695	Melanoma	+
A375	Melanoma	+
Hs852	Melanoma	+
A498	Kidney carcinoma	− (cond.)
A704	Kidney carcinoma	+
Hs755T	Wilms' tumor (kidney)	+
CEB001	Osteosarcoma	+
TE85	Osteosarcoma	+
Hs768	Osteosarcoma	+
Hs781	Osteosarcoma	+
A673	Rhabdomyosarcoma	+
A204	Rhabdomyosarcoma	+
A427	Lung carcinoma	−
A549	Lung carcinoma	+
A253	Epidermoid carcinoma, neck	−

Continued . . .

32 μM pretreatment saturated the recovery system. We have done this experiment using several strains of human fibroblasts, human embryonic kidney cells, and another kidney carcinoma cell strain, A704. In these cases, MNNG pretreatment gave no reduction in the survival of MNNG-treated adenovirus 5 even after a 32 μM MNNG pretreatment. We therefore believe that A498 cells may lack much of the repair capability of Mer^+ cells, and class the A498 strain as conditionally Mer^-. When A498 cells were infected with MNNG-treated adenovirus 5 for the normal 2-hr adsorption period, and then were treated with 32 μM MNNG, the depression in viral survival was the same as if the cells had been MNNG-treated 24 hr prior to infection. (The 32 μM MNNG had little direct killing effect on the

TABLE 1. Continued.

	Notes on origin	*Mer*$^+$/*Mer*$^-$
Tumor strains		
A431	Epidermoid carcinoma, vulva	+
A388	Epidermoid carcinoma	+
Hs703	Liver carcinoma	+
HuTu80	Stomach carcinoma	+
HT1080	Fibrosarcoma	+
Other transformed strains		
KDT	MNNG-tranformed KD	+
VA-13	SV40-transformed WI-38	−
SV80	SV40-transformed Fib.	−
Human fibroblast strains		
CRL1220	Apparently normal	+
CRL1187	Apparently normal	+
CRL1224	Apparently normal	+
KD (CRL1295)	Blood blister of lip	+
XP25RO	Xeroderma pigmentosum (A)	+
XP12BE	Xeroderma pigmentosum (A)	+
XP5BE	Xeroderma pigmentosum (D)	+
AT3BI	Ataxia telangiectasia	+
AT5BI	Ataxia telangiectasia	+
AT81CTO	Ataxia telangiectasia	+
AL1405	Acute myelogenous leuk.	+
AL639	Acute monomyel. leuk. (AML)	+
AL377	Brain tumor prone Fam.	+
AL2673	Brain tumor prone Fam.	+
AL1899	Brain tumor prone Fam.	+
AL1665	Familial AML	+
AL2642	Familial AML	+
AL409	Familial AML	+
AL2322	Familial ALL	+
WR001	Cockayne's syndrome	+
GM1598	Wiscott-Aldrich syndrome	+
GM449	Fanconi's anemia	+
Other cell strains		
HEK	Human embryonic kidney	+

intracellular virus, being 100-fold less concentrated than the lowest concentration (0.5 mg/ml) used to kill the virus.) This fact allowed us to measure the kinetics of intracellular recovery of the MNNG-damaged adenovirus in A498 cells. If, for example, intracellular repair of 30% of the viruses were complete, MNNG-treatment of the infected cells would be able to block the recovery of only 70% of the viruses (see Fig. 4). Recovery was observed to occur over a 12-hr period. Because we do not know if any incubation period after cellular MNNG treatment is required for loss of cellular ability to repair MNNG-treated viruses, we do not know that the kinetics as plotted reflect the precise time-course of repair of the virus. Because the true kinetics are likely to be related to those plotted, shifted only by the addition to the time scale of the (possibly) required incubation period, the rate of repair (given by the slope of the curve) should be a fairly accurate measure of the *in vivo* rate.

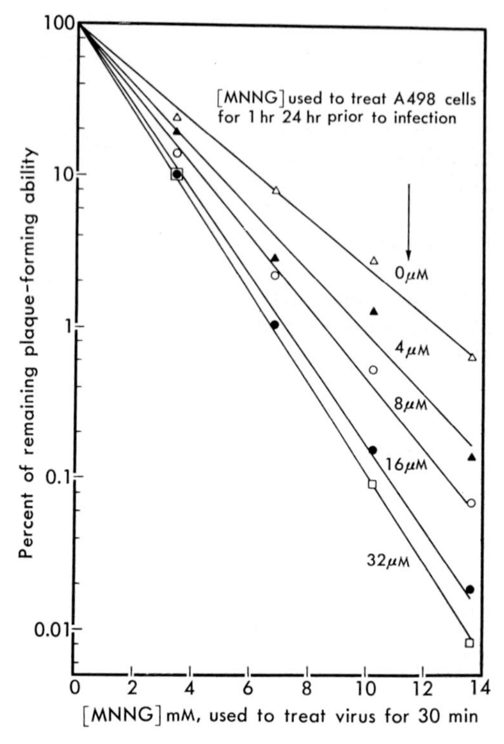

FIG. 3. Effect of cellular pretreatment on survival of MNNG-treated adenovirus in cells of the kidney carcinoma cell strain A498.

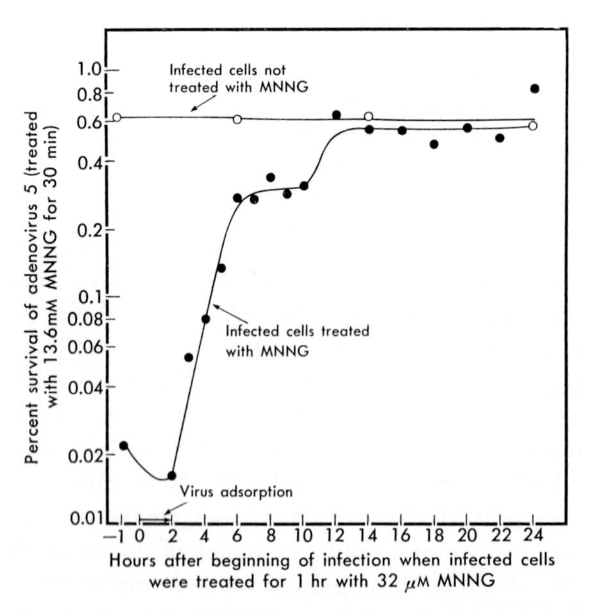

FIG. 4. Survival of MNNG-treated adenovirus 5 in cells treated for 1 hr before (−1 hr point) or after (0–24 hr points) infection with 32 μM MNNG.

4. Two SV40-transformed human lines are Mer⁻

On the suggestion of Dr. Kurt Kohn (NIH), we tested the SV40-transformed WI-38 line, VA-13, for its ability to support the growth of MNNG-damaged adenovirus 5. VA-13 was found to be *Mer*⁻. We have not yet tested WI-38 cells, but have found that another SV40-transformed human fibroblast line SV80, is also *Mer*⁻. (In collaboration with Dr. A. Lubinecki, we have determined that none of the *tumor* strains identified here as *Mer*⁻ synthesizes enough SV40 T-antigen to be detected in an indirect immunofluorescence assay. Therefore it is unlikely that the *Mer*⁻ strains were accidentally infected with SV40 during passage.)

5. Mer⁻ strains have abnormally low post-MNNG colony-forming ability

We have found that cells of four of the *Mer*⁻ strains, U-105MG, U-87MG, BE, and A427, are extremely sensitive to MNNG-produced loss of colony forming ability in comparison to the other strains tested. Figure 5 shows the results obtained for only two of these *Mer*⁻ strains. The *Mer*⁺ tumor cell strains, U-138MG and U-373MG, appear to be somewhat more sensitive to killing by MNNG than do the skin fibroblast strains, KD, CRL1187, and CRL1220. We believe, therefore, that the *Mer*⁻ phenotype is due to defects in a metabolic process that acts to restore both MNNG-damaged adenoviruses and to restore colony-forming ability to MNNG-damaged cells.

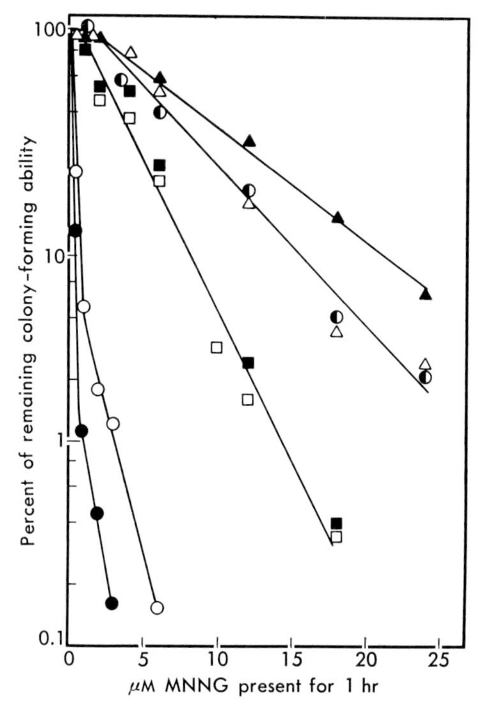

FIG. 5. Post-MNNG colony-forming ability of various cell strains. KD, CRL1220, and CRL1187 are human fibroblast strains, U-138MG and U-373MG are *Mer*⁺ astrocytoma strains, U-87MG and U-105MG are *Mer*⁻ astrocyoma strains. ▲ KD; △ CRL 1220; ◐ CRL 1187; ■ U-373MG; □ U-138MG; ● U-87MG; ○ U-105MG.

6. Mer⁻ strains repair UV damage normally

Cells of the U-105MG, U-87MG, and A172 *Mer*$^-$ strains support the growth of UV-damage adenoviruses normally (*5*). Furthermore, the post-UV colony-forming ability of the U-105MG and U-87MG strains is like that of the *Mer*$^+$ tumor or human skin fibroblasts (Fig. 6). These facts would appear to indicate that the *Mer*$^-$ strains are not defective in the way either XP or Cockayne's syndrome fibroblasts are defective (*1*, *7–9*) although XP cells have been reported to be defective in the excision of O^6-methylguanine residues from their MNNG-damaged DNA (*10*). In addition, XP cells do not show the *Mer*$^-$ phenotype (Table 1).

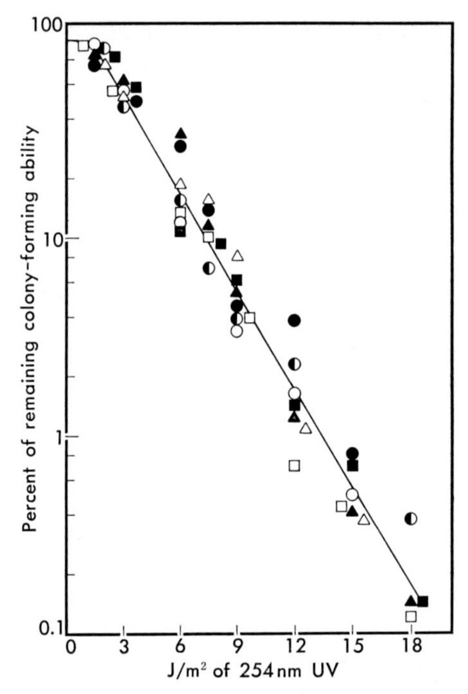

FIG. 6. Post-UV colony-forming ability of the strains in Fig. 5. ▲ KD; △ CRL 1220; ◐ CRL 1187; ■ U-373MG; □ U-138MG; ● U-87MG; ○ U-105MG.

7. MNNG-treated Mer⁻ strains show increased DNA repair-synthesis

We wished to determine whether the defect in *Mer*$^-$ strains could be detected by measuring the amount of MNNG-stimulated incorporation of ^{3}H-TdR into double-stranded DNA ("DNA repair-synthesis"). The *Mer*$^-$ strains U-105MG and U-87MG and the *Mer*$^+$ strains U-373MG, U138MG, and CRL1187 were blocked in semiconservative DNA synthesis for 30 min by 10 mM hydroxyurea (HU), and treated with selected concentrations of MNNG in complete medium containing 10 mM HU plus 5 μCi/ml ^{3}H-TdR for 60 min. From these cultures SDS lysates were prepared, treated with RNase and pronase, passed 5 times through a 20 g needle, and applied to benzoylated-naphtholyated-DEAE-cellulose columns to separate the fully double-stranded from fully or partially single stranded DNA. The "DNA repair-synthesis" (*i.e.*, the amount of label incorporated into fully

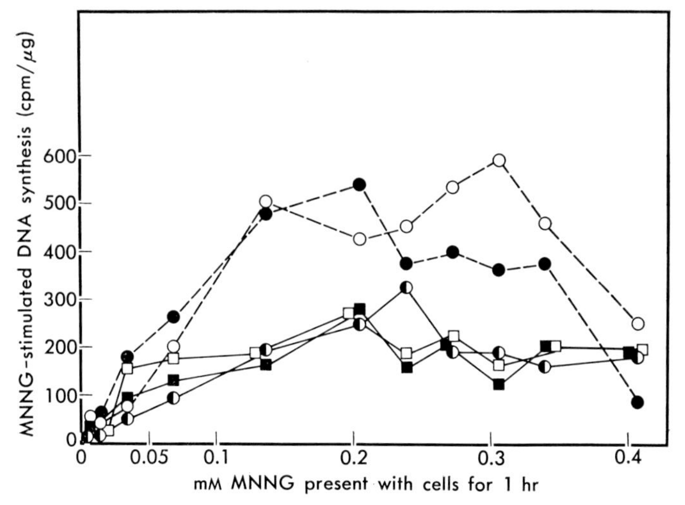

FIG. 7. Post-MNNG "DNA repair synthesis" by *Mer*$^+$ (CRL1187, U-138MG, and U-373 MG) and *Mer*$^-$ (U-87MG and U-105MG) strains. ◐ CRL 1187; ■ U-373MG; □ U-138MG; ● U-87MG; ○ U-105MG.

double-stranded DNA) was then determined (Fig. 7). Somewhat paradoxically, the MNNG-treated *Mer*$^-$ cells incorporated more ^{3}H-TdR into their DNA than did the *Mer*$^+$ strains. Further, the *Mer*$^+$ tumor strains showed the same level of incorporation as did the human fibroblasts (see also Refs. *11* and *12*).

8. Semiconservative DNA synthesis in Mer$^-$ *strains is sensitive to MNNG*

To measure the effect of MNNG on semiconservative DNA synthesis in *Mer*$^-$

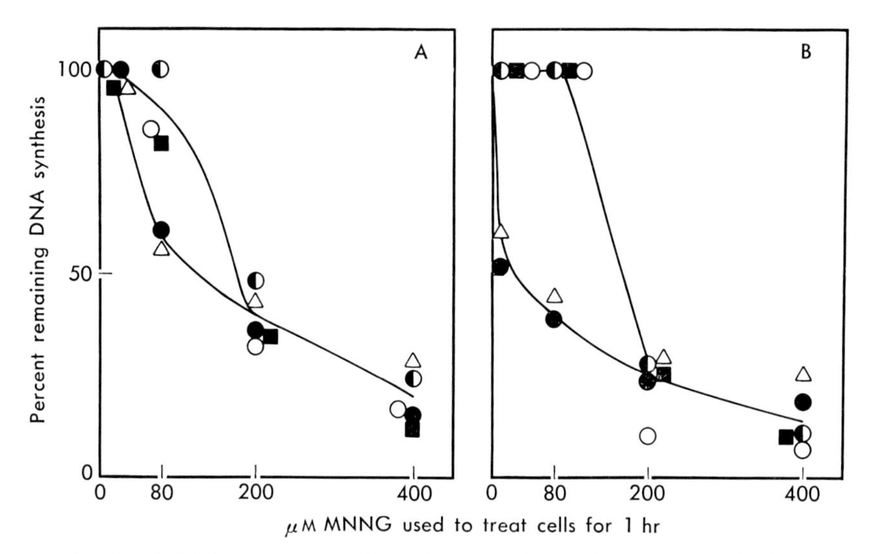

FIG. 8. Post-MNNG semi-conservative DNA synthesis by *Mer*$^+$ (CRL1187, U-373MG, and HT-29) and by *Mer*$^-$ strains (BE and U-87MG). (A) 0 time after MNNG treatment, (B) 24 hr after MNNG treatment. ◐ CRL1187; ■ U-373MG; ○ HT-29; △ BE; ● U-87MG.

and *Mer*$^+$ cells, cells were incubated 16–20 hr with 0.005 μCi/ml ^{14}C-TdR (50 mCi/mM), for 60 min in fresh medium without label, then for 60 min in fresh medium containing selected concentrations of MNNG. Immediately following such treatment, or 24 hr later, the cells were incubated with 100 μCi/ml ^{3}H-TdR (55 Ci/mM). The ^{3}H-incorporation for cultures treated at each dose was measured at 30, 60, 90, 120 min. The percent remaining DNA synthesis was then determined as follows (Fig. 8):

$$\%\text{DNA synthesis} = \frac{(^3\text{H cpm}/^{14}\text{C cpm})_{120\,\text{min},\,+\text{MNNG}}}{(^3\text{H cpm}/^{14}\text{C cpm})_{120\,\text{min},\,-\text{MNNG}}}$$

Two general conclusions can be made. The first is that DNA synthesis by the *Mer*$^-$ strains, BE and U-87MG, was more sensitive to the MNNG-produced depression than that by the *Mer*$^+$ strains U-373MG, HT-29, and CRL1187. Second, the major effect of the 24-hr incubation was to amplify the effects of the MNNG on the *Mer*$^-$ strains.

DISCUSSION

Eleven cell strains unable to support the growth of MNNG-damaged adenovirus 5 as well as either human fibroblasts (22 strains) or the majority of human tumor strains (30 strains) include both a minority of human tumor strains (9 strains, originating from tumors arising in the skin, colon, lung, brain, and kidney) and SV40-transformed human fibroblasts (2 strains). The phenotype (which we have called *Mer*$^-$) that is common to these 11 strains appears to be associated with (1) abnormally high sensitivity to MNNG-produced killing as determined by colony-forming ability, (2) abnormally high "DNA repair-synthesis" following MNNG treatment, (3) abnormally high susceptibility of semiconservative DNA synthesis to inhibition by MNNG, but with (4) normal ability to repair UV-produced damage.

We believe that these findings have possible implications in two areas: carcinogenesis and chemotherapy. It is possible that the tumors from which the *Mer*$^-$ strains arose were entirely composed of repair-deficient cells. Their repair deficiency either might have been caused as a result of a certain step in transformation (a step also required for SV40 to produce transformation), or might have facilitated the transformation of a repair deficient normal cell into a tumor cell (in which case it might also facilitate SV40 transformation). Further, if tumors that give rise to the *Mer*$^-$ cell strains are indeed composed of repair-deficient cells, then these might be the tumors that are most successfully treated by the chemotherapeutically used alkylating agents.

REFERENCES

1. Cleaver, J. E. and Bootsma, D. Xeroderma pigmentosum: Biochemical and genetic characteristics. Annu. Rev. Genet., *9*: 19–38, 1975.
2. Sasaki, M. Fanconi's anaemia: A condition possibly associated with a defective

DNA repair. *In*; P. C. Hanawalt, E. C. Friedberg, and C. F. Fox. (eds.), DNA Repair Mechanisms, pp. 675–684, Academic Press, New York, 1978.

3. Taylor, A.M.R., Harnden, D. G., Arlett, C. F., Harcourt, S. A., Lehmann, A. R., Stevens, S., and Bridges, B. A. Ataxia telangiectasia: A human mutation with abnormal radiation sensitivity. Nature, *258*: 427–429, 1975.
4. Paterson, M. C. Ataxia telangiectasia: A model disease linking deficient DNA repair with radiosensitivity and cancer proneness. *In*; P. C. Hanawalt, E. C. Friedberg, and C. F. Fox (eds.), DNA Repair Mechanisms, pp. 637–650, Academic Press, New York, 1978.
5. Day, R. S., III and Ziolkowski, C.H.J. Human brain tumour cell strains with deficient host-cell reactivation of *N*-methyl-*N'*-nitro-*N*-nitrosoguanidine-damaged adenovirus 5. Nature, *279*: 797–799, 1979.
6. Rupert, C. S. and Harm, W. Reactivation of biological damage. *In*; L. G. Augenstein, R. Mason, and M. Zelle (eds.), Advances in Radiation Biology, vol. 2, pp. 1–81, Academic Press, New York, 1966.
7. Day, R. S., III and Ziolkowski, C. Studies on UV-induced viral reversion, Cockayne's syndrome, and MNNG damage using adenovirus 5. *In*; P. C. Hanawalt, E. C. Friedberg, and C. F. Fox (eds.), DNA Repair Mechanisms, pp. 535–539, Academic Press, New York, 1978.
8. Schmickel, R. D., Chu, E.H.Y., Trosko, J. E., and Chang, C. C. Cockayne syndrome: A cellular sensitivity to ultraviolet light. Pediatrics, *60*: 135–139, 1977.
9. Day, R. S., III. Studies on repair of adenovirus 2 by human fibroblasts using normal, xeroderma pigmentosum, and xeroderma pigmentosum heterozygous strains. Cancer Res., *34*: 1965–1970, 1974.
10. Strauss, B., Bose, K., Altamirano, M., Sklar, R., and Tatsumi, K. Response of mammalian cells to chemical damage. *In*; P. C. Hanawalt, E. C. Friedberg, and C. F. Fox (eds.), DNA Repair Mechanisms, pp. 621–624, Academic Press, New York, 1978.
11. Scudiero, D. A. Repair deficiency in *N*-methyl-*N'*-nitro-*N*-nitrosoguanidine treated ataxia telangiectasia fibroblasts. *In*; P. C. Hanawalt, E. C. Friedberg, and C. F. Fox (eds.), DNA Repair Mechanisms, pp. 655–658, Academic Press, New York, 1978.
12. Scudiero, D. A. Decreased repair synthesis and defective colony-forming ability of ataxia telangiectasia fibroblast cell strains treated with *N*-methyl-*N'*-nitro-*N*-nitrosoguanidine. Cancer Res., *40*: 984–990, 1980.

GENETIC AND ENVIRONMENTAL FACTORS IN EXPERIMENTAL AND HUMAN CANCER,
H. V. GELBOIN ET AL. (EDS.), JAPAN SCI. SOC. PRESS, TOKYO, PP. 259–270, 1980

Genetic Aspects of Xeroderma Pigmentosum and Other Cancer-prone Diseases

Hiraku TAKEBE,*[1] Osamu NIKAIDO,*[1] Kanji ISHIZAKI,*[1] Takashi YAGI,*[1] Masao S. SASAKI,*[1] Mituo IKENAGA,*[2] Takehito KOZUKA,*[2] Yoshisada FUJIWARA,*[3] and Yoshiaki SATOH*[4]

Radiation Biology Center, Kyoto University, Kyoto, Japan,[1] Faculty of Medicine, Osaka University, Osaka, Japan,*[2] Kobe University School of Medicine, Kobe, Japan,*[3] and Tokyo Medical and Dental University, Tokyo, Japan*[4]*

Abstract: Xeroderma pigmentosum (XP) appears to be more common in Japan (approx. 1/40–100,000) than in other countries. Survey of 120 cases of XP patients for clinical characteristics and DNA repair capacities of their cells showed that clinical symptoms were apparently related to the level of repair deficiency. Skin cancers developed in approximately half of the patients, particularly among those having low repair capacities. Neurological abnormalities occurred mainly in those patients whose cells exhibited low repair capacities. Most of the patients over the age of 30 years had mild symptoms of XP and had no neurological disorders with considerably higher repair capacities than younger patients.

Genetic complementation tests of the cells revealed the existence of groups A, C, D, and F as well as variants in Japanese patients, the distribution of the groups being quite different from that in western countries. A high frequency of group A, which is usually associated with a low repair capacity may account for the presence of many young XP patients with severe symptoms and early development of cancers. With early diagnosis of the disease by the repair test and precaution against sun exposure, the prevention of cancers has been possible in many cases.

Examples and characteristics of other repair-deficient cancer-prone diseases in Japan, namely ataxia telangiectasia, Bloom's syndrome and Fanconi's anemia are briefly described.

A high incidence of xeroderma pigmentosum (XP) and the unique distribution of its genetic complementation groups in Japan have been reported by us (*1–5*). The majority of XP patients in Japan belong to the complementation group A characterized by very low DNA repair activities of their cells. The severe symptoms of these group A patients may account for the high ratio of patients under the age of 10 years. This report confirms the above findings of the survey and repair tests of 120 XP patients in Japan, and will discuss the genetic aspects of repair deficiency in relation to carcinogenesis. In addition, preliminary survey of other hereditary diseases with repair deficiency in Japan will be presented.

Clinical and Repair Characteristics of XP Patients in Japan

Table 1 gives the age distribution of XP patients with clinical and repair characteristics. The ages are at the time of biopsies and, therefore, do not reflect the age of onset or the first diagnosis. For most of the patients, the age represents the first visit to University Hospitals, from where the information for the table was obtained. The table may be summarized as follows:

1) Young patients, 0–9 years old, comprise 43%, most of them showing very low repair activities as measured by the relative amount of unscheduled DNA synthesis (UDS) after UV irradiation.
2) Approximately half of the patients developed skin cancers. There are patients who have not developed skin cancers in old age.
3) Mental retardation and other neurological abnormalities appeared predominantly in young patients and no patients over the age of 40 years showed any neurological disorders.
4) There is an apparent relationship between repair activities of the cells and the ages of the patients; the lower the repair activities, the younger the patients appear in the table.
5) The above tendency may have several exceptional cases, particularly in a group of older patietns with 9–25% of UDS.

The lack of patients in the lowest repair group over the age of 30 may mean the patients with low repair capacities of their cells have died by the age of 30. The possibility of having developed higher repair activity with age is unlikely since no change in repair activities was noted in any case and also because the patients in the same siblings always showed the same level of repair activity despite the differences in age. The distribution of the neurological abnormalities also does not support the possibility of development of repair activity at a greater age. Since the neurological abnormalities should be regarded as irreversible symptoms, the lack of such disorders in older patients denies the progressive change in repair capacities in XP patients.

Nearly 70% of the patients aged 10 years or more have developed skin cancers. The frequency is higher in the group with low repair activity (60% or less) than with near-normal levels of repair at these ages. Most of the children younger than

TABLE 1. Age Distribution, Clinical Characteristics, and DNA Repair of XP Patients in Japan

Ages	Number of patients (skin cancers)	Mental retardation and neurolog. abnorm.			UDS (% of normal) (skin cancers)			
		Yes	No	Unknown	≦5	9–25	30–60	60<
0–9	52 (13)	26	4	22	47 (12)	3 (1)	1	1
10–19	21 (15)	15	6	0	15 (11)	2 (2)	1 (1)	3 (1)
20–29	17 (12)	3	13	1	3 (3)	2 (2)	3 (1)	9 (6)
30–39	9 (5)	1	7	1	0	0	2 (2)	7 (3)
40–49	9 (5)	0	8	1	0	2 (1)	1 (1)	6 (3)
50≦	12 (9)	0	12	0	0	3 (2)	1 (1)	8 (6)
Total	120 (59)	45	50	25	65 (26)	12 (8)	9 (6)	34 (19)

10 years may still be too young to develop skin cancers, but are expected to develop the malignancy unless proper care is taken.

The age distribution in Table 1 appears to reflect the degree of severity of the symptoms. The younger patients should have been diagnosed as XP or XP suspects mainly because the symptoms appeared early, presumably due to the low repair capacities. On the contrary, the patients with intermediate and near-normal levels of repair capacity were not diagnosed as XP until their teens or twenties, possibly because the symptoms were very mild. Actually, most of the patients belonging to the intermediate and near-normal levels of repair showed very mild symptoms even when they developed skin cancers.

One of the exceptional cases was a female patient, 45 years old, with very mild symptoms and no neurological abnormalities. Her cells, however, showed only 10% of UDS after UV. Further analysis of her cells revealed that the cells should have a much higher repair activity as a whole and were found to belong to a new complementation group. A detailed description of this patient was published elsewhere (*6*) and will be mentioned briefly in the following sections.

Genetic Groups of XP Patients in Japan

The age and repair distribution of the patients in Table 1 seems to be quite different from the survey of XP patients in Europe, the United States, and Egypt compiled from the literature (*2*). The major differences noted were as follows:

1) Frequency of younger (0–9) patients was higher in Japan than in other countries.
2) There were fewer patients belonging to the intermediate repair level in Japan than in Europe or the United States. In particular, the patients with relatively low repair capacity but without neurological abnormalities comprise a considerable ratio in Europe and the United States, but were rarely found in Japan.
3) Relationship between levels of repair deficiency and age as well as the development of symptoms was not so clear in Europe or the United States as in Japan.

Introduction of genetic complementation tests in classification of XP patients (*7*) gave answers to some of these questions. Table 2 gives the comparison between the distribution of complementation groups in Japan and that in other countries.

TABLE 2. Genetic Groups of XP Patients

Area	Complementation group							Variant	Total
	A	B	C	D	E	F	G		
Japan	21	0	1	1	0	3	0	14[a]	40
Other countries[b]	18	1	24	7	2	0	2	7	61
UDS (%) (UV)	0–5[c]	3–7	5–31	10–55	40–60	10	2[d]	70–100	

[a] All tested for post-replication repair by Fujiwara. [b] Bootsma (personal communication), including 7 cases in Egypt. [c] Rare exceptions, *e.g.* 36%, are known to exist. [d] For 1 case (XP2BI) only (*8*) and reported to be 10–15% later (*10*).

Although we have not tested all the 120 cases listed in Table 1, the relative number of each complementation group may not change to a great extent judging from the distribution of UDS levels in Table 1.

The most noteworthy difference between the two areas is in group C patients. This group is represented predominatnly by the patients with relatively low repair activity in UDS, considerably severe symptoms but without neurological disorders. Often in textbooks of dermatology this group may correspond to a type described as "classical XP" in contrast to another group "de Sactis-Cacchione syndrome" with neurological abnormalities. Our survey of the complementation groups explains why we had only a few patients in Table 1 who may correspond to the former group. Another surprising feature of Table 2 is the high frequency of patients belonging to group A. The patients belonging to group A, with the exception of a few cases, showed very low repair activities of their cells (Photo A). Almost all those in Japan showed the symptoms very early in their life and all were diagnosed as XP before the age of 10. It should be noted that most of the patients in this group in Japan were not related to each other and came from different parts of the country. A new complementation group, F, was found in Japan from 2 families (XP23OS, XP2YO, and XP3YO). The characteristics of this group will be described in the next section.

Apparent high frequency of the "variants" whose cells showed a normal level of UDS but clear symptoms of XP may not reflect the exact ratio since Fujiwara has been extensively engaged in studies on XP variants and tried hard to test as many variants as possible. All variants in this table were proved to be defective to some extent in post-replication repair by Fujiwara. The only group C patient (XP2KA) was found recently (*9*). The symptoms of the patient were severe and she developed a squamous cell carcinoma of conjunctiva (Photos B and C). The only group D (XP10TO) patient (*2*) was found relatively early in the survey. The patient did not show any of the mental retardation or neurological disorders which accompany many group D patients in other countries. The presence of other patients in groups C and D in low frequencies may be expected from Table 1 provided the assignment which depends on levels of UDS is generally valid, but which may not be true.

Characteristics of Complementation Group F

Three patients, XP23OS, XP2YO, and XP3YO have so far been identified as belonging to the new complementation group, F. XP2YO (F, 65) and XP3YO (M, 29) are related, XP2YO being a sister of the grandfather of XP3YO. No other member of the family in three generations were affected according to the patients, except for the mother of XP3YO, who is related to XP2YO, who showed excessive freckles but has not been tested for repair activity yet. There were no consanguineous marriages in this family.

DNA repair characteristics of XP23OS cells recently were reported by us (*6*). Similar results were obtained with cells from XP2YO and XP3YO. All had low UDS activities, approximately 10% of the normal cells. The symptoms of

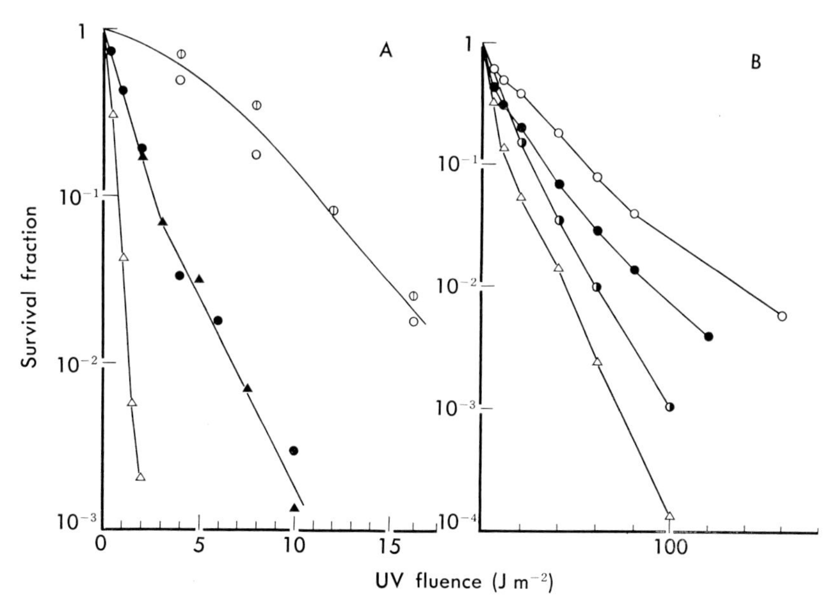

FIG. 1. A. UV inactivation of colony-forming ability in normal, XP2QS (complementation group A) and XP23OS (group F) cells. ○ ⦶ normal; ● ▲ 23OS; △ 2OS(A). B. Host-cell reactivation of UV-irradiated herpes simplex virus in normal, XP1OS (group A), XP5BE (group D) and XP 23OS cells. ○ normal; ● 23OS; ◑ 5BE (D); △ 1OS(A).

these patients, as seen in the Photographs (Photos D, E, and F), were milder than the patients in other complementation groups or those not indentified for the complementation group with approximately equal levels of UDS. XP3YO showed the most severe symptoms among the three group F patients and developed a squamous cell carcinoma on the lower lip which degenerated on treatment with 5-fluorouracil (5-FU). XP2YO was also reported to have developed skin cancers but a detailed record was not tracable. XP23OS showed very mild symptoms and has not developed cancers. None of these patients had neurological abnormalities.

DNA repair tests other than UDS revealed that the cells of group F patients should have efficient repair capacities as a whole as shown in Fig. 1. Both UV survival of the cells and the UV-irradiated herpes simplex virus using cells as hosts indicated that XP23OS should have an intermediate level of repair activity in comparison with normal cells and other XP cells belonging to complementation groups A and D. Similar results were obtained for XP3YO cells. Presence of the high repair activity in XP23OS cells was supported by the finding that these cells showed about 80% of the normal repair activity as measured by the removal of UV-endonuclease susceptible sites from the DNA during incubation after UV exposure (*10*). Such high overall repair activity may account for the mild symptoms of patient XP23OS. From these data, it is important to note that the use of UDS as a measure of excision repair activity should be limited to a certain extent and with some reservation.

Prevention of Cancers in XP Patients

The best and probably the only effective way to prevent the development of skin cancers in XP patients is to avoid sun exposure. This may be achieved not only by keeping the patients indoors, but also by the topical application of sun-screens. There are 2 types of sun-screen, light-scattering and light-absorbing. The former may be preferable since protection against the UV spectrum of sunlight is more effective than the light-absorbing compounds and also because there is less possibility of involvement of photodynamic activity which could take place.

Photograph G gives a good example of how the careful protection against sun exposure is effective in preventing the development of XP symptoms and skin cancer. The patient, XP8OS, was identified as belonging to complementation group A with very low activity of UDS after UV exposure. The photograph was taken when she was 6 years old. Earlier photographs of the same patient appeared in the Proceedings of the 6th Symposium of this series (*1*). She was first diagnosed as XP at 6 months, one of the earliest dates of diagnosis in all patients tested by us. Since then, extensive care was undertaken by her parents and doctors. Although she has developed some freckles on the face as can be seen from the photograph, essentially no part of her skin has become dry or deeply pigmented in 6 years. Although it may still be too early to conclude whether she could survive without developing skin cancers, the symptoms are far milder than any other patients in group A at her age.

Frequency of XP in the Japanese Population

Colleagues in other countries might have been surprised by the number of patients, 120, in our survey. Although we have had a good cooperative research group supported by government grants, we did not try too hard to look for patients. Apparently the incidence of XP in Japan is very high compared to other countries, with a possible exception of Egypt (*11*). Based on the frequency of consanguineous marriages among parents of patients and that in the general population, a rough estimate of the frequency of XP patients at birth may be 1 in 40,000 to 100,000 (*5*). This is higher than the estimated frequency in other countries, 1/250,000. The unique gene distribution as shown in Table 1 may account for the high incidence, since there could be a higher chance of producing homozygous XP genes of the same complementation group due to the predominant presence of group A. In addition, severe symptoms of group A patients appearing at a young age may increase the possibility of a definite identification as XP.

Other Hereditary Diseases with Repair Deficiency and a High Incidence of Cancer

1. Ataxia telangiectasia (AT)

Figure 2A gives X-ray survivals of 2 AT cell lines obtained from patients in Japan. Taylor *et al.* (*12*) reported that the most sensitive AT cells against X-rays were approximately 5 times more sensitive than normal cells; the cells in Fig. 2

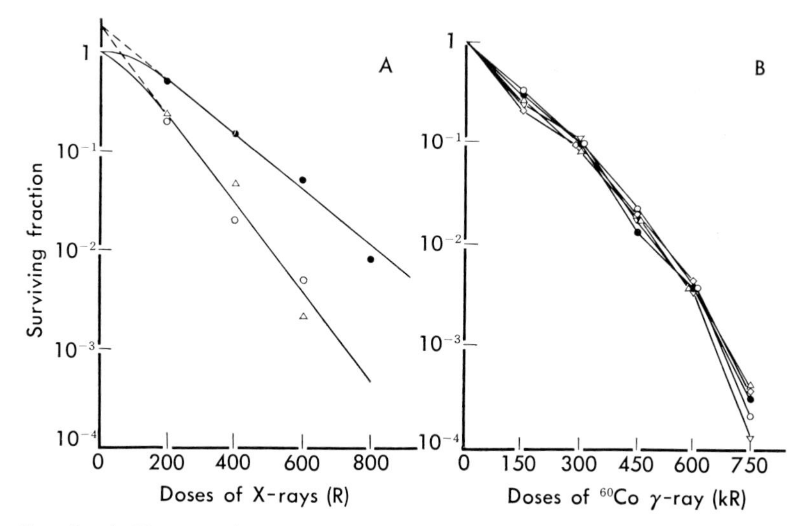

FIG. 2. A. X-ray survival curves of normal (●) and 2 AT cell lines, AT1SE (△) and AT1TO (○). B. Host-cell reactivation of ^{60}Co γ-irradiated herpes simplex virus in 2 normal and 3 AT cell lines. There is some doubt about the diagnosis of AT1KA. Detailed data will be published by Nikaido (in preparation). ○ ● control; △ AT1SE; ▽ AT1TO; ◇ AT1KA.

are not as sensitive as theirs. Host-cell reactivation of γ-irradiated herpes simplex virus did not show any difference between normal and AT cells (Fig. 2B), supporting the report by Rainbow (*13*). A systematic survey of AT patients in Japan has started but few data have been available to date. Apparently the frequency of AT is lower than that of XP, in contrast to Caucasean people in other countries where AT appears to be more frequent than XP. No comparative genetic tests between AT patients in Japan and other countries have been carried out. Some clinicians mentioned that the symptoms may be different in Japanese, particularly in telangiectasia which often does not show clearly in Japanese.

2. Bloom's syndrome (BS)

Three cases of patients with BS were confirmed by high frequency of spontaneous sister chromatid exchanges in Japan during the last 2 years (*14–16*). The first and second cases (Photos H and I) are included in Dr. German's BS registry as reported at this symposium (*17*). Detailed descriptions of both patients will be published elsewhere and the third case is being tested for further cell repair characteristics. As in AT patients in Japan, clinical features of the first 2 cases in Japan were not so clear as those in Jewish BS patients according to Dr. German who looked at both patients personally recently. The first case, 78AkSak, would not have been diagnosed as BS, as he was not too small and showed little telangiectasia, if sister chromatid exchange had not been examined. He had developed malignant lymphoma at the age of 14. The other 2 patients have not yet developed any malignancy. In previous literature, only one or two certain cases of BS in Japan have been reported, suggesting the frequency of the BS gene in Japan may be as low as in other countries.

3. Fanconi's anemia (FA)

Twenty cases of FA in Japan have been compiled by Sasaki (*18*). There were no cases of malignancy. Cytogenetic data clearly divided these patients into two groups; group A having a high frequency of chromosome aberrations (average: 55.7% of metaphases having spontaneous aberrations), and group B with a normal level of aberrations (5.9%). The latter group, although clinically diagnosed as FA, could represent other disease(s) of aplastic anemia coupled with growth retardation and/or malformations. Lack of high sensitivity to mitomycin C, prominent in group A, inducing chromosome aberrations in group B also suggests this possibility. All the patients in groups A and B were 11 years old or younger and could have been too young to develop a malignancy. The number of cases and their family data were not sufficient to estimate the gene frequency of FA in Japan.

Genetic Aspects of Repair Deficiency

1. Gene mapping

An effort to map the genes of repair-deficient diseases has not so far been successful. We tried to incorporate the chromosomes of normal human cells into XP cells using a microcell hybridization technique developed by Fournier and Ruddle (*19*). As shown in Fig. 3, XP cells became resistant to UV, presumably due to the incorporation of the chromosomes of normal cells. A detailed analysis is in progress. At least it may be reasonable to state that the gene(s) responsible for UV sensitivity, or XP gene(s), is located on a limited number of chromosome(s), since the microcells used for hybridization were assumed to have a small number of chromosomes.

2. Possible partial deficiency in heterozygotes and genetic high-risk group of cancer

In AT, Swift *et al.* (*20*) reported that the relatives of the AT patients were

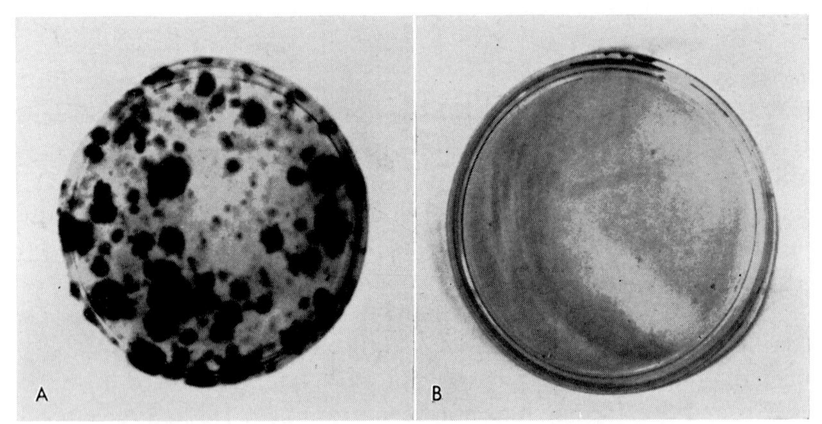

FIG. 3. A. A UV-resistant clone of XP2QS(SV) cells (complementation group A, transformed by SV40) which were hybridized by cell fusion mediated by HVJ (Sendai virus) with microcells of normal human fibroblast cells. B. XP cells mixed with the microcells, but without HVJ did not survive UV exposure. UV dose on both Petri dishes was 10 J/m². Details will be published by Ishizaki *et al.* (in preparation).

shown to have a higher incidence of malignancy than the general population. Similar results with blood relatives of xeroderma pigmentosum patients and spouse controls, the former showing a significantly higher incidence of nonmelanomatous skin cancers than the latter, were also reported by Swift and Chase (*21*). Our survey on XP patients and their families in Japan may not be able to contribute to this subject since the incidence of skin cancers in general is extremely low in Japanese, presumably due to their darker skin color, and no case of skin cancer has been encountered in families of XP patients in Japan.

Partial repair deficiency in AT heterozygotes was reported by Chen *et al.* (*22*) and Paterson *et al.* (*23*) and they suggested the presence of a genetically cancer-prone subpopulation of humans. An attempt to identify XP heterozygotes by partial repair deficiency has not been clearly successful, our trial so far being negative.

Persons carrying one AT gene were estimated to constitute approximately 1% of the whites in the United States (*20*). A similar estimate based on Hardy-Weinberg's law may be applied for XP heterozygotes in Japan which may comprise as much as 1% of the general population. Although little data have been available on the incidence of malignancies other than skin cancers in XP patients, it is possible that XP heterozygotes are also cancer-prone due to repair deficiency. Continued efforts are needed to detect the genetically cancer-prone high-risk group for these rare repair deficiency diseases and to evaluate the role of genetic factors in carcinogenesis.

ACKNOWLEDGMENTS

This work was supported by Grants-in-Aid from the Ministry of Education, Science and Culture of Japan, by the Subsidy for Cancer Research and by a Research Grant from the Intractable Disease Division, Public Health Bureau, Ministry of Health and Welfare of Japan. We are indebted to Drs. Kiyoji Tanaka, Seiji Arase, Masao Inoue, Hiroko Kawashima, and Toshio Kawaguchi for their cooperation and the supply of materials and photographs.

REFERENCES

1. Takebe, H. Genetic complementation tests of Japanese xeroderma pigmentosum patients and their skin cancers, and DNA repair characteristics. *In*; P. N. Magee, S. Takayama, T. Sugimura, and T. Matsushima (eds.), Fundamentals in Cancer Prevention (Proc. 6th Int. Symp. Princess Takamatsu Cancer Res. Fund.), pp. 383–395, Japan Scientific Societies Press, Tokyo, 1976.
2. Takebe, H., Miki, Y., Kozuka, T., Furuyama, J., Tanaka, K., Sasaki, M. S., Fujiwara, Y., and Akiba, H. DNA repair characteristics and skin cancers of xeroderma pigmentosum patients in Japan. Cancer Res., *37*: 490–495, 1977.
3. Takebe, H., Fujiwara, Y., Sasaki, M. S., Satoh, Y., Kozuka, T., Nikaido, O., Ishizaki, K., Arase, S., and Ikenaga, M. DNA repair and clinical characteristics of 96 xeroderma pigmentosum patients in Japan. *In*; P. C. Hanawalt, E. C. Friedberg, and C. F. Fox (eds.), DNA Repair Mechanisms, pp. 617–720, Academic Press, New York, 1978.

4. Takebe, H. Xeroderma pigmentosum: DNA repair defects and skin cancer. GANN Monogr. Cancer Res., *24*: 103–117, 1979.
5. Takebe, H. Genetic aspects of repair deficiency. *In*; S. Okada, M. Imamura, T. Terashima, and H. Yamaguchi (eds.), Proc. 6th Int. Congr. Radiat. Res., pp. 506–508, Jpn. Assoc. Radiat. Res., Tokyo, 1979.
6. Arase, S., Kozuka, T., Tanaka, K., Ikenaga, M., and Takebe, H. A sixth complementation group in xeroderma pigmentosum. Mutat. Res., *59*: 143–146, 1979.
7. de Weerd-Kastelein, E. A., Keijzer, W., and Bootsma, D. Genetic heterogeneity of xeroderma pigmentosum demonstrated by somatic cell hybridization. Nature New Biol., *238*: 80–83, 1972.
8. Keijzer, W., Jaspers, N.G.J., Abrahams, P.J., Taylor, A.M.R., Arlett, C. A., Zelle, B., Takebe, H., Kinmont, P.D.S., and Bootsma, D. A seventh complementation group in excision-deficient xeroderma pigmentosum. Mutat. Res., *62*: 183–190, 1979.
9. Inoue, M., Yokaichi, M., Tsukada, S., Saito, I., Sasaki, K., Yamasaki, H., Annen, Y., Hirone, T., Saito, T., Ishizaki, K., and Takebe, H. DNA repair studies on xeroderma pigmentosum patients at Kanazawa Medical University. J. Kanazawa Med. Univ., *4*: 65–70, 1079.
10. Zelle, B. and Lohman, P.H.M. Repair of UV-endonuclease-susceptible sites in the 7 complementation groups of xeroderma pigmentosum A through G. Mutat. Res., *62*: 363–368, 1979.
11. El-Hefnawi, H. and Smith, S. M. Xeroderma pigmentosum. A brief report on its genetic linkage with ABO blood groups in the United Arab Republic. Br. J. Dermatol., *77*: 35–41, 1965.
12. Taylor, A.M.R., Harnden, D. G., Arlett, C. F., Harcourt, S. A., Lehmann, A. R., Stevens, S., and Bridges, B. A. Ataxia telangiectasia: A human mutation with abnormal radiation sensitivity. Nature, *258*: 427–429, 1975.
13. Rainbow, A. J. Production of viral structural antigens by irradiated adenovirus as an assay for DNA repair in human fibroblasts. *In*; P. C. Hanawalt, E. C. Friedberg, and C. F. Fox (eds.), DNA Repair Mechanisms, pp. 541–545, Academic Press, New York, 1978.
14. Arase, S., Takahashi, O., Ishizaki, K., and Takebe, H. Bloom's syndrome in a Japanese boy with lymphoma. Clin. Genet., in press.
15. Kawashima, H., Sato, T., Taniguchi, N., Yagi, T., Ishizaki, T., and Takebe, H. Bloom's syndrome in a Japanese girl. Clin. Genet. *17*: 143–148, 1900.
16. Uchiya, S., Torigoe, K., Ohta, Y., Murai, T., Yaguchi, K., Sakai, K., Koshu, Y., and Saito, K. A case of Bloom's syndrome without a history of sun-sensitivity. Teratology, *20*: 180 (Abstr.), 1979.
17. German, J. and Schonberg, S. Bloom's syndrome; review of cytological and biochemical aspects. This volume, pp. 175–186.
18. Sasaki, M. S. Fanconi's anemia: A condition possibly associated with defective DNA repair. *In*; P. C. Hanawalt, E. C. Friedberg, and C. F. Fox (eds.), DNA Repair Mechanisms, pp. 675–684, Academic Press, New York, 1978.
19. Fournier, R.E.K. and Ruddle, F. H. Stable association of the human transgenome and host murine chromosomes demonstrated with trispecific microcell hybrids. Proc. Natl. Acad. Sci. U.S., *74*: 3937–3941, 1977.
20. Swift, M., Shloman, L., Perry, M., and Chase, C. Malignant neoplasms in the families of patients with ataxia telangiectasia. Cancer Res., *36*: 209–215, 1976.
21. Swift, M. and Chase, C. Cancer in families with xeroderma pigmentosum. J. Natl. Cancer Inst., *62*: 1415–1421, 1979.

22. Chen, P. S., Lavin, M. F., Kidson, C., and Moss, D. Identification of ataxia telangiectasia heterozygotes, a cancer-prone population. Nature, *274*: 484–486, 1978.
23. Paterson, M. C., Anderson, A. K., Smith, B. P., and Smith, P. J. Enhanced radiosensitivity of cultured fibroblasts from ataxia telangiectasia heterozygotes manifested by defective colony-forming ability and reduced DNA repair replication after hypoxic γ-irradiation. Cancer Res., *39*: 3725–3734, 1979.

EXPLANATION OF PHOTOS

PHOTO A. XP30S (F, 6 yr.), belonging to complementation group A and one of the most severely affected patients.

PHOTO B. XP2KA (F, 5yr.), the only patient belonging to complementation group C in Japan (Photo, courtesy of Dr. M. Inoue).

PHOTO C. Squamous cell carcinoma of conjunctiva of XP2KA.

PHOTO D. XP23OS (F. 45yr.), belonging to complementation group F.

PHOTO E. XP2YO (F. 65 Yr.), belonging to complementation group F (Photo, courtesy of Dr. T. Kawaguchi).

PHOTO F. XP3YO (M. 29 yr.), belonging to complementation group F (Photo, courtesy of Dr. T. Kawaguchi).

PHOTO G. XP8OS (F. 6 yr.), belonging to complementation group A and the best attended case.

PHOTO H. Bloom's syndrome, 78AkSak (M. 14 yr.), showing telangiectasia in an eye (Photo, courtesy of Dr. S. Arase).

PHOTO I. Bloom's syndrome, 86NoKi (F. 5 yr.) (Photo, courtesy of Dr. H. Kawashima).

Refer to text for detailed descriptions of each case.

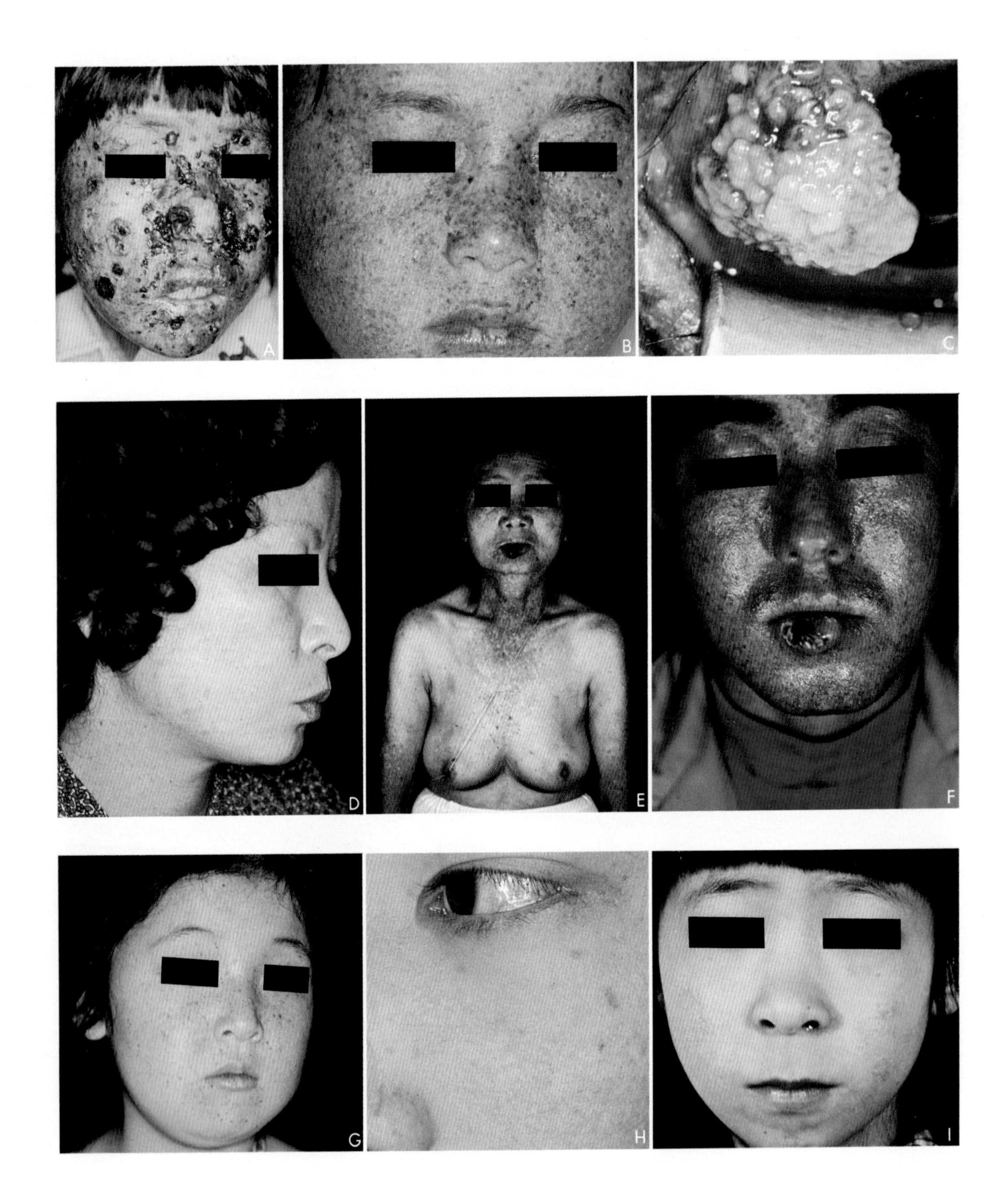

EPIDEMIOLOGY

GENETIC AND ENVIRONMENTAL FACTORS IN EXPERIMENTAL AND HUMAN CANCER,
H. V. GELBOIN ET AL. (EDS.), JAPAN SCI. SOC. PRESS, TOKYO, PP. 273–290, 1980

Cancer Risk and Lifestyle: Cancer among Mormons (1967–1975)

Joseph L. LYON, John W. GARDNER, and Dee W. WEST

Department of Family and Community Medicine, University of Utah College of Medicine, Salt Lake City, Utah 84132, U.S.A.

Abstract: Since 1950, Utah has had the lowest cancer mortality rates of the fifty states. Seventy-one percent of Utah residents are members of the Church of Jesus Christ of Latter-day Saints (Mormon or LDS), which proscribes tobacco, alcohol, coffee, and tea, and advocates strong family ties. We measure the contribution of these factors to Utah's low cancer incidence. The LDS had lower than expected rates for all cancers causally associated with tobacco and alcohol and had lower incidence for cancers of the stomach, colon, rectum, female breast, and uterine cervix. The non-LDS were not significantly different from the U.S. in overall cancer incidence. Much of the LDS difference was in the urban areas of the state.

Since 1950, the state of Utah has had the lowest cancer mortality of all the states (*1*). The area now comprising Utah was settled, beginning in 1847 by a religious group that actively advocates certain behaviors which have been found to have a direct bearing on the risk of developing cancer. This religious group (The Church of Jesus Christ of Latter-day Saints, commonly called Mormon or LDS) has proscribed the use of alcohol, tobacco, coffee, and tea in all forms, for health reasons, for at least the last 80 years. Additional Church teachings emphasize strong families (*2*), high educational attainment (*3*), and strict sexual mores proscribing premarital and extramarital sexual intercourse for both men and women (*4*).

About 70% of the state of Utah are presently members of the LDS Church. This presents the opportunity to study, in a defined geographic area, two populations sharing the same physical environment, but differing in a number of personal habits and behaviors that relate to risk of developing cancer.

METHODS

There were 20,379 cases of cancer diagnosed in Utah residents in 1967–1975. These cases were identified by the Utah Cancer Registry, a population based cancer registry whose operations have been described previously (*5, 6*). Table 1

TABLE 1. Utah Cancer Registry (1966-1975)——Percentage of Cases, by Site, Confirmed by Histology

ICD[a]	Site	%	ICD	Site	%
140	Lip	96	174	Breast (male)	94
141	Tongue	97	180	Cervix uteri	98
142	Salivary gland	95	181	Placenta	95
143-145	Mouth	97	182	Corpus uteri	98
146-149	Pharynx	97	183	Ovary-tube-ligament	97
150	Esophagus	91	184	Female genital	96
151	Stomach	86	185	Prostate	94
152	Small intestine	97	186	Testis	97
153	Colon	93	187	Male genital	98
154	Rectum-anus	96	188	Bladder	98
155	Liver	73	189	Kidney	93
156	Gall bladder	94	190	Eye	92
157	Pancreas	73	191-192	Brain and CNS	91
160	Nasal cavity-sinus	96	193	Thyroid	98
161	Larynx	97	194	Other endocrine glands	79
162	Lung	88		Ill defined sites	81
163	Pleura	95	199	Unknown primary	80
164	Mediastinum	95	200-202	Lymph, non-Hodgkins,	95
165	Respiratory tract	100		Hodgkins	98
170	Bone	96	204-205	Multiple myeloma	91
171	Connective tissue	97		Acute leukemia	95
173	Melanoma	98		Chronic leukemia	93
174	Breast	97		Other hemapoietic	83
				Total sites	94

Death certificate only cases, 1967-1975. All cases registered, 20,379; death certificate only, 652. [a] ICD: International Classification of Diseases.

shows the percentage of all cancer cases in this period, verified histologically, and the overall percentage for all cancer sites identified by death certificate only. To characterize each cancer case as to religion, we used the central membership files maintained by the LDS Church, which have been described elsewhere (*7*, *8*). Briefly, the matching procedure involved search of this file using the individual's name, birthdate, place of residence, and next of kin as matching variables.

Populations for the mid-point of the interval 1967–75 were obtained by linear interpolation of annual Church censuses. The age distribution was estimated from a 5% systematic sample of the Church population taken in August of 1971 (*7*, *8*). The non-LDS population was obtained by subtraction of the LDS from the total state population, which was derived by linear interpolation of the 1970 U.S. census and a 1975 population estimate from the State Bureau of Vital Statistics. The incidence of cancer in the LDS and non-LDS populations for the state was compared to the Third National Cancer Survey (TNCS) white rates, using standardized incidence ratios (SIRs) (*9*). The expected cases were calculated by applying the TNCS age-specific incidence rates to the respective Utah populations, using 5-year age groupings. Statistical significance of the difference for each population from the TNCS and U.S. were determined using the method of Bailar and Ederer

(*10*), and differences between LDS and non-LDS rates were tested using a modification of the Mantel-Haenszel procedure controlling for sex and age in 5-year age intervals (*7, 11*). Because of the large proportion of the Utah population residing in the urban areas (approximately 80% at the 1970 census), and the known increased risk to urban residents *vs.* rural residents, we also divided the state into urban and rural areas and calculated rates for each geographic area, using the procedure described above.

Beginning in 1977, we carried out three case-control studies in the state of Utah, using a random digit dialing technique to select control groups as representative of the state general population. The overall response rate from the control group was 85% with the majority of non-completions being due to individuals moving between the time of ascertainment through the telephone census and the time when they were approached by an interviewer.

The control groups, while being randomly drawn from the general population of the state, generally had age distributions reflecting a disease under study. Weighting factors were therefore necessary to adjust their responses to the general population of the state, and are given in Appendix Table 1.

RESULTS

Of the 20,379 cases, 439 (2.2%) were excluded because of inability to ascertain religion, and an additional 187 (0.9%) because no county of residence was available (63 LDS and 124 non-LDS cases). Of the 19,753 cases remaining, 12,112 were classified as LDS (9,002 urban and 3,110 rural), and 7,641 were classified as non-LDS (6,158 urban and 1,483 rural). Tables 2–5 present SIRs by cancer site for both LDS and non-LDS populations of the state. Figures 1–8 contain the same data, represented graphically.

TABLE 2. Observed and Expected Numbers and SIRs for Cancers by Site Occurring among LDS Males and Females (UTAH LDS, 1967–1975)

Primary site (MOTNAC[a]) Topography Code No.[b])	Male			Female		
	Observed number	Expected number	SIR	Observed number	Expected number	SIR
Total malignant neoplasms	6,070	8,347.6	73‡	6,105	8,050.0	76‡
Malignant neoplasm of:						
Lip, oral cavity, and pharynx	(434)	(413.2)	(105)	(91)	(158.9)	(57)
Lip (1400–1409)	294	100.4	294‡	26	7.9	329‡
Tongue (1410–1419)	31	76.0	41‡	13	32.2	40‡
Salivary glands (1420–1429)	28	30.7	91	19	27.0	70
Gum and mouth (1430–1459)	38	94.5	40‡	16	52.6	30‡
Nasopharynx (1471–1479)	14	17.6	80	9	7.2	125
Other and unsp. pharynx (1461–1469, 1480–1489, 1499)	29	93.8	31‡	8	29.6	27‡
Digestive organs and peritoneum	(1,326)	(2,137.1)	(62)‡	(1,191)	(1,844.9)	(65)‡
Esophagus (1500–1509)	42	111.6	38‡	7	39.1	18‡

Continued . . .

TABLE 2. Continued.

Primary site (MOTNAC[a]) Topography Code No.[b])	Male			Female		
	Observed number	Expected number	SIR	Observed number	Expected number	SIR
Stomach (1510–1519)	216	323.1	67‡	142	192.2	74‡
Small intestine (1520–1529)	24	23.3	103	22	19.0	116
Colon (1531–1539, 544[c)])	489	787.7	62‡	542	869.3	62‡
Rectum, rectosigmoid junction, and anal canal (1540–1542, 1549)	257	425.6	60‡	225	324.7	69‡
Liver (1550)	32	71.2	45‡	21	40.8	51‡
Gallbladder (1560)	22	26.4	83	53	64.6	82
Othe biliary (1551, 1561–1569)	28	38.3	73	31	30.3	102
Pancreas (1570–1579)	198	287.7	69‡	122	214.2	57‡
Retroperitoneum and peritoneum and unspecified digestive organs including abdomen (1580–1588, 1599)	18	40.1	45‡	26	48.3	54‡
Respiratory system	(745)	(1,924.5)	(39)‡	(166)	(461.3)	(36)‡
Larynx (1610–1619)	77	196.1	39‡	10	26.4	38‡
Lung (1620–1622, 623–624[c)])	639	1,691.8	38‡	137	414.7	33‡
Other respiratory organs including: pleura, mediastinum, intrathoracic site (1600–1609, 1630–1639, 649[c)])	29	36.6	79	19	19.5	97
Bones and joints (1700–1709)	41	28.4	144*	19	22.9	83
Connective subcutaneous and other soft tissues (1710–1719)	65	61.9	105	60	51.2	117
Melanomas—skin (1730–1739 for M-8723–8783)	150	126.3	119*	183	137.1	133‡
Breast (1740–1749)	17	17.4	98	1,786	2,184.4	82‡
Female genital system				(1,259)	(1,635.7)	(77)‡
Cervix-*in-situ*[d)] (1800–1809 for M-8—2)				517		
Cervix (1800–1809 excluding M-8—2)				252	455.1	55‡
Corpus uteri (1820)				609	602.0	101
Uterus, NOS (1829)				29	68.5	42‡
Ovary (1830)				308	419.0	74‡
Vagina (1840)				5	17.2	29‡
Vulva and clitoris (1842, 1843)				23	52.9	43‡
Other female genital (1810, 1832–1839, 1849)				33	19.6	168‡
Male genital system	(1,589)	(1,471.7)	(108)‡			
Prostate (1859)	1,453	1,332.5	109‡			
Testis (1869)	127	111.3	114‡			
Penis (1870)	7	22.8	31‡			
Scrotum and other male genital (1871–1874, 879[c)])	2	4.2	48			
Urinary system	(519)	(793.3)	(65)‡	(219)	(318.3)	(69)‡
Bladder (1889)	370	558.2	66‡	132	183.9	72‡
Kidney and renal pelvis (1890, 1891)	135	212.9	63‡	77	121.0	64‡

Continued . . .

TABLE 2. Continued.

Primary site (MOTNAC[a]) Topography Code No.[b])	Male			Female		
	Observed number	Expected number	SIR	Observed number	Expected number	SIR
Other urinary organs (1892–1898, 1899[c])	14	21.7	65	10	13.2	76
Eye (1900–1909)	27	23.3	116	29	24.3	119
Nervous system	(142)	(166.6)	(85)	(141)	(136.8)	(103)
Brain (1910–1919 except M-9533)						
Other nervous system (1920–1929 and 1910–1919 for M-9533)						
Endocrine system	(67)	(76.6)	(87)	(193)	(182.4)	(106)
Thyroid (1930–1931)	57	61.6	93	182	172.1	106
Other endocrine glands (1940–1949)	10	15.0	67	11	11.0	100
Lymphomas	(371)	(396.7)	(94)	(248)	(312.2)	(79)‡
Lymphosarcoma and reticulum cell sarcomas all sites[e]) for (M-9613, M-9623, M-9633, M-9643)	197	204.2	96	139	166.9	83*
Hodgkin's disease all sites[e]) for (M-9653–9683)	130	128.2	101	67	95.7	70‡
Other lymphomas including mycosis fungoides all sites[e]) for (M-9593, M-9603, M-9693–9723 M-9743, M-9753, T-1692, T-1690, T-1692, T-698,[c]) T-1699 except M-9803–9933)	44	64.6	68‡	42	50.2	84
Multiple myeloma——all sites[e]) with (M-9733)	78	93.3	84	65	80.9	80
Leukemia	(244)	(345.6)	(71)‡	(207)	(244.1)	(85)*
Lymphatic (lymphocytic) all sites[e]) for						
Acute (M-9825)	54	53.0	102	45	35.1	128
Chronic (M-9827)	47	86.7	54‡	28	54.7	51‡
Other (M-9823, M-9828–9829)	8	13.6	59‡	5	11.1	45
Other leukemia (granulocytic, monocytic, *etc.*) all sites[e]) for						
Acute (M-9805, M-9865, M-9895)	75	103.6	72‡	74	80.8	92
Chronic (M-9807, M-9867, M-9897)	38	51.5	74	34	33.8	101
Other (M-9803–9933 except those specified above under leukemia)	22	36.4	60*	21	28.9	73
Other and unknown primary (1543, 589[c]), 1730–1739[f]) except M-8723–8793 and M-8003, M-8013, M-8073 and M-8093, 1960–1969 except M-9593–9933, 1991)	255	271.8	94	248	255.3	97

[a]) Manual of Tumor Nomenclature and Coding. [b]) In certain places Morphologic code numbers must be taken in account. [c]) Third National Cancer Survey special code. [d]) If carcinoma-*in-situ* of cervix figures are available, include here, but exclude from subtotal in female genital system. [e]) All sites for the specific types of lymphomas are being placed here , even extra-nodal lymphomas (*e.g.*, lymphosarcoma of stomach). This is a different rule than was followed in TNCS. [f]) Exclude superficial skins (M-8003, M-8013, M-8073, and M-8093).

TABLE 3. Observed and Expected Numbers and SIRs for Cancers by Site Occurring among Males and Females (Utah Non-LDS, 1967–1975)

Primary site (MOTNAC[a]) Topography Code No.[b])	Male			Female		
	Observed number	Expected number	SIR	Observed number	Expected number	SIR
Total malignant neoplasms	3,943	3,708.5	(106)‡	3,822	3,314,3	115‡
Malignant neoplasm of:	(320)	(197.0)	(162)*	(96)	(65.8)	(146)‡
Lip, oral cavity, and pharynx						
Lip (1400–1409)	169	47.0	360‡	17	3.1	548‡
Tongue (1410–1419)	46	36.3	127	22	13.3	165*
Salivary glands (1420–1429)	20	13.6	147	13	10.7	121
Gum and mouth (1430–1459)	44	45.2	97	28	22.2	126
Nasopharynx (1471–1479)	5	8.7	57	5	3.0	167
Other and unsp. pharynx (1461–1469, 1480–1489, 1499)	36	46.1	78	11	12.5	88
Digestive organs and peritoneum	(838)	(943.1)	(89)‡	(745)	(739.5)	(101)
Esophagus (1500–1509)	57	50.2	114	13	15.8	82
Stomach (1510–1519)	153	139.1	110	68	76.4	89
Small intestine (1520–1529)	13	11.0	118	15	7.7	195*
Colon (1531–1539, 544[c]))	311	343.5	91	372	348.4	107
Rectum rectosigmoid junction, and anal canal (1540–1542, 1549)	144	191.9	75‡	122	131.4	93
Liver (1550)	18	31.9	56‡	10	16.1	62
Gallbladder (1560)	15	10.9	138	30	25.3	119
Other biliary (1551, 1561–1569)	10	17.3	58	11	12.2	90
Pancreas (1570–1579)	108	128.6	84	92	86.3	107
Retroperitoneum and peritoneum and unspecified digestive organs including abdomen (1580–1588, 1599)	8	7.4	108	8	5.8	138
Respiratory system	(829)	(924.6)	(90)‡	(163)	(193.9)	(84)*
Larynx (1610–1619)	118	96.7	122*	16	11.2	143
Lung (1620–1622, 623–624[c]))	684	811.5	84‡	129	174.6	74‡
Other respiratory organs including: pleura, mediastinum, intrathoracic site (1600–1609, 1630–1639, 1649[c]))	27	16.4	165*	18	7.7	234‡
Bones and joints (1700–1709)	19	11.0	173*	12	8.4	143
Connective subcutaneous and other soft tissues (1710–1719)	46	26.3	175‡	36	20.2	178‡
Melanomas——skin (1730–1739 for M-8723–8783)	89	56.7	157	111	55.6	200‡
Breast (1740–1749)	9	8.1	111‡	1,084	927.0	117‡
Female genital system				(799)	(685.8)	(117)‡
Cervix-*in-situ*[d]) (1800–1809 for M-8—2)	—	—	—	597		
Cervix (1800–1809 excluding M-8—2)	—	—	—	232	190.1	122‡
Corpus uteri (1820)	—	—	—	324	255.5	127‡
Uterus, NOS (1829)	—	—	—	11	28.6	38‡
Ovary (1830)	—	—	—	184	175.2	105
Vagina (1840)	—	—	—	9	6.7	134

Continued . . .

TABLE 3. Continued.

Primary site (MOTNAC[a]) Topography Code No.[b])	Male			Female		
	Observed number	Expected number	SIR	Observed number	Expected number	SIR
Vulva and clitoris (1842, 1843)	—	—	—	21	21.3	99
Other female genital (1810, 1832-1839, 1849)	—	—	—	18	7.6	237‡
Male genital system	(763)	(586.4)	(130)‡	—	—	—
Prostate (1859)	709	531.3	133‡	—	—	—
Testis (1869)	47	42.4	111	—	—	—
Penis (1870)	7	10.5	67	—	—	—
Scrotum and other male genital (1871-1874, 879[c])	0	1.0	0	—	—	—
Urinary system	(363)	(355.9)	(102)	(141)	(128.5)	(110)
Bladder (1889)	264	247.0	107	86	73.4	117
Kidney and renal pelvis (1890, 1891)	89	98.8	90	48	49.7	97
Other urinary organs (1892-1898, 899[c])	10	10.0	100	7	5.3	132
Eye (1900-1909)	15	10.5	143	13	9.8	133
Nervous system	(71)	(78.8)	(90)	(58)	(55.9)	(104)
Brain (1910-1919 except M-9533)						
Other nervous system (1920-1929 and 1910-1919 for M-9533)						
Endocrine system	(29)	(33.6)	(86)	(107)	(70.3)	(152)‡
Thyroid (1930-1931)	22	26.7	82	99	66.2	150‡
Other endocrine glands (1940-1949)	7	6.7	104	8	4.3	186
Lymphomas	(161)	(170.8)	(94)	(162)	(122.5)	(132)‡
Lymphosarcoma and reticulum cell sarcomas all sites[e] for (M-9613, M-9623, M-9633, M-9643)	94	91.6	103	74	68.2	109
Hodgkin's disease all sites[e] for (M-9653-9683)	49	50.4	97	53	34.5	154‡
Other lymphomas including mycosis fungoides all sites[e] for (M-9593, M-9603, M-9693-9723, M-9743, M-9753 T-1692, T-1690, T-698,[c] T-1699 except M-9803-9933)	18	28.8	63*	35	20.1	174‡
Multiple myeloma——all sites[e] with (M-9733)	44	42.0	105	39	32.4	120
Leukemia	(141)	(144.4)	(98)	(85)	(95.5)	(89)
Lympatic (lymphocytic) all sites[e] for						
Acute (M-9825)	23	22.7	101	21	13.6	154
Chronic (M-9827)	35	36.1	97	15	21.6	69
Other (M-9823, M-9828-9829)	7	5.3	132	2	4.2	48
Other leukemia (granulocytic, monocytic, *etc.*) all sites[e] for						
Acute (M-9805, M-9865, M-9895)	38	43.6	87	23	31.2	74
Chronic (M-9807, M-9867, M-9897)	24	21.0	114	18	13.4	134
Other (M-9803-9933 except those specified above under leukemia)	14	15.2	92	6	11.3	53

Continued . . .

TABLE 3. Continued.

Primary site (MOTNAC[a]) Topography Code No.[b])	Male			Female		
	Observed number	Expected number	SIR	Observed number	Expected number	SIR
Other and unknown primary (1543, 589,[c]) 1730–1739[f]) except M-8723–8793 and M-8003, M-8013, M-8073 and M-8093, 1960–1969 except M-9593–9933, 1991)	206	119.0	173‡	171	102.9	166‡

a–f) See the footnotes for Table 2.

TABLE 4. SIRs Compared to TNCS, with Comparison of Urban to Rural Residence

ICD	Site	Sex	Mormon			Non-Mormon		
			Urban[a)]	Rural[a)]	Difference[b)]	Urban[a)]	Rural[a)]	Difference[b)]
140–207	All sites	M	73†	72†	—	115†	76†	$P=0.0005$
		F	78†	67†	$P=0.02$	116†	105	—
Tobacco-related sites		M	44†	43†	—	106	59†	$P=10^{-5}$
		F	46†	33†	$P=0.01$	97	86	—
141, 143–9	Oral cavity and pharynx	M	39†	39†	—	108	54†	$P=0.0006$
		F	41†	30†	—	128	127	—
150	Esophagus	M	36†	38†	—	134	63	$P=0.02$
		F	21†	9†	—	79	97	—
161	Larynx	M	39†	37†	—	130*	90	$P=0.08$
		F	41†	28	—	157	86	—
162	Lung	M	37†	39†	—	96	54†	$P=0.00003$
		F	33†	32†	—	76†	64*	—
188	Bladder	M	69†	59†	—	122†	66†	$P=0.002$
		F	83	41†	$P=0.0006$	12	107	—
140	Lip	M	250†	370†	$P=0.004$	304†	406†	—
		F	296†	357†	—	431†	808†	—
151	Stomach	M	62†	80	—	120	81	$P=0.05$
		F	72†	78	—	95	62	—
153	Colon	M	67†	49†	$P=0.02$	104	54†	$P=0.0003$
		F	66†	53†	$P=0.09$	105	112	—
154	Rectum	M	62†	55†	—	78*	57†	—
		F	69†	68†	—	95	70	—
157	Pancreas	M	69†	69†	—	81	91	—
		F	60†	50†	—	98	144	—
173	Melanoma	M	125*	99	—	164†	129	—
		F	142†	108	—	194†	213†	—
174	Breast	F	84†	74†	$P=0.06$	121†	97	—
180	Cervix-*in-situ*[c)]	F			$P=0.05$			$P=0.04$
	invasive	F	54†	60†	—	120*	111	—
182	Corpus uteri	F	104	91	—	130†	107	—
183.0	Ovary	F	76†	65†	—	108	92	—
185	Prostate	M	108†	111†	—	145†	98	$P=0.03$
189.0–1	Kidney	M	62†	68†	—	92	85	—
		F	68†	52†	—	93	109	—

a) * $P<0.05$, † $P<0.01$ indicates significance of difference from TNCS. b) P values from modified Mantel-Haenszel chi-square test of difference by urban-rural residence. c) Not included in "all sites" category.

TABLE 5. SIRs Compared to TNCS, with Comparison of Mormons to Non-Mormons

ICD	Site	Sex	Urban			Rural		
			LDS[a]	Non-LDS[a]	Difference[b]	LDS[a]	Non-LDS[a]	Difference[b]
140–207	All sites	M	73†	115†	$P<10^{-5}$	72†	76†	—
		F	78†	116†	$P<10^{-5}$	67†	105	$P=0.00003$
Tobacco-related sites		M	44†	106	$P<10^{-5}$	43†	59†	$P=0.009$
		F	46†	97	$P<10^{-5}$	33†	86	$P=10^{-5}$
141, 143–9	Oral cavity and pharynx	M	39†	108	$P<10^{-5}$	39†	54†	—
		F	41†	128	$P<10^{-5}$	30†	127	$P=0.0002$
150	Esophagus	M	36†	134	$P<10^{-5}$	38†	63	—
		F	21†	79	$P=0.0004$	9†	97	$P<10^{-5}$
161	Larynx	M	39†	130*	$P<10^{-5}$	37†	90	$P=0.006$
		F	41†	157	$P=0.0003$	28	86	—
162	Lung	M	37†	96	$P<10^{-5}$	39†	54†	$P=0.03$
		F	33†	76†	$P<10^{-5}$	32†	64*	$P=0.009$
188	Bladder	M	69†	122†	$P<10^{-5}$	59†	66†	—
		F	83	120	$P<10^{-5}$	41†	107	$P=0.02$
153	Colon	M	67†	104	$P=0.0001$	49†	54†	—
		F	66†	105	$P<10^{-5}$	53†	112	$P=0.0004$
154	Rectum	M	62†	78*	—	55†	57†	—
		F	69†	95	$P=0.04$	68†	70	—
174	Breast	F	84†	121†	$P<10^{-5}$	74†	97	$P=0.03$
180	Cervix-*in-situ*[c]	F			$P<10^{-5}$			$P<10^{-5}$
	invasive	F	54†	120*	$P<10^{-5}$	60†	111	$P=0.008$
140	Lip	M	250†	304†	—	370†	406†	—
		F	296†	431†	—	357†	808†	—
151	Stomach	M	62†	120	$P=0.05$	80	81	—
		F	72†	95	—	78	62	—
157	Pancreas	M	69†	81	—	69†	91	—
		F	60†	98	$P=0.005$	50†	144	$P=0.002$
173	Melanoma	M	125*	164†	$P=0.06$	99	129	—
		F	142†	194†	$P=0.04$	108	213	$P=0.01$
182.0	Corpus uteri	F	104	130†	$P=0.02$	91	107	—
183.0	Ovary	F	76†	108	$P=0.003$	65†	92	—
185	Prostate	M	108†	145†	$P=0.01$	111*	98	—
189.0–1	Kidney	M	62†	92	—	68*	85	—
		F	68†	93	—	52†	109	$P=0.07$

a) * $P<0.05$, † $P<0.01$ indicates significance of difference from TNCS. b) P values from modified Mantel-Haenszel chi-square test of difference in religious status. c) Not included in "all sites" category.

Utah's comparative advantage in cancer incidence, compared to the U.S., is clearly reflected and is contributed exclusively by the LDS portion of the state population. The non-LDS have cancer rates slightly higher than that expected from the TNCS. The incidence of cancer for urban and rural areas of the state shows the expected gradient (*12–16*), but only for non-LDS men and women. A slight gradient was observed for LDS women, but virtually none for LDS men.

The cancer sites associated with cigarette use in prior studies (*17*) (lung, larynx, tongue, gum and mouth, esophagus, and bladder), demonstrate the largest

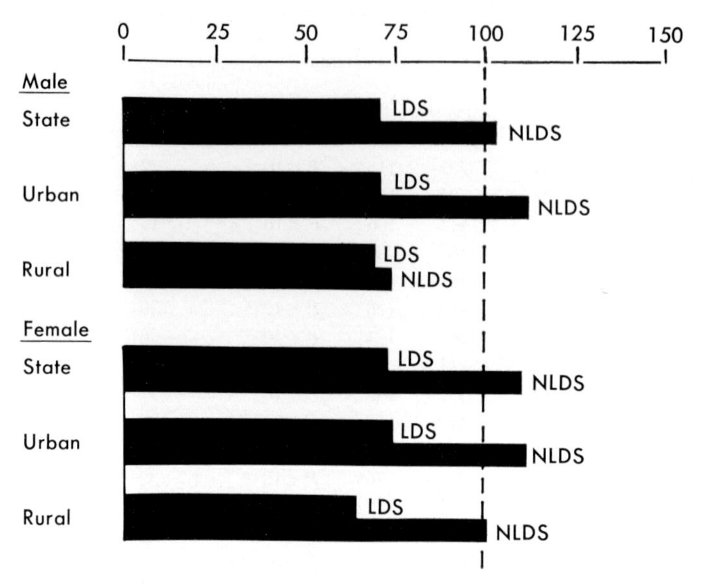

FIG. 1. SIRs for men and women by religion and residence—all sites.

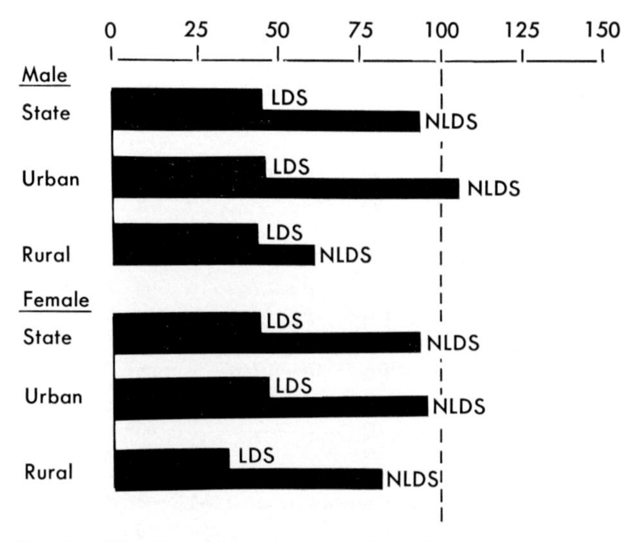

FIG. 2. SIRs for men and women by religion and residence smoking—mouth, pharynx, lung, esophogus, bladder.

difference, with the incidence in LDS men and women about 55% less than that of the U.S. Non-LDS men and women have SIRs about 7% below that of the United States; primarily, because of a deficit of lung cancer cases.

For LDS men, there was no difference in cancer risk for the tobacco related sites by place of residence, but for LDS women there was a small (8%) but significant difference. For the non-LDS, large urban-rural gradients were present for all cancers related to tobacco.

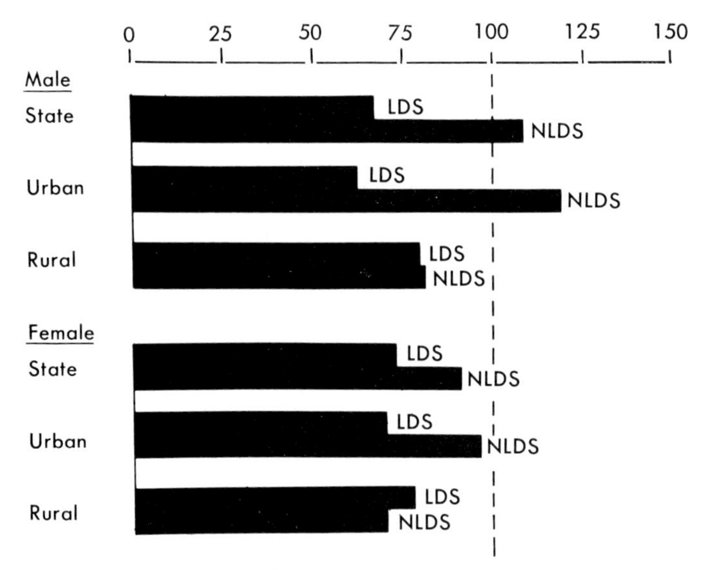

FIG. 3. SIRs for men and women by religion and residence—stomach.

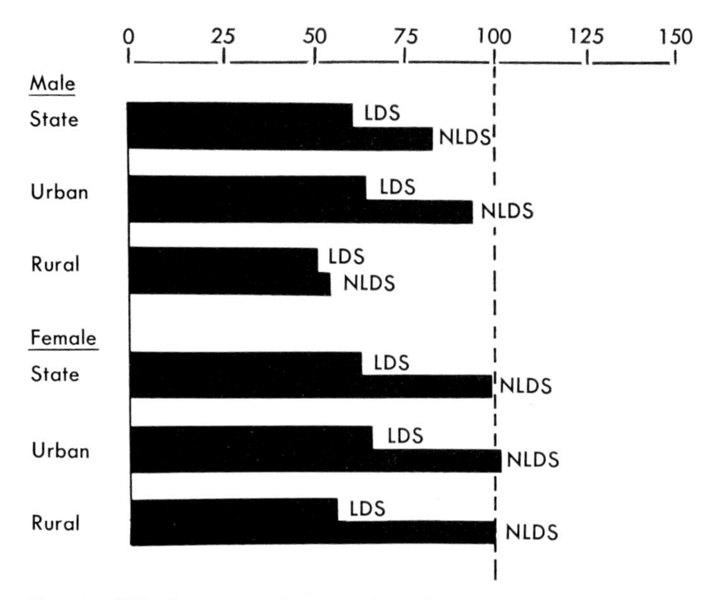

FIG. 4. SIRs for men and women by religion and residence—colon-rectum.

For digestive tract sites of stomach, pancreas, and colon and rectum, significant differences between the two religious groups were present, with the LDS substantially lower than the non-LDS. Again, these same sites (particularly, colon) demonstrated an urban-rural difference for the two religious groups, with the exception of non-LDS women. Interestingly, among the LDS we found significantly lower rates of cancers of the colon and stomach: a pattern not generally seen elsewhere in the world (*18*).

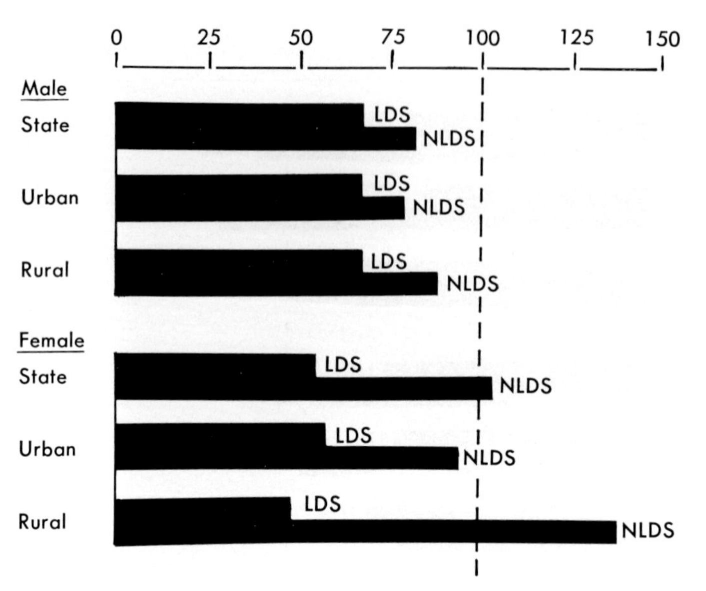

FIG. 5. SIRs for men and women by religion and residence—pancreas.

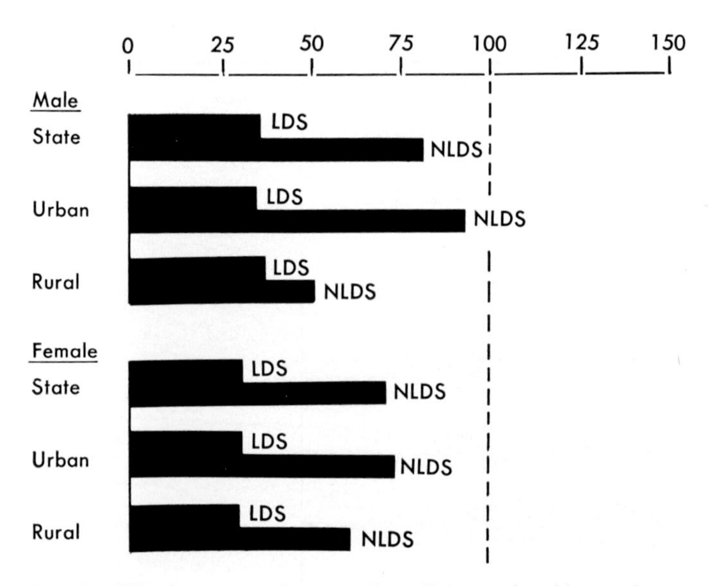

FIG. 6. SIRs for men and women by religion and residence—lung.

Cancer of the female breast showed a significant difference between the two religious groups and demonstrated an urban-rural gradient in both groups. Cancer of the uterine cervix demonstrated the largest difference in a non-tobacco related site and, for LDS women, had slightly lower urban incidence than rural incidence. The reverse was not, however, found for the non-LDS.

Finally, cancers of the prostate and lip and malignant melanoma both demonstrated rates above expectation. Tables 6–9 contain information concerning different

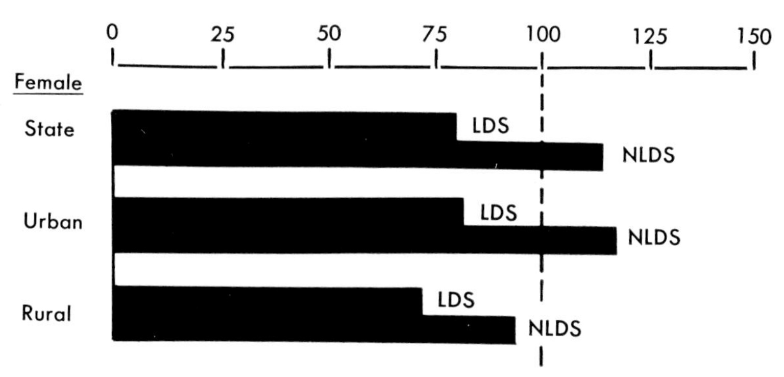

FIG. 7. SIRs for women by religion and residence—breast.

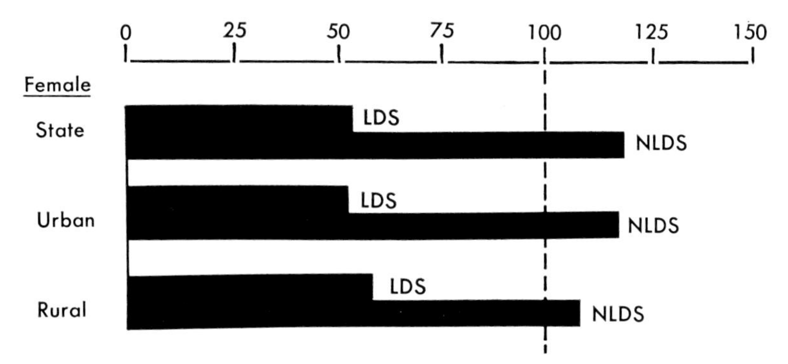

FIG. 8. SIRs for women by religion and residence—cervix uteri (invasive).

TABLE 6. Educational and Occupational Factors

Factors	Male		Female	
	LDS	Non-LDS	LDS	Non-LDS
Education				
% high school				
Colon	16	23	13	12
Cervical			18	17
Total	16	23	15	14
% high school graduate				
Colon	21	26	35	44
Cervical			37	27
Total	21	26	36	38
% some college				
Colon	35	30	37	22
Cervical			35	39
Total	35	30	36	28
% college graduate or professional				
Colon	27	21	16	22
Cervical			15	17
Total	27	21	16	20
Occupation				
$\bar{X}$ (scale 1 (high) to 12 (low))				
Colon/total	4.3	4.5	4.9	5.4

TABLE 7. Use of Tobacco, Coffee, Tea, and Alcohol among Control Groups in Utah by Sex and Religion

Factors	Male		Female	
	LDS	Non-LDS	LDS	Non-LDS
Smoking				
% current smokers				
Colon	10	35	5	38
Cervical			10	45
Ovarian			7	45
Total	10	35	7	41
$\bar{X}$ age started smoking				
Colon/total	20	18	22	19
$\bar{X}$ years smoked				
Colon	18	20	12	17
Cervical			24	18
Total	18	20	17	17
% smoked for at least 1 yr.				
Colon/total				
$\bar{X}$ cigarettes smoked in 1 day				
Colon/total	23	26	12	20
Coffee				
% drank since age 30				
Cervical/total			42	87
Tea				
% drank after age 30			21	59
Drinking				
% currently drink				
Colon/total	16	64	17	60
% drank liquor single age 30				
Cervical/total			18	52
% drank beer since age 30				
Cervical/total			18	30
% drank wine since age 30				
Cervical/total			15	48

risk factors in the LDS population. These tables contain a proportion of respondents reporting Church membership, and a proportion of the smokers of each of the control groups, the information on educational level, marital status, reproductive and sexual factors.

Significant differences between the two religious groups and their reported use of alcohol, tobacco, coffee, and tea are striking, as are differences in marital status and reproductive factors.

DISCUSSION

The validity of the data in the Cancer Registry has been discussed in a number of prior publications (*7*, *8*, *19*).

TABLE 8. Sexual and Reproductive Factors

Factors	Female	
	LDS	Non-LDS
$\bar{X}$ age at first intercourse		
Cervical/total	20	20
% with intercourse with more than one partner		
Cervical/total	25	61
$\bar{X}$ Number of pregnancies		
Colon	3.6	2.9
Cervical	4.5	2.7
Ovarian	4.0	3.4
Total	4.0	3.0
$\bar{X}$ age at birth of first child		
Colon	21.7	23.1
Cervical	22.0	22.5
Ovarian	21.7	20.3
Total	21.8	22.3
% Ever used birth control pills		
Colon	29	30
Cervical	48	71
Ovarian	42	72
Total	37	50
$\bar{X}$ Age began using birth control pills		
Cervical	25	27
Ovarian	27	27
Total	26	27

TABLE 9. Reproductive Factors

Factors	Female	
	LDS	Non-LDS
% miscarriage or stillbirth		
Ovarian/total	11	25
$\bar{X}$ age of menopause		
Ovarian/total	44	39
% with hysterectomy		
Ovarian/total	17	40
% who examine breast at least once a month		
Ovarian/total	63	19

1. Cancers Associated with Cigarette and Alcohol Use

Much of the difference in cancer incidence between Utah and the remainder of the United States occurs because of the strikingly lower rates in the LDS population at these cancer sites. Using the formulas for attributing an etiologic fraction to an incidence rate (*20*), and using the state incidence rate and the control group data on smoking to estimate the incidence in the unexposed, we find that for lung cancer the predicted incidence explains 96% of the difference observed between the LDS and the state rates. However, applying the same formula to cancer of the

APPENDIX TABLE 1. Utah Population, Sample Size, Weights, and Weighted Sample Size for Colon, Cervical, and Ovarian Studies

Age	Utah population	%	Colon			Cervical			Ovarian		
			N	Wt. factor	*N* (wt.)	*N*	Wt. factor	*N* (wt.)	*N*	Wt. factor	*N* (wt.)
24–34	129,961	27.2	84	5.86	492	128	1.23	158	6	14.17	85
35–39	53,558	11.2	48	4.23	203	65	1.00	65	3	11.67	35
40–44	53,990	11.3	46	4.44	204	37	1.76	65	3	11.67	35
45–49	52,511	11.0	48	4.15	199	17	3.77	64	18	1.94	35
50–54	48,472	10.2	51	3.59	183	19	3.11	59	9	3.56	32
55–59	41,226	8.6	60	2.60	156	9	5.56	50	23	1.17	27
60–64	34,979	7.3	72	1.83	132	12	3.50	42	23	1.00	23
65–69	27,620	5.7	69	1.48	102	5	6.60	33	8	2.25	18
70–80	35,681	7.5	135	1.00	135	13	3.31	43	16	1.50	24
Total	477,998	100.0	613		1,806	305		579	109		314

APPENDIX TABLE 2. Utah Population 1970

	Male	Female	Male and female
Utah	540,575	553,722	1,094,297
Utah LDS	380,821	400,914	781,735
% LDS	70.4	72.4	71.4
LDS birthrate approximately 30/1,000			

larynx leaves a residual of about 22% (LDS rates are 22% less than predicted if smoking were the single risk factor involved and using a relative risk of 5). This finding, however, is compatible with the reported synergistic association between alcohol and tobacco for cancers of the oral cavity, larynx, and esophagus (*21*).

The urban-rural cancer risk reported in previous studies does not hold for Mormon men, but is quite striking in non-Mormon men and women. Again, past explanations have attempted to implicate air pollution as causal (*22*, *23*). The paucity of any gradient in Mormons, particularly males, is counter to this explanation. Unfortunately, our case-control studies did not contain an adequate number of rural residents to permit precise estimates of smoking for Mormon and non-Mormon by rural areas.

The issue of whether smoking-related cancer explains the differences between Utah and the remainder of the United States can partly be answered by removing these sites from consideration and computing SIRs from remaining observed and expected numbers of cases. For Mormon men, an SIR of 87.7 is obtained, and for LDS women, 79.5, if the tobacco-related sites are removed. If we further remove the two sites which were found to be significantly higher for Mormon men (lip and prostate), an SIR of 75.7 is obtained. This contrasts with SIRs for non-Mormon men and women of 113.6 and 117.6. If we remove cancers of the lip and prostate from consideration in non-Mormon men, we obtain an SIR of 101.3. These differences are not explained by tobacco, but must have some other causal factor or factors.

2. Cancers not Associated with Smoking

Differences between the LDS and non-LDS populations persist for cancers at a number of sites not associated with cigarette smoking. These include cancers of the stomach, colon, breast, ovary, and uterine cervix. For cancer of the uterine cervix, a more complex pattern emerges between urban and rural risks, with rural LDS higher than urban. This may be due to the possible effect of better acceptance of screening in the urban areas. Again, applying the formula for attributable risk in a population to cervical cancer data, some interesting results emerge. If we assume that sexual intercourse with multiple sexual partners is a risk factor (with a risk of 4.7 from our control group), and apply these data to the LDS and non-LDS populations, and then predict the expected LDS incidence from the observed frequency of intercourse with more than one sexual partner, we find that 92% of the LDS rates would be explained by this factor. However, if we used the smaller risk ratio of 2, then only about 77% of the LDS rates would be explained by this factor.

For cancer of the female breast, significant differences were present between LDS and non-LDS. This disease does show an urban-rural gradient for both the LDS and non-LDS in the state.

Finally, cancer of the lip and prostate and melanoma all show incidence significantly above U.S. expectation. What the reasons for these excesses are, is not clear.

Thus, factors in lifestyle relating to use of alcohol, tobacco, and sexual behavior, appear to have a significant modifying behavior on the risk of cancer for a subset of the United States population either compared to other populations in the United States or to a similar group sharing the same environment in the state of Utah.

REFERENCES

1. Mason, T. J. and McKay, F. W. U.S. Cancer Mortality 1950–1969, DHEW Publication No. (NIH) 74–615, Government Printing Office, Washington, D. C., 1974.
2. Spicer, J. O. and Gustavus, S. O. Mormon fertility through half a century: Another test of the Americanization. Soc. Biol., *21*: 70–76, 1974.
3. Hardy, D. R. Social origins of American scientists and scholars. Science, *185*: 497–506, 1974.
4. Gardner J. W. and Lyon, J. L. Low incidence of cervical cancer in Utah. Gynecol. Oncol., *5*: 68–80, 1977.
5. Young, J. L., Jr., Asire, A. J., and Pollack, E. S. (eds.). Cancer Incidence and Mortality in the U.S. 1973–1976, DHEW Publication No. (NIH) 78–1837, U.S. Dept. of HEW, National Cancer Institute, Bethesda, Md., 1978.
6. Lyon, J. L., Gardner, J. W., Klauber, M. R., and Smart, C. R. Low cancer incidence and mortality in Utah. Cancer, *39*: 2608–2618, 1977.
7. Lyon, J. L., Wetzler, H. P., Gardner, J. W., Klauber, M. R., and Williams, R. R. Cardiovascular mortality in Mormons and non-Mormons in Utah, 1969–1971. Am. J. Epidemiol., *108*: 357–366, 1978.
8. Lyon, J. L., Klauber, M. R., Gardner, J. W., and Smart, C. R. Cancer incidence

in Mormons and non-Mormons in Utah, 1966–1970. N. Engl. J. Med., *294*: 129–133, 1976.
9. Cutler, S. J. and Young, J. L. (eds.). Third National Cancer Survey: Incidence Data, DHEW Publication No. (NIH) 75–787, 1975.
10. Bailer, J. C., III and Ederer, F. Significance factors for the ratio of a Poisson variable to its expectation. Biometrics, *20*: 639–643, 1964.
11. Mantel, N. Chi-square tests with one degree of freedom: Extensions of the Mantel-Haenszel procedure. J. Am. Stat. Assoc., *58*: 690–700, 1963.
12. Haenszel, W., Marcus, S. C., and Zimmerer, E. T. Cancer Morbidity in Urban and Rural Iowa, Public Health Monograph No. 37, PHS Publication No. 426, Government Printing Office, Washington, D.C., 1956.
13. Clemmsen, J. and Neilsen, A. Comparison of age-adjusted cancer incidence rates in Denmark and the United States. J. Natl. Cancer Inst., *19*: 989–998, 1957.
14. Levin, M. L., Haenszel, W., and Carroll, B. E. Cancer incidence in urban and rural areas of New York State. J. Natl. Cancer Inst., *24*: 1243–1257, 1960.
15. Haenszel W. and Dawson, E. A. A note on mortality from cancer of the colon and rectum in the United States. Cancer, *18*: 265–272, 1972.
16. Buell, P. and Dunn, J. E. The relative impact of smoking and air pollution on lung cancer. Arch. Environ. Health, *15*: 291–297, 1967.
17. The Health Consequences of Smoking, DHEW publication No. (HMS) 73–8704, Government Printing Office, Washington, D. C., 1973.
18. Waterhouse, J., Muir, C., Correa, P., and Powell, J. (eds.), Cancer Incidence in Five Continents, vol. III, IARC Scientific Publication No. 15, International Agency for Research on Cancer, Lyons, 1976.
19. Lyon, J. L., Gardner, J. W., and West, D. W. Cancer in Utah, risk by religion and place of residence. Proceedings of the Conference on Cancer in Low Risk Populations.
20. MacMahon, B. and Pugh, T. Epidemiology Principles and Methods, pp. 274–275, Little Brown and Company, Boston, 1970.
21. Rothman, K. J. Alcohol. *In*; J. F. Fraumeni, Jr. (ed.), Persons at High Risk of Cancer, pp. 139–150, Academic Press, New York, 1975.
22. Shy, C. M. Lung cancer. *In*; A. J. Finkle and W. C. Duel (eds.), Urban Environment Review and Clinical Applications of Pollution Research, pp. 3–35, Publishing Sciences Group Incorporated, Acton, Mass., 1976.
23. Dean, G. Lung cancer and bronchitis in Northern Ireland 1960–61. Br. J. Med., *1*: 1506–1514, 1966.

GENETIC AND ENVIRONMENTAL FACTORS IN EXPERIMENTAL AND HUMAN CANCER,
H. V. GELBOIN ET AL. (EDS.), JAPAN SCI. SOC. PRESS, TOKYO, PP. 291–302, 1980

Genetics of Neoplasia in a Human Isolate

Alice O. MARTIN,*[1] Judith K. DUNN,*[2] Joe Leigh SIMPSON,*[1] Sherman ELIAS,*[1] Gloria E. SARTO,*[1] Bion SMALLEY,*[2] Carolyn L. OLSEN,*[3] Sam KEMEL,*[4] Michael GRACE,*[5] and Arthur G. STEINBERG*[6]

Department of Obstetrics and Gynecology,[1] and Cancer Center,*[2] Northwestern University Medical School, Chicago, Ill. 60611, U.S.A., Albany Medical College, Albany, N.Y., U.S.A.,*[3] Manitoba Cancer Foundation, Winnipeg, Manitoba, Canada,*[4] W. W. Cross Cancer Institute, Edmonton, Alberta, Canada,*[5] and Department of Biology, Case Western Reserve University, Cleveland, Ohio, U.S.A.*[6]*

Abstract: Investigations in human isolates represent one approach to the delineation of the complex interactions of genetic and environmental factors in oncogenesis. We have ascertained approximately 200 cases of cancer in the Hutterites, a large religious isolate which originated in Europe in 1528 and currently resides in the United States and Canada. This population was selected for genetic studies because it is inbred, available pedigree information extends to the 1700's, a small number of founders account for the current gene pool, family sizes are large, and all members share similar environments because they live on communal farms (colonies).

Our objectives were to: (1) compare cancer patterns of Hutterites with those in non-inbred control populations; (2) search for recessive gene control over oncogenesis by comparing inbreeding levels of individuals with cancer to controls; (3) detect and analyze familial aggregates of cancer, and (4) delineate genetic from environmental factors affecting cancer risk.

Results of our analysis indicate: (1) the Hutterites are a " low risk " population based on cancer related mortality 1965–1977, compared to 1970 U.S. Whites; this appears to be primarily due to a deficit of lung cancer, consistent with expectations in a non-smoking population; (2) recessive allele control may exist over some forms of cancer, notably childhood leukemia; and (3) familial aggregation most frequently consists of cancers of the digestive system and breast, usually adenocarcinomas.

Genetic factors, environmental factors, and genetic-environmental interactions may be invoked to explain differences in cancer rates and risks among human populations. However, establishment of etiological links between any of these factors and cancer is arduous.

As we have heard in previous sessions of this conference, impressive associations have been uncovered between rare genetic diseases produced by the action of recessive alleles and predisposition to cancer (*1–3*). In addition, effects of recessive alleles on cancer are well established in experimental organisms. Another method of investigating the role of recessive alleles in oncogenesis is to examine the association between inbreeding levels and cancer patterns, in inbred human populations (*4*). In these groups, marriage between relatives (consanguineous matings) increases the chance that two alleles in the same condition are united in an individual. We say that an individual is homozygous by descent, that is, both alleles (conditions of a gene) are the same at a particular genetic locus because they originated in a common ancestor. Homozygosity allows the expression of recessive alleles, and consanguinity increases the chance of homozygosity.

Several difficulties accompany studies in human inbred populations: (1) levels of inbreeding are often low compared to those generated in experimental populations; (2) population sizes are often small, consequently there are few cases of cancer to analyze; and (3) cultural factors peculiar to these populations may complicate such analyses.

Counterbalancing these problems is the availability of extensive family and medical histories of members of some inbred populations, and a range of inbreeding levels for many individuals. This is particularly true of the Hutterites, a large religious isolate we are investigating (*5*).

Other indications that genetic factors play a role in oncogenesis are the occurrence of different cancer patterns among racial and ethnic groups, and longitudinal transmission of some forms of cancer over several generations in certain families. These patterns could be caused by dominant alleles or environmental factors common to family members. Because the Hutterite population as a whole shares a relatively uniform environment and lives communally, familial aggregates detected in this population are more likely to be due to genes shared by family members, than in the general population where families differ from each other by both genetic and environmental background.

In summary, the objectives of our investigation are: (1) to compare the patterns of cancer in the Hutterites with those in various other populations, specifically with regard to incidence and mortality; (2) to assess the role of recessive alleles in oncogenesis; and (3) to look for familial aggregates of cancer.

Although correlations with environmental factors were not sought directly, the types and frequencies of cancer were interpreted relative to Hutterite lifestyle.

This project is still in progress; therefore, complete results are yet forthcoming.

The Population

The Hutterite population is a large religious isolate established in 1528 in Moravia. Religious tenets of the group prohibited participation in any type of military activity. This led to persecution and necessitated migrations. Consequently, population size fluctuated during the first 200 years of the group's existence. The current population may be traced to the late 1700's when the Hutterites resided in Russia. The oldest geneological records available to us extend to that period (*5–7*).

Between 1874 and 1877 approximately 900 members of the sect immigrated

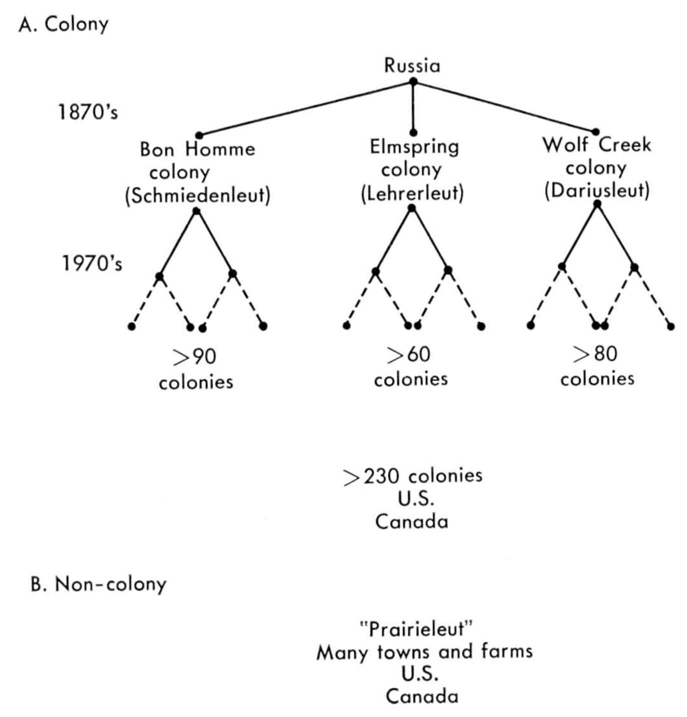

FIG. 1. Hutterite population structure.

from Russia to the U.S. to an area which is now southeastern South Dakota. It appears that less than half of these settled in three colonies (communal farms). The remainder took advantage of the Homestead Act and became independent farmers. They are referred to as the "Prairieleut" and have not yet been studied in detail. Descendents of the three original colonies comprise three subpopulations, the Schmiedenleut (S-leut), Lehrerleut (L-leut), and Dariusleut (D-leut) (Fig. 1). More than 200 colonies now exist in the United States and Canada, a result of a pattern of recurrent formation of daughter colonies from preexisting colonies. The total Hutterite population now numbers more than 25,000 individuals. However, this represents the multiplication of only a small number of independent genomes. For 2 of the 3 subgroups (referred to as "leut"), the S- and L-leut combined, there were only a maximum of 90 founders or minimum ancestors (*8*). These ancestral genomes have been multiplied extensively, resulting in approximately 15,000 contemporary members (as of 1977). Further reflection of this paucity of independent genomes is the existence of only 15 surnames and a relatively high coefficient of inbreeding (*5, 7*). The mean coefficients of inbreeding ($\bar{F}$) are greater than that expected for the offspring of second cousins, but less than that of the offspring of first cousins once removed ($\bar{F}=0.0211$ for the S-leut and $\bar{F}=0.0255$ for the L-leut).

The Hutterites are one of the most inbred large populations of European origins with accurate records because of the meticulous care with which colony preachers have recorded information on births, marriages, and deaths. Genetic analysis is also facilitated by the occurrence of large family sizes. Individuals are usually married between 20 and 25 years of age, and divorce is not practiced. All forms of birth control have traditionally been rejected by the Hutterites. Prenuptial intercourse is thought to be rare because of religious prohibition.

Population distributions computed or estimated at yearly intervals show that approximately 50% of the population is less than age 15 (*5*, *9*).

Particularly relevant to the analysis of traits that may have both a genetic and environmental component are observations that the Hutterites share a relatively uniform environment, particularly within a given colony, the basic social and economic subunit. The only economic base is agriculture; colony life being directed towards maintenance of the farm and family. Each colony is lead by a preacher. Other positions include a farm foreman ("Boss"), a head cook, a German teacher, and other specialized positions, for example, carpenter, poultry supervisor, and shoemaker. Modern farm methods and machinery are used. Newer or remodeled colonies have indoor sanitary facilities; however, outhouses are not uncommon. Very little time is spent outside the colony. Excursions are confined to expeditions for shopping, medical care, or social visits to other colonies. Children attend school on the colonies until minimal legal educational requirements are fulfilled, after which they assume adult responsibilities at age 15.

Food is prepared in a common kitchen and individuals usually do not eat meals in family units, that is, men and women usually sit at separate tables. Children either eat together in a dining room of their own or if such localities are not available, they eat in the main dining room at a separate time. These eating patterns lessen the possibility that familial aggregates of cancer result from familial dietary habits rather than common genes. However, genetic-environmental interaction may exist because family members tend to live on the same colonies, and there may be intercolony environmental differences.

Based upon field trip observations, Hutterite diet appears to be varied and reflects both their Germanic origins and agricultural lifestyle. Frequent foods include meat or poultry, usually fried or boiled; potatoes; soups with noodles; bread or rolls; butter; unpasteurized milk; coffee and dessert. Until recently, food was produced and prepared almost exclusively on the colony, but commercially produced items such as peanut butter, ice cream, cookies, and donuts are now consumed. Obesity is common, particularly among adult women.

Consumption of alcohol is generally permitted, but appears to be moderate. Beer and homemade wine are served during celebrations, but generally not at routine meals. Smoking is actively discouraged, and hence, appears to be rare. Members of our group have never observed smoking during many colony visits.

Case Ascertainment and Confirmation

Two procedures were used to ascertain cases of cancer: (1) record searches to detect individuals with one of the 15 Hutterite surnames; (2) field trips to colonies in order to obtain family and medical histories. ICDA-8 (*10*) was used as the anatomic classification system; and histologic classifications were made according to the systemized nomenclature of pathology (SNOP) (*11*). Because the population resides in the U.S. and Canada, data were obtained from several registries, many hospitals and death certificates. Except where noted, only cases that were confirmed to be cancer and that occurred in communal Hutterites are included in the analysis.

RESULTS

Our first objective was to compare cancer patterns in the Hutterites with those in other populations.

The total number of cases ascertained in Hutterites is small (177).* However, the size of the population surveyed is also small. The total population size of the S- and L-leut in Manitoba, South Dakota, and Alberta included in our survey is approximately 12,562. Furthermore, as previously mentioned, 50% of the population is less than 15 years old, that is, not in a high cancer risk age bracket.

The sites most frequently involved in all subgroups of the population are organs of the digestive system, notably stomach cancer. The numbers of cases of colon cancer are approximately equal to those of rectal cancer.

Breast cancer and leukemia are the next most frequent types. For cases for which diagnosis dates were available, it was noted that most breast cancer occurred in women over age 44.

The leukemias appear to fall into two groups based on age at death: (1) 18 years or younger (childhood leukemias); (2) greater than 50 years (adult leukemias). The possible increased risk of childhood leukemia may be related to inbreeding, as discussed in a subsequent section.

Cancers of the male and female genital systems were also found (prostate, uterus, ovary). Very few cases were found at sites not described above.

1. *Mortality patterns*

Only deaths occurring in the interval beginning with 1965 were used to compute standardized mortality ratios. Use of this interval seemed appropriate for comparison with the standard, 1970 U.S. Whites. Moreover, registry data and death certificate information prior to 1965 were incomplete. Therefore, cases ascertained prior to that date are not included in the mortality analysis.

Compared to expectations using 1970 U.S. Whites as a standard, there were significantly fewer deaths observed than expected ($P \leq 0.01$) (*12*), due principally to fewer deaths due to lung cancer among males than expected (one observed compared to 13 expected). This deficit is consistent with expectations in a nonsmoking population, although this hypothesis was not tested directly.

Childhood leukemia mortality showed significant increases over expectation in several categories (classified by leut$\supset$sex$\supset$region$\supset$type). An excess of rectal cancer was significant only for L-leut males.

Deaths related to uterine cancer are significantly elevated in the S-leut, but

* 124 excluding D-leut.

not in the L-leut. Only a single case of cervical cancer (adenocarcinoma) was identified, and that woman is still alive.

Because Hutterite women are obese, endometrial cancer might be increased, but this factor may be counteracted by the high parity and infrequent use of exogenous estrogens. The lack of apparent increase in breast cancer mortality may be due to counteracting effects of high parity which according to some reports is believed to decrease risk, and relatively older age at marriage, which is believed to increase risk. Breast feeding appears to be routine, at least for the first 3–4 months after birth. The paucity of cervical cancer is consistent with current etiological hypothesis showing relationships between cervical cancer and early age at first intercourse and multiple sexual partners. Neither factor is characteristic of this population.

In summary, cancer mortality patterns suggest that the Hutterites are a "low risk" population, and that this decrease may be explained by the lifestyle of the group.

2. *Effects of inbreeding*

Our second objective was to assess the role of recessive alleles in oncogenesis. The coefficient of inbreeding (F) has been calculated for individual Hutterites by computerized pedigree analysis. (Examples of numerical values of F would be 0 in a non-inbred population, $F \cong 0.06$ for the offspring of a first cousin mating, and $F=0.25$ for the offspring of parent-child and sib-sib matings.)

The level of inbreeding of individuals with cancer has been compared with two sets of control frequencies; total population values and individually matched controls. First, the mean population coefficient of inbreeding was computed separately by sex within each anatomic and histologic site,* and compared with the mean coefficient of inbreeding of Hutterites of the same leut and sex born within the same decade (from 1900 to 1970). Second, each case was compared to values for 2 sets of control individuals matched for leut, sex, birth date ± 2 years, and last recorded residence. We are in the process of sampling more sets of individually matched control samples.

Comparison with control means from the population suggests random fluctuation and no strong overall inbreeding effect (Tables 1 and 2). Considering only categories with several cases, and for which there is consistency between leut and sexes, only childhood leukemias exhibit increased levels of inbreeding. Generalizations for classifications with few or single cases seem unwarranted; however, an extremely high inbreeding value ($F \cong 0.1$) associated with dermatofibrosarcoma protuberans, a rare childhood cancer, is worth noting.

Comparisons of the mean coefficients of inbreeding of the individually matched controls and the cases show the overall mean of cases to be slightly higher than that for one of the two sets of controls thus far drawn, and identical with that of the other ($\bar{F}$ cases$=0.015$, $\bar{F}$ control $1=0.010$, $\bar{F}$ control $2=0.015$). Analyzing the

* And weighted by per cent completeness of pedigrees (*7*).

TABLE 1. Inbreeding Analysis: S-leut (Manitoba and South Dakota)

	Sex	Case $\bar{F}$	Case N	Population $\bar{F}$	Population N
Total digestive	M	0.0139	14	0.0193	1,889
	F	0.0152	15	0.0175	1,177
Pancreas	M	0.0242	3	0.0157	331
Stomach	M	0.0134	8	0.0174	1,156
	F	0.0082	4	0.0121	335
Colon	F	0.0259	7	0.0193	842
Rectal	M	0.0101	3	0.0143	258
	F	0.0061	2	0.0138	280
Colo-rectal	F	0.0215	9	0.0180	1,122
Liver-gallbladder	M	0.0000	1	0.0165	79
Small intestine	F	0.0019	1	0.0139	192
Breast[a]	F	0.0109	14	0.0081	1,347
<45	F	0.0239	4	0.0182	753
$\geqq 45$	F	0.0091	8	0.0150	727
Leukemia—total	M	0.0230	4	0.0218	783
	F	0.0328	3	0.0221	1,724
Leukemia ($\leqq 18$)	M	0.0375	2	0.0230	654
	F	0.0328	3	0.0221	1,724
Leukemia (>50)	M	0.0077	2	0.0157	129
Male genital (prostate)	M	0.0072	4	0.0148	337
Female genital					
Uterus	F	0.0168	4	0.0172	671
Ovary	F	0.0429	1	0.0200	724
Cervix uteri	F	0.0083	1	0.0183	479
Hodgkins	M	0.0105	2	0.0194	467
Connective[b]	M	0.0976	1	0.0221	1,102
Skin	M	0.0082	3	0.0160	208
	F	0.0371	1	0.0087	55
Urinary (kidney)	M	0.0039	1	0.0165	79
	F	0.0410	1	0.0226	614
Oral cavity	M	0.0000	1	0.0156	202
Larynx	F	0.0039	1	0.0137	8
Trachea, bronchus, lungs	M	0.0234	2	0.0190	546
	F	0.0058	1	0.0087	55
Reticulum cell sarcoma and lymphosarcoma	M	0.0176	2	0.0213	1,569
Unknown	M	0.0321	2	0.0211	1,304
	F	0.0012	2	0.0017	133
Melanoma	F	0.0190	1	0.0176	302
Brain	M	0.0234	1	0.0157	129
	F	0.0253	1	0.0200	724
Other parts—nervous system	M	0.0122	1	0.0195	728

[a] Age at diagnosis is not known for 2 cases. [b] Dermatofibrosarcoma protuberans.

leut separately, the case mean fell between that of controls for the L-leut, and was slightly higher than that of the controls for S-leut. The only significant difference between means for cases within sites and their respective controls was for leukemics

TABLE 2. Inbreeding Analysis: L-leut (Alberta)

Cancer type	Sex	Case $\bar{F}$	Case N	Population $\bar{F}$	Population N
Total digestive	M	0.0059	8	0.0141	317
	F	0.0178	4	0.0149	324
Stomach	M	0.0044	3	0.0122	163
	F	0.0146	2	0.0152	222
Colon	M	0.0009	1	0.0121	69
	F	0.0048	1	0.0143	102
Rectal	M	0.0080	4	0.0148	223
Colo-rectal	M	0.0078	5	0.0148	223
Liver-gallbladder	F	0.0332	1	0.0129	69
Small intestine	F	0.0078	1	0.0129	69
Breast (≧45)	F	0.0070	5	0.0147	362
Leukemia (≦18)	F	0.0205	1	0.0274	656
	M	0.0468	2	0.0269	763
Leukemia (>50)	M	0.0096	2	0.0177	390
Male genital (prostate)	M	0.0041	4	0.0123	192
Female genital					
Uterus	F	0.0019	2	0.0225	474
Dysgerminoma	F	0.0205	1	0.0182	324
Cervix uteri	F	0.0019	1	0.0191	243
Hodgkins	F	0.0126	1	0.0286	687
Connective	M	0.0200	1	0.0160	154
Skin	M	0.0000	1	0.0130	29
	F	0.0190	4	0.0179	669
Urinary (bladder)	M	0.0058	1	0.0121	69
Oral cavity	M	0.0195	1	0.0188	236
	F	0.0078	1	0.0124	38

(childhood plus adult, S- and L-leut combined) and control 1. However, the two control groups in this category also differed significantly, complicating the interpretation of this result (t-tests were used for these comparisons).

An additional significant difference emerged when the S-leut was separated

TABLE 3. Inbreeding Analysis: Childhood Leukemics

Age at diagnosis	Coefficient of inbreeding (F)	F case/F control[b]	
		1	2
1	0.0605	4.45	2.59
8	0.0468	2.28	16.14
17	0.0468	16.14	0.69
18	0.0439	5.63	2.81
3	0.0312	1.28	0.86
9[a]	0.0234	6.00	0.75
4	0.0205	2.11	0.44
14[a]	0.0127	1.63	0.33

[a] For 2 cases, only age at death is known. [b] 11/16 ratios are >1; cases are greater than all of control set 1.

by region (South Dakota, Manitoba). Breast cancer cases in South Dakota showed significantly higher F's than either control group. Whereas the numbers are small, it is intriguing that the breast cancer cases in South Dakota are all premenopausal, in fact, are the only such in the total sample. A pair of sibs is included in this group. Furthermore, in individual comparisons of cases of premenopausal breast cancer ($N=4$) with each of 2 controls, the inbreeding levels for cases are always higher.

Ratios of inbreeding levels for cases of childhood leukemia compared to each of their two controls are shown in Table 3. These results suggest an increased risk of childhood leukemia associated with inbreeding. This association is being investigated further.

Because the F distribution is skewed, we are also investigating distribution of cases and controls in the "tails" of the F distribution. Nine per cent of cases have F's $\geqq 0.04$ compared to 2% in control 1 and 5% in control 2. Cases in this "tail" are mostly childhood cancers (leukemia, dermatofibrosarcoma protruberans, kidney), or characterized by unusually early age of onset.

In summary, there appears to be little effect, or possibly a protective effect, of genetic homozygosity (as measured by the coefficient of inbreeding) and overall cancer risk. Thus, the relatively low risk of cancer many not be solely due to Hutterite lifestyle, but also to the genetic structure of the population.

The only consistent inbreeding effect appears to be on the risk of childhood leukemia, and possibly on premenopausal breast cancer. The leukemia results are consistent with the effects of consanguinity detected in Japan (*13*). As will be seen in the following section, some of the types of cancer, which exhibit vertical familial transmission suggestive of dominant inheritance (digestive system, postmenopausal breast cancer) are generally those categories for which mean coefficients of inbreeding were lower for cases than for controls, rendering further support for the role of dominant alleles in producing these cancers.

3. Familial aggregates

An example of a family exhibiting many affected members in sequential generations is shown in Fig. 2 to illustrate the nature of our results. This D-leut pedigree is notable for its aggregation of stomach cancer. For example, in generation 2, 6 of the 11 are affected, and 5 of these 6 sibs have stomach cancer (2 of these are confirmed by medical records, 3 are anecdotal.)

A list of sibship, parent-sibship, and parent-child aggregates may be found in

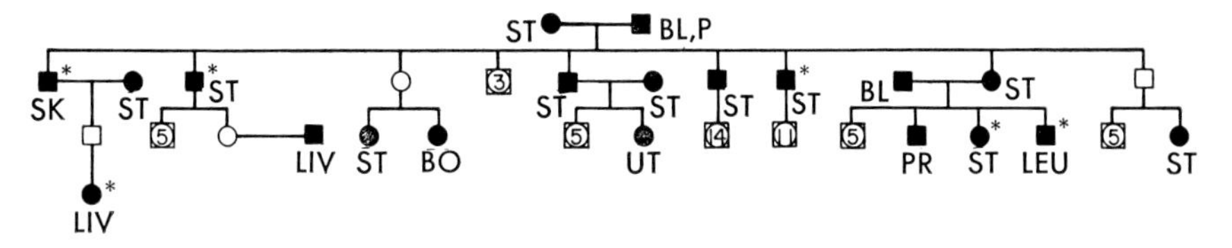

FIG. 2. Partial pedigree, D-leut, Alberta. ● female with cancer (anecdotal); ■* male with cancer (confirmed); ST, stomach; SK, skin; LIV, liver; BL, urinary bladder; P, prostate; BO, bone; UT, uterus; LEU, leukemia.

TABLE 4a. Familial Aggregates, S- and L-leut

Sibships		
S-leut: 12 sibships account for 28/84[a] cases		
1. Lung	Prostate	
2. Breast, carcinoma	Endometrium, adenocarcinoma	
3. Digestive or abdominal	Breast, carcinoma	Ear, squamous cell carcinoma
4. Unknown primary	Rectum, adenocarcinoma	Small intestine, adenocarcinoma
5. Breast, adenocarcinoma (premenopausal)	Breast, ductal adenocarcinoma (premenopausal)	
6. Stomach, carcinoma	Stomach, adenocarcinoma	
7. Stomach, mucin producing adenocarcinoma	Stomach, adenocarcinoma	
8. Stomach, mucin producing adenocarcinoma	Pancreas, adenocarcinoma	
9. Small intestine, adenocarcinoma	Rectum, adenocarcinoma	Unknown primary
10. Breast, adenocarcinoma	Breast, ductal adenocarcinoma	
11. Liver, adenocarcinoma	Breast, adenocarcinoma	Unknown primary
12. Larynx, carcinoma	Ear, squamous cell carcinoma	

[a] 51% of cases occur in sibship, child-parent or sibship-parent aggregates.

TABLE 4b. Familial Aggregates, S- and L-leut

Child-parent	
S-leut: 3 families account for 6/84[a] cases	
1. Ovary, adenocarcinoma	Rectum, adenocarcinoma
2. Stomach, adenocarcinoma	Acute lymphatic leukemia
3. Chondrosarcoma, site unknown	Breast, carcinoma simplex
Sibship-parent	
S-leut: 3 families account for 9/84[a] cases	
1. Stomach, carcinoma Leukemia	Breast, carcinoma
2. Stomach, carcinoma *in situ* Rectum, adenocarcinoma	Stomach, carcinoma
3. Rectum, adenocarcinoma Stomach, carcinoma *in situ*	Stomach, carcinoma

[a] 51% of cases occur in sibship, child-parent, or sibship-parent aggregates.

Table 4a-c. Fifty per cent of all 124* cases occur in familial aggregates. Types of cancers occurring in aggregates are generally those already reported to show clustering, for example, cancers of the digestive system, and of the breast. Most are adenocarcinomas, a form already reported (*14*) to be transmitted in a fashion consistent with dominant inheritance. Currently these aggregates are being combined into more complex pedigrees, and the relationships of the minimum ancestors (founders of the population) to cases and controls are being examined. Thus, we

* S- and L-leut.

TABLE 4c. Familial aggregates

Child-parent	
L-leut: 5 families account for 11/40[a] cases	
1. Stomach, carcinoma	Endometrium, carcinoma
2. Stomach, carcinoma	Breast, carcinoma
3. Skin, carcinoma	Breast, carcinoma
4. Ovary, dysgerminoma	Adenocarcinoma
5. Chronic lymphatic leukemia	Gallbladder, adenocarcinoma[b]
	Stomach, adenocarcinoma
Sibships	
L-leut: 3 sibships account for 6/40[a] cases	
1. Stomach	Colon, adenocarcinoma
2. Kaposi's sarcoma	Skin, carcinoma
3. Breast, carcinoma	Prostate, adenocarcinoma

a) 43% of cases occur in sibship or child-parent aggregates. b) Both parents affected.

are in the process of determining whether familial aggregation in the Hutterites is most likely due to genetic or environmental factors, or chance.

In summary, analysis of cancer patterns in the Hutterite population thus far has provided evidence for: (1) "low mortality risk" primarily due to a deficit of lung cancer; (2) recessive allele control over some forms, notably childhood leukemia; (3) familial aggregation of cancers of the digestive system and breast, usually adenocarcinomas. Analyses in progress include: (1) comparison of mortality rates of Hutterites to Canadian populations; (2) comparison of familial aggregates of cancer with aggregations of colonies related by historical splits; and (3) comparisons of the relationships of founders of the population to cases and controls. The main objective of the analysis in progess is to delineate genetic from environmental factors in oncogenesis.

ACKNOWLEDGMENT

This work was supported in part by U.S. PHS-NIH 3 RO1 CA 19822-01, National Cancer Institute.

REFERENCES

1. German, J. L. and Schonberg, S. This volume, pp. 175–186.
2. Takebe, H., Nikaido, O., Ishizaki, K., Yagi, T., Sasaki, M. S., Ikenaga, M., Kozuka, T., Fujiwara, Y., and Satoh, Y. This volume, pp. 259–270.
3. Hecht, T., and McCaw, B. K. Chromosome instability syndromes. *In*; J. J. Mulvihill, R. W. Miller, and J. F. Fraumeni (eds.), Genetics of Human Cancer, pp. 105–123, Raven Press, New York, 1977.
4. Schull, W. J. Cancer and inbreeding. *In*; J. J. Mulvihill, R. W. Miller, and J. F. Fraumeni (eds.), Genetics of Human Cancer, pp. 15–17, Raven Press, New York, 1977.
5. Steinberg, A. G., Bleibtreu, H. K., Kurczynski, T. W., Martin, A. O., and Kurczynski, E. M. Genetic studies on an inbred human isolate. *In*; J. F. Crow and

J. V. Neel (eds.), Proceedings of the Third International Congress of Human Genetics, pp. 267–290, Johns Hopkins Press, Baltimore, 1967.
6. Bleibtreu, H. K. Marriage and residence patterns in a genetic isolate. Ph.D. Thesis, Harvard University, 1964.
7. Mange, A. P. Growth and inbeeding of a human isolate. Human Biol., *36*: 104–133, 1964.
8. Martin, A. O. The founder effect in a human isolate: Evolutionary implications. Am. J. Phys. Anthropol., *32*: 351–367, 1970.
9. Martin, A. O., Dunn, J. K., Simpson, J. L., Olsen, C. L., Kemel, S., Grace, M., Elias, S., Sarto, G. E., Smalley, B., and Steinberg, A. G. Cancer mortality in a human isolate. Natl. Cancer Inst., in press, 1980.
10. U.S. Department of Health, Education and Welfare. Public Health Service National Center for Health Statistics, PHS Publication, No. 1693, 1968.
11. Percy, C. L., Berg, J. W., and Thomas, L. B. Manual of Tumor Nomenclature and Coding, American Cancer Society, 1968.
12. Bailar, J. C. and Ederer, F. Significance factors for the ratio of a Poisson variable to its expectation. Biometrics, *20*: 639–643, 1964.
13. Kurita, S., Kamei, Y., and Ota, K. Genetic studies on familial leukemia. Cancer, *34*: 1098–1101, 1974.
14. Fraumeni, J. F., Jr. Clinical patterns of familial cancer. *In*; J. J. Mulvihill, R. W. Miller, and J. F. Fraumeni (eds.), Genetics of Human Cancer, pp. 223–233, Raven Press, New York, 1977.

GENETIC AND ENVIRONMENTAL FACTORS IN EXPERIMENTAL AND HUMAN CANCER,
H. V. GELBOIN ET AL. (EDS.), JAPAN SCI. SOC. PRESS, TOKYO, PP. 303–312, 1980

A Population-based Study on Familial Aggregation of Breast Cancer in Iceland, Taking Some Other Risk Factors into Account

Hrafn TULINIUS,*[1] Nicholas E. DAY,*[2] Helgi SIGVALDASON,*[1] Ólafur BJARNASON,*[1] Guðmundur JÓHANNESSON,*[1] Maria A. LICEAGA DE GONZALEZ,*[2] Kristín GRÍMSDÓTTIR,*[1] and Guðrún BJARNADÓTTIR*[1]

Icelandic Cancer Society, Reykjavik, Iceland,[1] and International Agency for Research on Cancer, Lyons, France*[2]*

Abstract: Previous publications on the rise of cohort specific incidence of breast cancer in Iceland, and on the effect of reproductive factors in the Icelandic population are reviewed. The study concerns 1,740 breast cancer cases diagnosed since 1910 and up to the present. For the 1,740 breast cancer cases, genealogical information has been collected for little over 550 and the computerized file makes use of 477 families. Of these 130 have been completely traced for 1st-, 2nd-, and 3rd-degree relatives, except for great-grand parents and grand parents' sibs. The remaining 347 families have been traced three generations backwards, but the offspring have not been traced yet. All genealogies also include family members by wedding. The family file contains a total of 29,513 names of which 10,601 are female relatives and 3,788 are females related by marriage. When the breast cancer case file is matched to the family information there are 2.3% or 249 of the female relatives that also exist on the breast cancer case file, whereas 1.8% or 69 of the females related by marriage. When the types of relatives that are either purely maternal, on the one hand, or purely paternal or mixed maternal and paternal, on the other, are compared, it is found that the percentage of matching is slightly higher for the mixed maternal/paternal than the pure maternal lines, which confirms what has previously been stated, that the familiality of breast cancer does not favour the pure maternal line.

The relative risk of sisters compared to the general population is 2.91 or roughly three fold. When reproductive factors are taken into account the standardized relative risk is altered only to 2.74.

When the 1st-degree relatives are compared with the 2nd-degree relatives and the 3rd-degree relatives, we found a correlation between closeness of skinship and risk of breast cancer, this is, however so small, that there is very little difference in risk between 3rd-degree relatives and relatives by marriage, who in genetic terms could be taken as representatives of the general population.

It is concluded that it is possible to compute the risk of breast cancer contributed by having a close relative with the disease on a large population sample.

Genealogical information is of a high quality in Iceland and so has been the information on health for the last few score years. The Icelandic population is willing to participate in surveys and screening operations (*1*, *2*). There is no need, nor is there time, to review carefully what has been published (*3*) on the familial risk of breast cancer. Most, but not all, authors find the relative risk has increased. The increase in risk found is usually small, 2–4-fold, however, it has been shown that under particular circumstances this relative risk can be considerably higher, approaching 10-fold (*4–6*). This is when cancer is bilateral and one of the cancers of the two breasts has a premenopausal onset.

This excellent genealogical information in Iceland, and the general acceptance of familiality as a portion of the risk for breast cancer, prompted us to start investigations aimed at showing whether the familial risk is important in the Icelandic population and whether it can be measured accurately.

We felt obliged first to investigate some other known risk factors, so that whatever we found on the familiality, could be put in perspective with other risk factors.

When speaking of familial increase in risk, one has to take into account that this increase, if found, can be genetic, environmental, or both.

We first showed that the increase in breast cancer risk with time is very marked in Iceland (*7*). This was possible because there were complete records of all breast cancer cases diagnosed since 1910, now numbering 1,740. The incidence has steadily risen during this period. When we separated the population into birth decades, that is, we looked at all those born between 1840 and 1849, those born between 1940 and 1949, and all the intervening decades, we could plot an age distribution curve for each decade of birth cohort. When doing so, we found that the cohort-specific incidence rises very rapidly. Over this span of 110 years, the incidence has increased more than 10-fold (Table 1). We also found that the increase in incidence affects not only the higher age groups, the postmenopausal, as had been postulated

TABLE 1. Risk for Breast Cancer in Iceland Associated with Various Individual Characteristics

Adjusted standardized morbidity ratio by cohort: decade of birth

1840–	1850–	1860–	1870–	1880–	1890–	1900–	1910–	1920–	1930–	1940–
0.252	0.345	0.558	0.886	0.995	1.067	1.257	1.568	1.541	2.392	4.350

Adjusted age-specific incidence rates per 100,000 person-years: age

20–	25–	30–	35–	40–	45–	50–	55–	60–	65–	70–	75–	80–
1.0	2.6	8.2	21.6	45.7	71.4	76.1	88.8	94.8	146.1	116.3	175.3	238.1

Rel. risk associated with age at first birth (in years)

<20	20–24	25–29	30–34	35+
1.00	1.53	2.27	2.00	2.91

Rel. risk associated with parity

1–2	3–4	5+
1.78	1.25	1.00

Risk of nulliparous, relative to parity 5+ and age at first birth <20 =4.55.

by other researchers (*8–10*), but it equally affects the younger age groups. This means that whatever factor is changing in the population it must determine the breast cancer risk when the patients are relatively young.

In the same manner, we investigated reproductive factors by linking the breast cancer file with records collected by the Cervix Cancer Detection Clinic, which has operated in Iceland since 1964 (*1*). By so doing we could confirm the previous finding of the importance of age at first pregnancy, or first live birth, in determining the risk of breast cancer (*11*) but we could also show that the risk was substantially and significantly altered by parity, another reproductive factor, which previously had been found to be of no importance (*12*). Having done this, we feel capable of analyzing the information we have on the actual familial risk.

MATERIAL AND METHODS

The material on which we based this investigation was collected partly by genealogists who have worked for this study using interviews of family members and published information, series A, and partly by the office of the Genetical Committee of the University of Iceland, from the records already collected by that committee, series B. For the two groups uniform rules have been used throughout although the extent of the genealogical tree collected has differed. For the 1,740 cases we have on file, we have collected genealogical information for little over 550 of which we have computerized records for 477 proposity. Of this, 130 were collected by the genealogists and 347 by the office of the Genetical Committee. In both sets we have collected information on male and female members and we have also collected information on the spouses of the family members so that we have a groups of inwed non-relatives in the same file. Those can be used for internal control. The families collected by the genealogists, series A, have been collected for all 1st-, 2nd-, and 3rd-degree relatives, except for the great-grandparents and the grandparent sibs. Series B we have confined to parents, sibs, parents' sibs, grandparents, and their spouses.

1. Selection of proband cases

The Icelandic Cancer Registry which has been in operation since 1955 (*13–15*) has supplied a complete list of breast cancer cases diagnosed since that date. Prior to 1955, a list of all breast cancer cases diagnosed in Iceland has been established independently (*16*).

Series A consists of every 8th case diagnosed between 1955 to 1972, subsequently supplemented with 30 cases born between 1900 and 1916 and 5 cases born in 1864 or 1865 and 5 cases born in 1875. These cases are 130.

Series B consists of all 347 cases born in 1916 or later, diagnosed before the end of 1977.

(The two series have some members in common, but can be treated as distinct).

2. Construction of family trees

a) Series A

The family trees for series A include parents, sibs, children, parents' sibs, parents' parents, sibs' children, and sibs' childrens' children. The family trees include inmarried, not blood-related, individuals. The records of the Genetical Committee, other informative documents including the numerous published family histories and direct interviewing were used to establish the families.

b) Series B

The family trees for series B include parents, sibs, parents' sibs, parents' parents, and children. Inmarried individuals are included. The initial information came from the files of the Genetical Committee later supplemented by information from other sources as described for series A. Estimation of the degree of incompleteness of the Genetical Committee records will be an important subsidiary finding.

For both series, date of birth and date of death, if dead, were recorded. Whether death had occurred was ascertained from the complete national death records, those not appearing in the population roster being sought for in the death records to verify their status.

3. Ascertainment of breast cancer cases among families

Matching was initially performed by the computer, using date of birth and first 4 letters of forename and surname, a considerable margin of error was allowed. The possible matches were verified by personal examination of one of the genealogists (K.G.) and a member of the staff of the Cancer Registry (G.B.).

4. Factors included in the risk estimation for each female family member

The factors to be taken into consideration are decade of birth, age, age at first birth, and parity. Each factor is taken to have an independent, multiplicative effect, as shown previously (*12*). The risk associated with decade of birth and age come from Bjarnason *et al.* (*7*), as later corrected by Breslow and Day (*17*). The risk associated with age at first birth and with parity are from Tulinius *et al.* (*12*) (Table 1). For sisters and aunts from series B whose children were not included in the ascertained family trees, information on age at 1st birth and parity were obtained, when available, from the records of the Icelandic Cervix Cancer Detection Clinic (*1, 12*).

RESULTS

The total number on the case file is 1,740 and the number of cases for whom family information is used is 477. The total number of female relatives 10,601, containing 249 cases of breast cancer or 2.3% whereas the 3,788 females related by marriage on the family file contain 69 cases or 1.8% breast cancer cases (Table 2). This small difference between 2.3% and 1.8% would increase if the fact were taken into account that the family file contains a large number of children, whereas to be among the females related by marriage the person has to be married, see later.

TABLE 2. Female Breast Cancer on File

Total No. of female breast cancer on file		1,740	
No. of families traced, series A	130		
No. of families traced, series B	347	477	
		Matched	%
Female relatives	10,601	249	2.3
Male relatives	11,118		
Females related by marriage	3,788	69	1.8
Males related by marriage	4,006		
	29,513	318	

In Table 3 the number of each kind of relative is given and the number and % of relatives with breast cancer. For the 1st-degree relatives (Table 3a), the total % is 4.9, with mothers contributing 5.8%, sisters 5.0%, and daughters only 1.6%. This difference among the 1st-degree relatives reflects the different age distribution and will be allowed for by standardization. For the 2nd-degree relatives the total per cent is 2.2% or half that of the 1st-degree relatives. The greatest number come from the maternal and paternal aunts, 3% being breast cancer cases. This is comparable to the mothers among the 1st-degree relatives, being the same generation, and therefore of similar age distribution as the aunts. It is interesting to note that the percentage is practically the same on the maternal and paternal side. For the 3rd-degree relatives the total percentage is 1.4, the greatest number of these are 1st cousins who give 3.2% matched on the maternal side and 2.5% matched on the not-pure maternal side. Cousins are the same generation as cases and their sisters and this percentage of matching for 1st cousins should be compared to 5% for sisters.

TABLE 3a. Female Relatives on File

	No.	Matched	%
1st degree			
Mothers	468	27	5.8
Sisters	972	49	5.0
Daughters	187	3	1.6
Total 1st degree	1,627	79	4.9

	Maternal			Paternal		
	No.	Matched	%	No.	Matched	%
2nd degree						
Grandmothers	461	4	0.9	461	10	2.2
Aunts	1,260	38	3.0	1,278	39	3.1
Half-sisters	70	0		114	2	1.8
Nieces	371	5	1.3	509	11	2.2
Granddaughters	163	0		209	0	
Total	2,325	47	2.0	2,571	62	2.4
Total 2nd degree				4,896	109	2.2

TABLE 3b. Female Relatives on File (3rd Degree)

	Maternal			Not-pure maternal		
	No.	Matched	%	No.	Matched	%
Grandparent's sister[a]				6	0	
Parent's half-sister[b]	80	3	3.8	295	5	1.7
First cousin[c]	472	15	3.2	1,244	31	2.5
Half-sib's daughter[d]	24	0		172	2	1.2
Sib's granddaughter[e]	367	0		1,225	3	0.2
Great granddaughter[f]	36	0		156	0	
Total	979	18	1.8	3,098	41	1.3
Total 3rd-degree female relatives				4,077	59	1.4

[a] Father's father's sister and father's mother's sister. [b] Paternal, father's father's daughter, father's mother's daughter, and mother's father's daughter. [c] Paternal, father's brother's daughter, father's sister's daughter, mother's brother's daughter. [d] Paternal, father's son's daughter, father's daughter's daughter, mother's son's daughter. [e] Paternal, brother's son's daughter, brother's daughter's daughter, sister's son's daughter. [f] Paternal, son's son's daughter, son's daughter's daughter, daughter's son's daughter.

As mentioned earlier the group of inwed females on the file cannot be compared to the blood relatives directly since in order to be married to someone one has to be of marriage age. For that reason we have prepared Table 4a and 4b where only those 25 years and older are counted. In Table 4a we see again the same kind of gradient in risk when we compare within generations, mothers with 5.8% and

TABLE 4a. Relatives, 25 Years and Older, by Generations

Gen.	1st degree		Matched	%	2nd degree		Matched	%	3rd degree		Matched	%
−2					Grandmother	905	14	1.5				
−1	Mother	464	26	5.6	Aunt	1,803	77	4.3	Parent's half-sisters	272	8	2.9
0	Sisters	831	49	5.9	Half-sister	152	2	1.3	Cousins	1,396	46	3.3
1	Daughter	141	3	2.1	Niece	696	16	2.3	Daughter's of half-sisters	136	2	1.5
2									Sib's grand-daughter	517	3	0.6

TABLE 4b. Wives of Relatives, 25 Years and Older, by Generations[a]

Gen.	1st degree		Matched	%	2nd degree		Matched	%	3rd degree		Matched	%
−2					Grandfather	155	1	0.6				
−1	Father	102	2	2.0	Uncle	433	5	1.2	Parent's half-brother	87	2	2.3
0	Brother	313	10	3.2	Half-brother	63	3	4.8	Cousins	1,207	36	3.0
1	Son	147	2	1.4	Brother's son	198	3	1.0	Sons of half-sibs	114	1	0.9
2									Sib's grand-son	376	1	0.3

[a] Wives includes all mothers of the relatives offspring *e.g.* single mothers.

aunts with 4.3%, or in the generation of the case, sisters with 5.9%, and cousins, or 3rd-degree relatives 3.3%. Table 4b also gives all mothers of the offspring of the male relatives. Here this gradient between 1st-, 2nd-, and 3rd- degree is lost, for example the brothers' wives having 3.2% matching and the cousins' wives having 3.0% matching. The father's wives have 2.0% matching, the uncles' wives having 1.2% matching. Comparing the tables, one finds very little difference in the 3rd-degree relatives where the blood-related cousins have 3.3% matching, 46 out of 1,396 and the wives of the male cousins have 3.0% matching 36 out of 1,207.

In order to answer the question whether the reproductive factors, which we have previously shown to have an important risk contribution (*12*), influence the risk contributed by familial aggregation, we looked for information on age at 1st birth and parity for certain relatives of breast cancer cases. The information was obtained from the Cervix Cancer Detection Clinic of the Icelandic Cancer Society (*1*). Table 5 gives some of this information, 5a on age at 1st birth, and 5b on parity. In general one can see that the reproductive habits of sisters resemble those of the cases to a certain degree.

Table 6 shows observed and expected rates of certain family members and a relative risk. Account has been taken of the contribution made by decade of birth and age in all the computations, but only for sisters can we measure the

TABLE 5a. Reproductive Experience of Different Relative Types, and of the Detection Clinic Population, by Decade of Birth—1. Age at First Birth: % of Women in Each Category

Decade of birth	Sisters				Aunts				Detection clinic population			
	NP	<20	20–29	>30	NP	<20	20–29	>30	NP	<20	20–29	>30
<1840					22.0	0	51.2	26.8				
1840–59					17.7	1.3	48.1	32.9				
1860–79	27.8	11.1	55.5	5.6	22.9	1.1	42.5	33.5				
1880–99	23.9	2.8	59.2	14.1	27.9	4.3	49.5	18.3				
1900–09	29.2	0	50.8	20.0	13.3	0	66.7	20.0	12.3	4.3	65.8	17.9
1910–19	11.1	8.9	55.6	24.4	11.1	0	88.9	0	10.5	8.2	66.5	14.8
1920–29	21.4	14.3	50.0	14.3	0	0	100.0	0	7.7	14.0	69.9	8.4
1930–39	20.0	20.0	50.0	10.0					6.3	25.7	65.8	2.2

TABLE 5b. Reproductive Experience of Different Relative Types, and of the Detection Clinic Population, by Decade of Birth—2. Parity: % of Women in Each Category

Decade of birth	Sisters				Aunts				Detection clinic population			
	NP	<3	3–4	>5	NP	<3	3–4	>5	NP	<3	3–4	>5
<1840					22.0	14.6	14.6	48.8				
1840–59					17.7	31.6	12.7	38.0				
1860–79	27.8	5.6	5.5	61.1	22.9	26.3	20.7	30.2				
1880–99	23.9	22.5	29.7	23.9	28.0	22.6	24.7	24.7				
1900–09	29.2	30.8	26.2	13.8	15.1	20.0	40.0	26.7	12.3	29.1	34.4	24.2
1910–19	11.1	37.8	22.2	28.9	11.1	22.2	55.6	11.1	10.5	24.9	36.9	27.7
1920–29	21.4	35.7	14.3	28.6	0	50.0	50.0	0	7.7	21.1	40.9	30.3
1930–39	20.0	50.0	20.0	10.0					6.3	33.5	44.6	15.6

TABLE 6a. Observed and Expected Risk of Breast Cancer Cases by Relative Type

Type of relative	Observed	Expected	Rel. risk	*P*
Mothers	24	12.48	1.92	0.01
Grandmothers	14	12.33	1.14	NS
Maternal grandmothers	4	6.63		
Paternal grandmothers	10	5.70		
Aunts	67	43.15	1.55	0.01
Maternal aunts	31	21.13		0.05
Paternal aunts	36	22.02		0.01
Sisters	30	9.92	3.02	0.01
Daughters	3	1.74	1.72	NS
Firist cousins				
2246 series A	13	9.14	1.42	NS
series B	2	0.88	2.27	NS
2236 series A	9	8.10	1.11	NS
series B	0	0.92		NS
2146 series A	10	8.69	1.15	NS
series B	2	0.91	2.20	NS
2136 series A	9	7.55	1.19	NS
series B	1	0.39	2.56	NS
First cousins total	46	36.58	1.26	NS (<0.08)

Corrected for decade of birth and age.

TABLE 6b. Observed and Expected Risk of Breast Cancer Cases by Relative Type (Series A)

Type of relative	Observed	Expected	Rel. risk
Corrected for decade of birth and age			
Mothers	5	2.65	1.93 NS
Aunts	12	9.40	1.28 NS
Maternal aunts	7	5.04	
Paternal aunts	5	4.36	
Sisters	20	6.87	2.91 S
Excluding those born before 1890, and corrected for decade of birth, age, and reproductive history			
Mothers	1	0.39	
Aunts	2	2.77	
Sisters	18	6.74	2.74 S

contribution by reproductive history. This contribution appears to be minimal, or its effect is to reduce the relative risk of sisters from 2.91 to 2.74.

CONCLUSIONS

The preliminary conclusions that may be drawn from this are the following:

1. The frequency of breast cancer among relatives by blood, or by marriage, to breast cancer cases, can be measured in the data bank collected for that purpose.

2. When the risk of breast cancer is compared between degrees of relatedness one can see that there is a constant gradient, so that the nearest relatives have a greater risk than those more distantly related. In this manner it can be seen that there is an increase in risk of breast cancer among close family members of breast cancer cases.
3. No consistent difference was found in the likelihood of matching for relatives depending on whether they were related to the case through a pure maternal line or the relationship was, at least partly, paternal.
4. In trying to evaluate how much this risk is above that of the general population we have compared the percentage of breast cancer cases between comparable groups of blood relatives and relatives by marriage. From that, one can see that in the relatives by marriage there is no gradient between 1st-, 2nd-, and 3rd- degree relatives; however, the total risk in the relatives by marriage seems to be of a comparable size to that of the 3rd-degree relatives. One may, therefore, conclude that the small difference in risk between close relatives and the general population is confined to 1st- and 2nd-degree relatives and is lost when the genetic dilution has come down to one-eighth common genes, or 3rd-degree relatives.
5. Little is known about the familiality of reproductive factors. We have looked at reproductive factors among sisters of breast cancer cases and found that their reproductive habits resemble those of the cases to a certain degree, however, when this information is used for internal standardization of the relative risk, this contribution appears to be minimal or altering the relative risk of sisters from 2.91 to 2.74.

ACKNOWLEDGMENTS

This work was partially supported by a collaborative research agreement between the Unit of Epidemiology and Biostatistics, IARC, Lyons, France, and the Department of Pathology, University of Iceland, under contract No. N01-CB-43973 with the Breast Cancer Epidemiology Program of National Cancer Institute, National Institutes of Health, U.S.A., and from Grant No. GM22370 from the Institute of General Medical Research, National Institute of Health, U.S.A.

REFERENCES

1. Jóhannesson, G., Geirsson, G., and Day, N. The effect of mass screening in Iceland, 1965–1974, on the incidence and mortality of cervical carcinoma. Int. J. Cancer, *21*: 418–425, 1978.
2. Davidsson, D., Sigfússon, N., Ólafsson, Ó., Björnsson, O. J., and Thorsteinsson, Th. Report A III. Betalipoprotein, cholesterol and triglycerides in Icelandic males aged 34–61. Acta Med. Scand., *616* (Suppl): 1–150, 1977.
3. Anderson, D. E. Familial and genetic predisposition. *In*; B. A. Stoll (ed.), New Aspects of Breast Cancer, vol. 2, pp. 3–24, William Heinemann Medical Books Ltd., 1976.
4. Anderson, D. E. Some characteristics of familial breast cancer. Cancer, *28*: 1500, 1971.

5. Anderson, D. E. A genetic study of human breast cancer. J. Natl. Cancer Inst., *48*: 1029, 1972.
6. Anderson, D. E. Genetic study of breast cancer: Identification of a high risk group. Cancer, *34*: 1090, 1974.
7. Bjarnason, O., Day, N., Snædal, G., and Tulinius, H. The effect of year of birth on the breast cancer age-incidence curve in Iceland. Int. J. Cancer, *13*: 689–696, 1974.
8. DeWaard, F. The epidemiology of breast cancer; review and prospects. Int. J. Cancer, *4*: 577–586, 1969.
9. DeWaard, F., Cornelis, J. P., Aoki, K., and Yoshioa, M. Breast cancer incidence according to weight and height in two cities of the Netherlands and in Aichi prefecture, Japan. Cancer, *40*: 1969–1975, 1977.
10. DeWaard, F. Relationships between the epidemiology and endocrinology of mammary carcinoma. Front. Hormone Res., *5*: 1945–1954, 1978.
11. MacMahon, F., Cole, P., Lon, M., Lowe, C. R., Mirra, A. P., Ravnihar, B., Salber, E. J., Valaoras, V. G., and Yuasa, S. Age at first birth and breast cancer risk. Bull. WHO, *42*: 209–221, 1970.
12. Tulinius, H., Day, N. E., Jóhannesson, G., Bjarnason, Ó., and Conzales, M. Reproductive factors and risk for breast cancer in Iceland. Int. J. Cancer, *21*: 724–730, 1978.
13. Bjarnason, Ó. Iceland. *In*; R. Doll *et al.* (eds.), Cancer Incidence in Five Continents, pp. 168–178, Springer-Verlag, Berlin-Heidelberg-New York, 1966.
14. Bjarnason, Ó. *In*; N. Ringertz (ed.), Cancer incidence in Finalnd, Iceland, Norway and Sweden: A comparative study. Acta Pathol. Microbiol. Scand., 224 (Suppl.): 1971.
15. Bjarnason, Ó. Iceland. *In*; J. Waterhouse *et al.* (eds.), Cancer Incidence in Five Continents, vol. III, IARC Scientific Publication No. 15, International Agency for Research on Cancer, Lyons, 1974.
16. Snædal, G. Cancer of the breast. A clinical study of treated and untreated patients in Iceland, 1911–1955. Acta Chir. Scand., *338* (Suppl.): 1964.
17. Breslow, N. E. and Day, N. Indirect standardization and multiplicative models for rates, with reference to the age adjustment of cancer incidence and relative frequency data. J. Chron. Dis., *28*: 289–303, 1975.

GENETIC AND ENVIRONMENTAL FACTORS IN EXPERIMENTAL AND HUMAN CANCER,
H. V. GELBOIN ET AL. (EDS.), JAPAN SCI. SOC. PRESS, TOKYO, PP. 313–326, 1980

Radiation Carcinogenesis: The Hiroshima and Nagasaki Experiences

William J. SCHULL, Toranosuke ISHIMARU, Hiroo KATO, and Toshiro WAKABAYASHI

Radiation Effects Research Foundation, Hiroshima, Japan

Abstract: Thirty years ago, persuasive evidence had accumulated that leukemia occurred more frequently among survivors of the atomic bombings of Hiroshima and Nagasaki. Efforts to characterize the dose-response relationship and the effects of age at the time of bombing and calendar time on this phenomenon were hampered, however, by the failure to define a manageable cohort for study and the absence of a systematic method of case detection as well as other factors. Subsequently, with the establishment of the Life-Span Study sample, a cohort of 109,000 individuals, A-bomb survivors and their controls, and a Leukemia Registry, our understanding improved materially. We now know, for example, that the larger the exposure dose and the younger the age at the time of the bombing (ATB), the greater was the effect in the early period and the more rapid was the decline in risk in subsequent years, and that chronic granulocytic leukemia contributed substantially, especially in Hiroshima, to the total leukemogenic effect initially, but made little contribution after 1955. The leukemogenic effect among those of older age ATB occurred later and decreased more slowly. Leukemia has not been significantly increased in those individuals exposed *in utero*, nor among the offspring of exposed persons. Today, mortality from leukemia has returned to "normal" values in all age groups of survivors save that of the 35–49 years group ATB.

Subsequent studies of radiation-induced carcinogenesis have focussed primarily on the Life-Span Study cohort and have used death certificates, the findings of Tissue and Tumor Registries, and the records of local hospitals. It is now clear, for example, that mortality from cancers of the lung, breast and stomach increases with increasing dose and, recently, the suggestion of an increase in mortality from cancer of the colon has appeared. Multiple myeloma also increases significantly with radiation dose. As yet, an increase in mortality from cancer of the esophagus, urinary tract and lymphoma remains uncertain. The increase in mortality from cancers other than leukemia becomes significant, generally, when individuals reach the usual ages of onset for a given cancer and the distribution of times from

A-bombing to death does not differ significantly by radiation dose for solid tumors, but it does depend upon the age ATB. Both the relative and the absolute risk for cancers other than leukemia are higher for younger age ATB cohorts at the same age at risk, that is, age at death.

Numerous important issues remain unclear. Among these are *first* is the carcinogenic effect a general one, affecting all tissues and histologic types? Each successive year at risk of death in the Life-Span Study cohort would suggest so, but the data are not as yet compelling. *Second*, can reliable estimates of the relative biological effectiveness (RBE) of neutrons be obtained, given the paucity of the Nagasaki data? Our estimates are derived from the difference in radiation induction in Nagasaki (largely gamma exposure) and Hiroshima (mixed gamma and neutron), and are as a consequence no more reliable than the weaker set of data. *Third*, does ionizing radiation enhance the effect of other known carcinogens, for example, smoking? *Fourth*, can something further be done to clarify whether the period of time from exposure to cancer death changes systematically with increasing ionizing radiation. Our mortality data suggest not, but death certificate information is not equally reliable for all solid tumors.

Thirty years ago, Yamawaki and Borges accumulated persuasive evidence that leukemia was increased among the survivors of the atomic bombings of Hiroshima and Nagasaki (*1*). Much effort in the next several years was devoted to the study of the early hematologic and preclinical phases, and the effects of age at the time of bombing and calendar time on the incidences and types of radiation-related cases of leukemia (*2–4*). It was soon apparent that the peak annual incidence occurred in the early 50s and has declined steadily thereafter (*3*, *5*). These studies to which we allude were hampered, however, by the failure to define a manageable cohort for scrutiny, the absence of a systematic method of case detection, and the inability then to assign to individuals estimates of their exposures. The occurrence of leukemia could be functionally related only to the distances from the hypocenter of the exposed individuals and the presence or absence of a history of severe radiation complaints, that is, epilation, oropharyngeal lesions, or purpura (*6*). Subsequently with the establishment in 1955 of the Unified Study Program with its emphasis on fixed samples, the development of a Leukemia Registry (1958) and the evolution of means to assign to individuals doses of gamma and neutron radiation (the T65 dosimetry system; see Ref. *7*) our understanding of the leukemogenic effect of A-bomb exposure improved materially. We shall return to a presentation of our current knowledge shortly, but certain background remarks seem appropriate first.

The Unified Study Program itself embraces four major elements, namely, (1) a study of mortality in a cohort of 109,000 individuals, 82,000 A-bomb survivors and their controls, the so-called Extended Life-Span Study sample; (2) the biennial medical evaluation of approximately 20,000 members initially of the Life-Span Study cohort, the Adult Health Study; (3) an Autopsy Program, and (4) a series of genetic studies (see Ref. *8* for a full description of the major samples). Recent investigations of radiation-related malignancies have focused primarily on the Life-Span Study sample and have used death certificates, the results of the periodic examinations of members of the Adult Health Study sample, routine autopsy diagnoses and surgical pathology reports, the findings of the Tissue and Tumor Registries, and the records of the local hospitals. The Tumor Registries, the first continuous such registries in Japan, were begun in 1957 in Hiroshima and 1958 in Nagasaki under the sponsorship of the City Medical Associations, with operational support from the Atomic Bomb Casualty Commission (ABCC), the predecessor of the present Radiation Effects Research Foundation. The Tissue Registries are more recent; Hiroshima's was established in 1973 and Nagasaki's in the following year. The Autopsy Program to which reference has been made has centered upon members of the Life-Span Study sample. Some 4,900 of the approximately 16,000 Life-Span Study subjects who died between 1961 and 1975 have been autopsied under this program (*9*). The results of these latter examinations have served as the bases for a series of successive assessments of the accuracy of death certificate diagnoses of cancer (for the most recent see Ref. *9*). These have invariably shown the confirmation rate at autopsy for all cancers reported on death certificates to be generally high. Rates vary, of course, as a function of site, the age of the individual at death, and whether death occurred at home or in a hospital. Both confirmation and detection rates appear essentially independent of radiation dose.

We shall now summarize briefly the findings of the past 30 years. For convenience, the latter will be grouped under the broad 3-digit rubrics of the International Classification of Disease, 9th revision (*10*).

Malignant Neoplasms of the Lip, Oral Cavity, and Pharynx (140–149)

A recent review of all malignancies under this rubric seen in the years 1957 through 1976 among members of the Life-Span Study sample identified 85 possible cases, exclusive of salivary gland tumors, of which 63 were considered definite, that is, had been histologically confirmed (*11*). Save for cancers of the tongue the number of malignancies for each specific site (lip; gum; mouth other than lip, gum or tongue; tonsil; nasopharynx; hypopharynx and pharynx) is invariably small, and no one of these tumors is significantly associated with dose. Patently, given the infrequency of most of these malignancies, our ability to detect any but the most pronounced radiation effects is extremely small.

Several years ago Belsky and his colleagues (*12*, *13*) examined the relationship of salivary gland tumors to exposure. Although the number of cases available for study was not large (30), among those individuals exposed to 300 rads or more the number with tumors was significantly greater than expected. Indeed, the relative risk was almost 22 times that which obtains for individuals who were either not present in these cities ATB or if present, received less than 1 rad. They emphasize, of course, the need for caution in drawing conclusions under these circumstances, especially so when only three cases were observed in the highest dose group, that is, among those individuals receiving 300 rads or more.

Malignancies of the Digestive Organs and Peritoneum (*150–159*)

At least three cancers here seem radiation related, namely, those of the esophagus, stomach, and large bowel excluding the rectum. Associations have been sought for the liver (*14*, *15*) and the gall bladder, bile ducts and Vater's ampulla (*16*) but none have as yet been found. Some 90 cases of primary carcinoma of the pancreas were diagnosed in the 15-year period, 1956–1970, in the ABCC autopsy series, but no clear association with exposure emerged (*17*).

To return to the three significantly associated malignancies, the effect of ionizing radiation on the frequency of deaths attributed to an esophageal malignancy appears, retrospectively, to have been first evidenced in the years 1954–1958 (*18*, *19*). At the present time, the estimated excess mortality attributable to this tumor per million person years per rad (PYR) is 0.15. A conspicuous elevation in the frequency of this tumor is only seen among survivors who were 50 years of age or more ATB, and while the effects in the two cities cannot be shown to be significantly different, only the one in Hiroshima is significant by conventional standards.

The situation with respect to malignancies of the stomach remains enigmatic. Nakamura (*20*, *21*), using death certificates for the period 1950–1973, reported a consistent increase in mortality with increasing radiation dose in Hiroshima, the rate being highest among those survivors who had received 400 or more rads, whereas the evidence of a radiation effect in Nagasaki was very weak. An excess was found only at doses above 500 rads. Beebe *et al.* (*18*) reported that the evidence of an association of tumors of this organ with radiation derives from a steady accumulation of excess deaths rather than some recent change. This assertion was based on an examination of successive 4-year time increments beginning in 1950 and terminating in 1974; they observed none for which the excess in mortality was of more than borderline significance. Kato and Schull (*19*), however, in the most recent analysis find the estimated mortality PYR to be significantly elevated in the years 1975–1978. Again, although the cities do not differ significantly one from another as judged in terms of estimated mortality per million PYR only the effect in Hiroshima is a statistically significant one.

Malignant neoplasms of the large intestine, excluding the rectum have in the past not appeared to be radiation related. Indeed, as recently as 1978 Beebe, Kato and Land asserted that "there is little to suggest a relationship with radiation" although they did note that their analysis was not entirely negative in that tests on Hiroshima females revealed discrepancies between expectation and observation with very low probabilities. Kato and Schull (*19*), however, now find the estimated mortality per million PYR for this tumor when cities, sexes, and all ages are combined to be significantly elevated. It is 0.29 or approximately twice the risk associated with esophageal tumors. This change in events is attributable mainly to a marked increase in the number of deaths ascribed to this malignancy in the years 1975–1978. It is seen in both cities, more strikingly so in Hiroshima than in Nagasaki, however, and more conspicuously affects two age groups, namely, those individuals 10–19 or 35–49 ATB.

Malignancies of the Respiratory and Intrathoracic Organs (160–165)

Harada and Ishida (*22*), in an analysis of the Hiroshima City Tumor Registry, were the first to suggest that the incidence of lung cancer was significantly higher among those who were exposed within 1,500 m of ground zero. Ciocco (*23*) observed that mortality from lung cancer among male survivors exposed within 1,500 m of the hypocenter and more than 50 years of age in 1950 was 2–3 times higher than expected. Similarly, Beebe and his associates (*24*) in a review of autopsies in the Life-Span Study sample noted that pulmonary cancer was about twice that expected in subjects who were exposed to 90 or more rads and had died in 1961–1965. Wanebo *et al.* (*25*) were the first, however, to attempt an incidence study based on the Life-Span Study cohort utilizing all possible methods of ascertainment. They observed that the number of deaths ascribed to lung cancer among survivors was approximately twice that expected, but noted no significant differences between more heavily exposed survivors and other subjects with lung cancer when comparisons were made as to age of onset of the disease, presence of other primary neoplasms, histologic type, or other epidemiological factors commonly associated with increased risks of lung cancer. Cihak *et al.* (*26*), reviewing the autopsy findings in the years 1961–1970, reported that small cell anaplastic carcinomas were definitely more common in irradiated persons compared to controls (relative risk 3.9), and that epidermoid and bronchogenic adenocarcinomas, the two most commonly seen malignancies of the lung, were also increased but not significantly so. They too failed to find evidence of an influence of either smoking or occupational exposure to which the apparent radiation effect might be ascribed (*27*). Subsequent studies (*18*, *19*) have been limited to mortality ascribed to cancer of the lung among the Life-Span Study sample. The carcinogenic effect continues to be seen only in the oldest age groups, that is among individuals 35 or more years of age ATB. Beebe *et al.* (*18*) observed that "although both relative and absolute risk measures for Hiroshima are higher than those for Nagasaki, the values for the two cities do not differ significantly; only those for Hiroshima meet the usual criteria for statistical significance." This observation continues to hold in the more recent analysis (*19*) which spans the years 1950–1978.

Malignancies of Bone, Connective Tissue, and Breast (170–175)

Yamamoto and Wakabayashi (*28*) in a study of benign and malignant bone tumors found at autopsy or in surgical specimens in Hiroshima and Nagasaki in the years 1950–1965 were unable to demonstrate an association of these tumors and exposure distances. With the exception of osteosarcomas, however, the number of malignant bone tumors of a specified kind was too small to permit consideration according to the histologic type of the tumor. Death certificate data have not been particularly informative here, for the number of deaths ascribed to malignant tumors of the bone are so few. Under this rubric, the malignancy most clearly radiation related is cancer of the breast. Wanebo *et al.* (*29*) were the first to establish this. Their study was limited, however, to women who were members of the Adult Health Study group. Subsequent investigations have focused on members of the Life-Span Study sample, some 63,000 females in all. The most recent of these (*30*) finds that (1) the distribution of histologic types of mammary cancers does not vary significantly with radiation dose, (2) among all women who received at least 10 rads, those irradiated before age 20 will experience the highest rates of breast cancer throughout their lifetimes, and (3) the effects of neutrons and gamma radiation appear to be linearly related to dose and about equal in the induction of cancer of the breast. McGregor and his colleagues (*31*) had earlier reported much these same findings on a somewhat smaller number of cases. Nakamura *et al.* (*32*) have examined the relationship between exposure and other epdiemiologic risk factors in the development of breast cancer among the survivors. Specifically, they examined the association of such risk factors as family history of malignant neoplasm; education; menstrual, marital, and reproductive history; and history of breast feeding and conclude that the risk increases in proportion to the number of other risk factors involved.

Malignancies of the Genitourinary System (*179–189*)

Efforts to relate malignancies of this system to exposure to the A-bombs have involved both incidence studies and the continued mortality surveillance of the Life-Span Study sample. Insofar as the former are concerned, Bean and his colleagues (*33*) were unable to demonstrate an association with exposure to an atomic bomb of the occurrence of prostatic adenocarcinoma in 1,357 male members of the Life-Span Study sample who were 50 years of age or older at death and were autopsied at ABCC-RERF betwen 1961 and 1969. Nor could they demonstrate an effect of exposure on histologic type or biologic activity. Sawada (*34*) in a much earlier study attempted to evaluate the effect of A-bomb radiation on the occurrence of gynecologic tumors as seen in a series of routine examinations of 1,785 exposed women and 1,802 control subjects, but unfortunately, almost half of the women of this series would not accept an examination of their pelves. Carcinoma of the cervix was the most frequently encountered tumor; however, its occurrence could not be shown to depend on exposure nor was the age of onset significantly different between control and irradiated individuals. The most recent examination of deaths in the Life-Span Study sample (*19*) also failed to find a significant association. This latter study identified 443 deaths attributable to a malignancy of the cervix and uterus, 93 to the uterus alone and 139 deaths attributed to the other urinary organs.

Other Malignancies (*190–199*)

Malignancies of the eye, brain, "other" and unspecified parts of the nervous system; the thyroid gland and other endocrine glands and their related structures fall within this rubric. Included too are malignant neoplasms of "other and ill-defined" sites, malignancies without specification of site, and secondary malignant neoplasms of the lymph nodes, respiratory and digestive system, as well as other sites. We are, of course, here concerned only with primary tumors and of those within this rubric, only malignant neoplasms of the thyroid have been significantly associated with exposure to A-bomb radiation. Hollingsworth *et al.* (*35*) first noted, in a study of thyroid disease in Hiroshima, that carcinoma of the thyroid constituted some 7% of the total number of individuals with thyroid disorders they saw and that a greater number of these malignancies were seen among the most heavily exposed individuals. They cautiously noted, however, that differences among distance groups were not statistically significant. Subsequent studies have removed any doubt of the association of this tumor with exposure. Indeed, Socolow *et al.* (*36*) were soon to show that among 15,369 Adult Health Study subjects examined in Hiroshima and Nagasaki there was a highly significant correlation between proximal exposure to the atomic bombs and the eventual development of thyroid carcinoma. In their study, biopsies were obtained in approximately 50% of the patients with thyroid nodules, without knowledge of the patient's radiation history, and of these only 2 of the 21 cases of malignancy occurred among nonexposed individuals. Sixteen cases were seen in females and 5 in males; among the 5 male patients with carcinoma 4 were exposed within 1,400 m of ground zero whereas 10 of 16 females were similarly exposed. Wood and his colleagues (*37*) suggested that age at exposure altered the subsequent proclivity to develop carcinoma of the thyroid, but Sampson *et al.* (*38*, *39*), in a study of 525 individuals with papillary carcinomas of the thyroid seen among 3,067 consecutive autopsies failed to find any indication that dose-prevalence relationships were different in the youngest and oldest age groups. More recently, Parker *et al.* (*40*) summarized the results of the continuing, periodic examinations of the Adult Health Study group in the years 1958–1971. In all, 74 cases of thyroid cancer in the Adult Health Study were seen and histologically confirmed in these years. They found thyroid carcinoma to be commoner in women and significantly more prevalent in persons exposed to 50 or more rads of atomic radiation. They also found carcinoma of the thyroid to be diagnosed during life more commonly in persons who were less than 20 years ATB although they failed to confirm the previously reported radiosensitivity in childhood. An increased risk was found even among persons who were 50 years of age or older at the time of radiation exposure. It warrants note here that this is one of the few malignancies where a significant relationship with ionizing radiation has been demonstrable largely on clinical grounds. No effect has emerged through the continuing mortality surveillance of the Life-Span Study sample but few of the individuals with thyroid malignancies, confirmed either at autopsy or on histologic section, have had their deaths ascribed to these malignancies.

Malignancies of the Lymphatic and Hematopoietic Tissue (200–208)

As has been previously indicated, the first radiation-related malignancy in this rubric to emerge was leukemia, and the findings of the Foundation and its predecessor ABCC have been extensively reported (most of the relevant publications will be found in Ref. *41–44*). As succinctly put as possible, these studies suggest, first, the incidence of leukemia (for all cell types) to be directly proportional to the dose of neutrons and to the square of the dose of gamma rays. It must be noted, however, that other models such as linear-quadratic in gamma and linear in neutrons or a linear-linear can not be excluded; indeed, they fit the data almost equally well. Second, annual incidence of leukemia among the survivors reached a peak, as previously stated in 1952 and has declined steadily since. It had not, however, completely disappeared as recently as 1971. Third, when incidence by dose is examined as a simultaneous function of the age ATB and calendar time of onset, it is clear that the larger the exposure dose and the younger the individual at the time of the exposure, the greater was the radiation effect in the early period, *i.e.* prior to October 1955, and the more rapid was the decline in risk in subsequent years. The leukemogenic effect among individuals who were older ATB occurred later and decreased more slowly. Finally, insofar as the incidence of specific leukemia is concerned, the radiation related risk of acute lymphocytic leukemia as well as "the other types" of acute leukemia seems higher among people, exposed at younger ages whereas the frequency of chronic granulocytic leukemia as well as the risk of acute granulocytic and lymphocytic leukemia is greater among individuals who were older at the time of their exposure.

Nishiyama and his colleagues (*45*) were the first to report an increased prevalence of malignant lymphoma and multiple myeloma in survivors of the atomic bomb in Hiroshima exposed to 100 rads or more (a similar relationship was not seen in Nagasaki). Most of the effect they saw was, however, attributable to malignant lymphoma rather than myeloma; they identified only six cases of the latter. Earlier reports had suggested that the frequency of myelofibrosis with myeloid metaplasia was increased in proximally exposed survivors in Hiroshima (*46*, *47*). Ichimaru and his colleagues (*48*) have recently reviewed all of the cases of multiple myeloma seen in the period from October 1950 through December 1976 in the Life-Span Study sample. They find the risk of myeloma to increase consistently and significantly with radiation exposure. Indeed, the standardized relative risk is about five times greater among individuals exposed to 100 rads or more than in the control group.

Finally, it should be noted that as yet it has not been possible to demonstrate an increased frequency of malignancies in either those individuals exposed *in utero* or born to individuals who were exposed, that is, among the F_1 (*49*, *50*).

Figure 1 is an effort to summarize some of the foregoing. It sets forth the relative risks of specific malignancies among individuals who received 200 rads or more exposure as contrasted with individuals who received zero rad. Note that these risks range from 1.5 times or so to almost 15 times in the case of leukemia. It should be noted, however, that these figures can be somewhat misleading when the impact on these cities is considered if one does not bear in mind the exposure of the average survivor, less than 50 rads. A more meaningful assessment of the public health implications might be the increase in cancer deaths which are radiation related in the years 1950–1978. Some 5,135 exposed members of the Life-Span Study sample died in these years of a malignant neoplasm; of this number approximately 10%, that is, about 500 deaths can be ascribed to the fact of exposure.

Figure 2 gives the estimated excess mortality per million PYR for the tumors

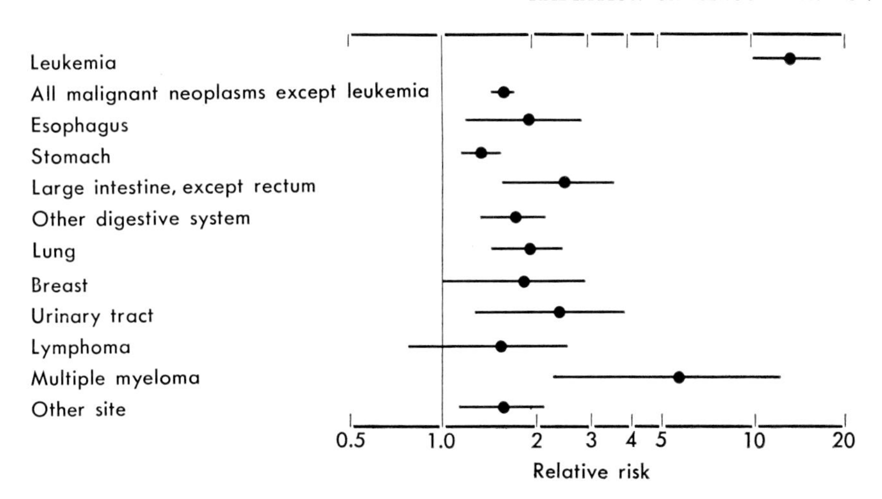

FIG. 1. Relative risk and 80% confidence intervals for malignant neoplasms, 200+ rads *vs.* 0 rad, 1950–1978.

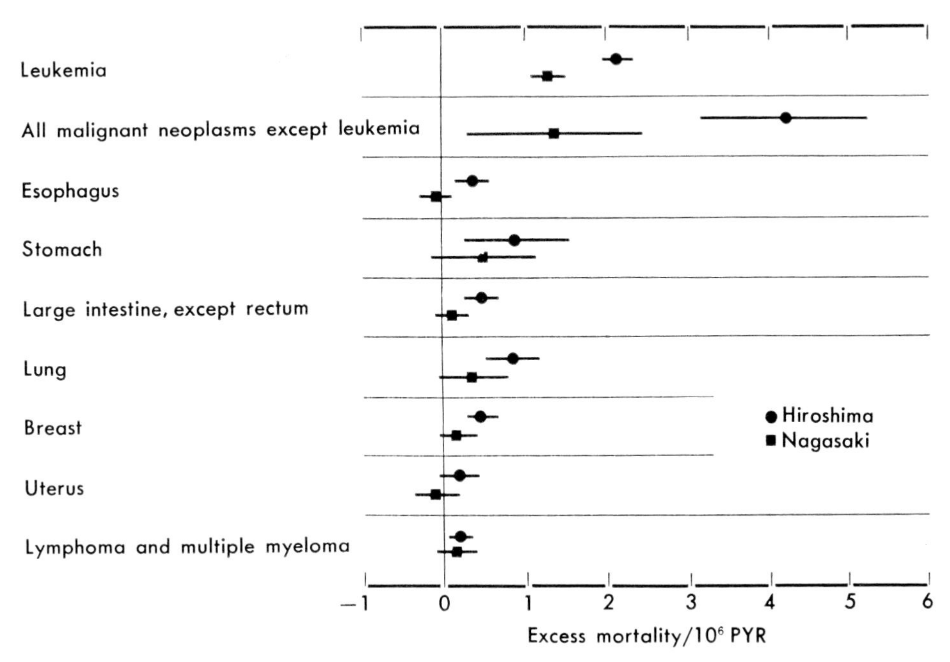

FIG. 2. Estimated excess mortality per million PYR for selected sites with 90% confidence limits.

shown in Fig. 1 for the cities separately. Note that this excess is consistently less in Nagasaki than in Hiroshima, indeed, the Nagasaki values are not significantly different from zero (but also not different from those for Hiroshima). It should be recalled that the qualities of the radiation from these two nuclear devices were different. Exposures in Nagasaki were largely to gamma radiation whereas a significant neutron component existed in Hiroshima. Given these differences and those seen in Fig. 2 it is natural to consider the biological effectiveness of gamma radiation as contrasted with exposure to neutrons. Among the two best studied malignancies, namely, breast and leukemia, it appears that the risk of radiation induced malignancy increases approximately linearly with dose in the former instance but quite possibly not in the latter (*30*, *51*). Gamma and neutron radiation appear approximately equally effective in inducing malignancies of the breast whereas the two types of radiation may have dissimilar effects on the induction of leukemia. The incidence of leukemia increases linearly with neutrons but appears to have a non-negligible quadratic component in the case of gamma exposure. As yet we do not have equally comprehensive data on most other malignancies; however, Fujita *et al.* (*52*) have examined the effects of gamma and neutron exposure as reflected in mortality ascribable to cancer in the Life-Span Study sample. Their results underscore both the need for further studies of the effects of radiation quality and the difficultness to be expected in distinguishing between different sorts of dose responses.

Table 1 illustrates still another unresolved problem. Here we set forth the relative risk of death ascribable to either leukemia or other malignancies prior to a particular time as a function of age ATB. Note these risks can vary greatly although it appears the young ATB are especially vulnerable. As yet our data are often too sparse to permit us to generate similar figures for specific malignancies other than leukemia in which much confidence would be placed. Patently here is an important issue. Why do these differences in vulnerability exist? Unfortunately, we still know so little about how radiation exerts its effect that we can not formulate meaningful biochemical or physiological models to guide us in our inquiries.

We need too further investigation of the effect of radiation in concert with exposure to other carcinogens. We've cited some preliminary studies of this nature, such as that of Ishimaru *et al.* (*53*) on leukemia or Nakamura *et al.* (*32*) on breast cancer. But frequently the number of pairs of cases and controls has been modest (164 pairs in the breast cancer study cited) and the power in the statistical sense, to detect certain reasonable differences small. There is here a need for innovative

TABLE 1. Relative Risk for Leukemia and Nonleukemic Malignancies by Age ATB, 100+rads *vs.*, 0 rad, 1950–1978

Age at time of bomb		Age at death					
		<30	30–	40–	50–	60–	70+
<10	Leukemia	23.8	16.5	—			
	Nonleukemia	25.0	4.8	7.6			
10–19	Leukemia	—	4.0	0.6	—		
	Nonleukemia	0.7	2.7	2.3	9.3		
20–34	Leukemia	8.4	6.6	—	7.5	0.0	
	Nonleukemia	—	1.9	1.9	2.0	1.7	
35–49	Leukemia			9.0	8.6	1.6	1.9
	Nonleukemia			1.1	1.1	1.3	1.4
50+	Leukemia				—	9.7	—
	Nonleukemia				2.3	1.0	1.4

statistical and epidemiological strategies to advance our understanding to the next plateau. The opportunities to implement such developments seem especially noteworthy in these cities. The populations of survivors are still large; they spanned the full spectrum of ages at the time of exposure; their exposures are reasonably well known and are not confounded with inherent disease or disability. The survivors are cooperative and concerned. Moreover, their experiences are clearly relevant to the "outside world" for the results of ABCC-RERF studies tend to agree with studies of other exposed populations, mostly medically or occupationally exposed. As others have observed, their experiences can confer structure on our risks generally, *e.g.*, age dependencies, shape of dose-response curve, and the like. Indeed, their experiences constitute the cornerstone of what is presently known about radiation carcinogenesis. It seems reasonable to presume, therefore, that future insights into this process will be no less dependent upon careful observations on the experiences of these survivors.

REFERENCES

1. Folley, J. H., Borges, W., and Yamawaki, T. Incidence of leukemia in survivors of the atomic bombs in Hiroshima and Nagasaki, Japan. Am. J. Med., *13*: 11–21, 1952.
2. Lange, R. D., Moloney, W. C., and Yamawaki, T. Leukemia in atomic bomb survivors. I. General observations. Blood, *9*: 574–585, 1954.
3. Moloney, W. C. Leukemia in survivors of atomic bombing. N. Engl. J. Med., *253*: 88–90, 1955.
4. Moloney, W. C. and Lange, R. D. Leukemia in atomic bomb survivors. II. Observations on early phases of leukemia. Blood, *9*: 663–685, 1954.
5. Tomonaga, M. Statistical investigation of leukemia in Japan. N. Z. Med. J. Haematol. *65* (Suppl.): 863–869, 1966.
6. Moloney, W. C. and Kastenbaum, M. A. Leukemogenic effects of ionizing radiation on radiation on atomic bomb survivors in Hiroshima city. Science, *121*: 308–309, 1955.
7. Milton, R. C. and Shohoji, T. Tentative 1965 radiation dose (T65D) estimation for atomic bomb survivors, Hiroshima-Nagasaki. ABCC Tech. Rep., *1–68*: 1–43, 1968.
8. Beebe, G. W. and Usagawa, M. The Major ABCC Samples. ABCC Tech. Rep., *12–68*: 1–66, 1968.
9. Yamamoto, T., Moriyama, I. M., Asano, M., and Guralnick, L. The Autopsy Program and the Life Span Study, January 1961-December 1975. RERF Tech. Rep., *18–78*: 1–35, 1978.
10. World Health Organization. International Classification of Diseases, Ninth Revision. WHO, Geneva, vol. 1, pp. xxxiii and 773, 1977.
11. Pinkston, J. A., Wakabayashi, T., Yamamoto, T., Asano, M., Harada, Y., Kumagami, H., and Takeichi, M. Cancer of the head and neck in atomic bomb survivors, Hiroshima and Nagasaki, 1957–76. Unpublished, 1979.
12. Belsky, J. L., Tachikawa, K., Cihak, R. W., and Yamamoto, T. Salivary gland tumors in atomic bomb survivors, Hiroshima-Nagasaki, 1957–1970. J. Am. Med. Assoc., *219*: 864–868, 1972.
13. Belsky, J. L., Takeichi, N., Yamamoto, T., Cihak, R. W., Hirose, F., Ezaki, H.,

Inouye, S., and Blot, W. J. Salivary gland neoplasms following atomic radiation: Additional cases and reanalysis of combined data in a fixed population, 1957–1970. Cancer, *35*: 555–559, 1975.

14. Schreiber, W. M., Kato, H., and Robertson, J. D. Primary carcinoma of the liver in Hiroshima and Nagasaki, Japan. Cancer, *26*: 69–75, 1970.
15. Asano, M., Seyama, S., Itakura, H., Hamada, T., Iijima, S., and Kato, H. Relationship between incidence of liver cancer and exposure to A-bomb. Trans. Soc. Pathol. Jpn., *67*: 143–144 (Abstr.), 1978 (in Japanese).
16. Robertson, J. D., Kato, H., and Schreiber, W. M. Carcinoma of the gallbladder, bile ducts and Vater's ampulla Hiroshima and Nagasaki. ABCC Tech. Rep., *7–70*: 1–13, 1970.
17. Cihak, R. W., Kawashima, T., and Steer, A. Adenoacanthoma (adenosquamous carcinoma) of the pancreas. Cancer, *29*: 1133–1140, 1972.
18. Beebe, G. W., Kato, H., and Land, C. E. Studies of the mortality of A-bomb survivors. 6. Mortality and radiation dose, 1950–1974. Radiat. Res., *25*: 138–201, 1978.
19. Kato, H. and Schull, W. J. Studies of the mortality of A-bomb survivors. 7. Mortality and radiation dose, 1950–1978. Unpublished, 1979.
20. Nakamura, K. Stomach cancer in atomic bomb survivors, 1950–73. RERF Tech. Rep., *8–77*: 1–8, 1977.
21. Nakamura, K. Stomach cancer in atomic bomb survivors. Lancet, *ii*: 866–867, 1977.
22. Harada, T. and Ishida, M. Neoplasms among A-bomb survivors in Hiroshima: First Report of the Research Committee on Tumor Statistics, Hiroshima City Medical Association, Hiroshima, Japan. J. Natl. Cancer Inst., *25*: 1253–1264, 1960.
23. Ciocco, A. JNIH-ABCC Life Span Study and ABBCC-JNIH Adult Health Study: Mortality 1950–1964 and diseases and survivorship, 1958–1964 among sample members aged 50 or older, October 1, 1950. ABCC Tech. Rep., *18–65*: 1–49, 1965.
24. Beebe, G. W., Yamamoto, T., Matsumoto, Y. S., and Gould, S. E. ABCC-JNIH Pathology Studies, Hiroshima and Nagasaki. Report 2. ABCC Tech. Rep., *8–67*: 1–73, 1967.
25. Wanebo, C. K., Johnson, K. G., Sato, K., and Thorslund, T. W. Lung cancer following atomic radiation. Am. Rev. Respir. Dis., *98*: 778–787, 1968.
26. Cihak, R. W., Ishimaru, T., Steer, A., and Yamada, A. Lung cancer at autopsy in A-bomb survivors and controls, Hiroshima and Nagasaki, 1961–1970. I. Autopsy findings and relation to radiation. Cancer, *33*: 1580–1588, 1974.
27. Ishimaru, T., Cihak, R. W., Land, C. E., Steer, A., and Yamada, A. Lung cancer at autopsy in A-bomb survivors and controls, Hiroshima and Nagasaki, 1961–1970. II. Smoking, occupation, and A-bomb exposure. Cancer, *36*: 1723–1728, 1975.
28. Yamamoto, T. and Wakabayashi, T. Bone tumors among the atomic bomb survivors of Hiroshima and Nagasaki. Acta Pathol. Jpn., *19*: 201–212, 1969.
29. Wanebo, C. K., Johnson, K. G., Sato, K., and Thorslund, T. W. Breast cancer in the ABCC-JNIH Adult Health Study, Hiroshima-Nagasaki, 1950–66. ABCC Tech. Rep., *13–67*: 1–12, 1967.
30. Tokunaga, M., Norman, J. E., Jr., Asano, M., Tokuoka, S., Ezaki, H., Nishimori, I., and Tsuji, Y. Malignant breast tumors among atomic bomb survivors, Hiroshima and Nagasaki, 1959–74. J. Natl. Cancer Inst., *62*: 1347–1359, 1979.
31. McGregor, D. H., Land, C. E., Choi, K., Tokuoka, S., Liu, P. I., Wakabayashi, T., and Beebe, G. W. Breast cancer incidence among atomic bomb survivors, Hiroshima and Nagasaki, 1950–1969. J. Natl. Cancer Inst., *59*: 799–811, 1977.

32. Nakamura, K., McGregor, D. H., Kato, H., and Wakabayashi, T. Epidemiologic study of breast cancer in A-bomb survivors. RERF Tech. Rep., *9–77*: 1–31, 1977.
33. Bean, M. A., Yatani, R., Liu, P. I., Fukazawa, K., Ashley, F. W., and Fujita, S. Prostatic carcinoma at autopsy in Hiroshima and Nagasaki Japanese. Cancer, *32*: 498–506, 1973.
34. Sawada, H. Evaluation of gynecological tumors in the atomic bomb survivors. ABCC Tech. Rep., *6–59*: 1–13, 1959.
35. Hollingsworth, D. R., Hamilton, H. B., Tamagaki, H., and Beebe, G. W. Thyroid disease: A study in Hiroshima, Japan. Medicine, *42*: 47–71, 1963.
36. Socolow, E. L., Hashizume, A., Neriishi, S., and Niitani, R. Thyroid carcinoma in man after exposure to ionizing radiation: A summary of the findings in Hiroshima and Nagasaki. N. Engl. J. Med., *268*: 406–410, 1963.
37. Wood, J. W., Tamagaki, H., Neriishi, S., Sato, T., Sheldon, W. F., Archer, P. G., Hamilton, H. B., and Johnson, K. G. Thyroid carcinoma in atomic bomb survivors Hiroshima and Nagasaki. Am. J. Epidemiol., *89*: 4–14, 1969.
38. Sampson, R. J., Key, C. R., Buncher, C. R., and Iijima, S. The age factor in radiation carcinogenesis of the human thyroid: A study of 536 cases of thyroid carcinoma, Hiroshima-Nagasaki. ABCC Tech. Rep., *7–69*: 1–6, 1969.
39. Sampson, R. J., Key, C. R., Buncher, C. R., and Iijima, S. Thyroid carcinomas in Hiroshima and Nagasaki. I. Prevalence of thyroid carcinoma at autopsy. J. Am. Med. Assoc., *209*: 65–70, 1969.
40. Parker, L. N., Belsky, J. L., Yamamoto, T., Kawamoto, S., and Keehn, R. J. Thyroid carcinoma after exposure to atomic radiation: A continuing survey of a population, Hiroshima and Nagasaki, 1958–1971. Ann. Intern. Med., *80*: 600–604. 1974.
41. Hoshino, T., Kato, H., Finch, S. C., and Hrubec, Z. Leukemia in offspring of atomic bomb survivors. Blood, *30*: 719–730, 1967.
42. Ichimaru, M., Ishimaru, T., and Belsky, J. L. Incidence of leukemia in atomic bomb survivors belonging to a fixed cohort in Hiroshima and Nagasaki, 1950–71. J. Radiat. Res., *19*: 262–282, 1978.
43. Liu, P. I., Ishimaru, T., McGregor, D. H., Yamamoto, T., and Steer, A. Autopsy study of leukemia in atomic bomb survivors, Hiroshima-Nagasaki 1949–1969. Cancer, *31*: 1315–1327, 1973.
44. Okada, H., Tomiyasu, T., Ishimaru, T., Hoshino, T., and Ichimaru, M. Risk of leukemia in offspring of atomic bomb survivors, Hiroshima and Nagasaki, May 1946–June 1959. ABCC Tech. Rep., *30–72*: 1–14, 1972.
45. Nishiyama, H., Anderson, R. E., Ishimaru, T., Ishida, K., Ii, Y., and Okabe, N. The incidence of malignant lymphoma and multiple myeloma in Hiroshima and Nagasaki atomic bomb survivors, 1945–1965. Cancer, *32*: 1301–1309, 1973.
46. Anderson, R. E., Hoshino, T., and Yamamoto, T. Myelofibrosis with myeloid metaplasia in survivors of the atomic bomb in Hiroshima. Ann. Intern. Med., *60*: 1–18, 1964.
47. Anderson, R. E. and Yamamoto, T. Myeloproliferative disorders in atomic bomb survivors. *In*; W. J. Clarke, E. B. Howard, and P. L. Hackett (eds.), Myeloproliferative Disorders of Animals and Man, USAEC Proc. 8th Annu. Hanford Biol. Symp., Richland, Washington, 1968.
48. Ichimaru, M., Ishimaru, T., Mikami, M., and Matsunaga, M. Multiple myeloma among atomic bomb survivors in Hiroshima and Nagasaki, 1950–1976. RERF Tech. Rep., *9–79*: 1–16, 1979.

49. Jablon, S. and Kato, H. Childhood cancer in relation to prenatal exposure to atomic-bomb radiation. Lancet, *ii*: 1000–1003, 1970.
50. Kato, H. Mortality of *in-utero* children exposed to the A-bomb and of offspring of A-bomb survivors. *In*; Late Biological Effects of Ionizing Radiation, vol. 1, pp. 49–60, International Atomic Energy Agency, Vienna, 1978.
51. Ishimaru, T., Otake, M., and Ichimaru, M. Dose-response relationship of neutrons and gamma rays to leukemia incidence among atomic bomb survivors in Hiroshima and Nagasaki by type of leukemia, 1950–1971. Radiat. Res., *77*: 377–394, 1979.
52. Fujita, S., Shimizu, Y., Masaki, K., Yoshimoto, Y., and Kato, H. RBE of neutrons in cancer mortality among atomic bomb survivors, Hiroshima and Nagasaki, 1950–1979. Unpublished.
53. Ishimaru, T., Okada, H., Tomiyasu, T., Tsuchimoto, T., Hoshino, T., and Ichimaru, M. Occupational factors in the epidemiology of leukemia in Hiroshima and Nagasaki. Am. J. Epidemiol, *93*: 157–165, 1971.

GENETIC AND ENVIRONMENTAL FACTORS IN EXPERIMENTAL AND HUMAN CANCER,
H. V. GELBOIN ET AL. (EDS.), JAPAN SCI. SOC. PRESS, TOKYO, PP. 327–340, 1980

Cancer Patterns among Migrant and Native-born Japanese in Hawaii in Relation to Smoking, Drinking, and Dietary Habits

Laurence N. KOLONEL, M. Ward HINDS, and Jean H. HANKIN
Epidemiology Program, Cancer Center of Hawaii, University of Hawaii, Honolulu, Hawaii 96813, U.S.A.

Abstract: Cancer incidence rates for major sites were compared among Japanese in Japan, first generation Japanese migrants to Hawaii (Issei), and subsequent generations of Japanese born in Hawaii (Nisei, *etc.*). In general, Hawaii rates were higher than rates in Japan; notable exceptions were the rates for esophageal, gastric, and cervical cancer. Nisei rates were lower than Issei rates for lung, prostate, esophagus, stomach, and cervix uteri; Nisei rates exceeded Issei rates for corpus uteri and female breast cancer; Nisei rates were about the same as Issei rates for colon, rectum, liver, bladder, thyroid, and lymphomas. There was notable consistency between males and females in the various site-specific incidence patterns. The age-specific breast cancer incidence curves for both Nisei and Issei women resembled the curve for women in Japan, including the absence of a "menopausal hook" and decreasing rates in the older ages.

For the Japanese 45 years of age and older in Hawaii, we compared the cancer incidence rates with correponding smoking, drinking, and dietary habits by nativity and sex. Lung cancer incidence patterns correlated well with rates of cigarette smoking for both sexes. Similar relationships were seen for esophageal and rectal cancer incidence and alcohol use. Dietary fat intake corresponded with colon cancer incidence patterns for both sexes, but not with the patterns for breast or prostate cancer. Within each sex, stomach cancer incidence showed positive associations with dried/salted fish and carbohydrate intake, and negative associations with preserved meat and vitamin C intake. These findings are discussed in relation to prevailing hypotheses and other published reports.

The study of migrant populations has yielded much useful information in epidemiological research. In the field of cancer, particularly, the comparison of disease rates among first generation migrants, their parent populations, and their descendants has been very informative (*1–5*). Most of these studies have utilized mortality data, since cancer incidence statistics are generally less readily available, especially for selected subgroups of the population like migrants.

Data on Japanese have been prominent in studies of this kind. Migrant studies among Japanese have shown, for example, decreasing stomach cancer mortality rates and increasing colon cancer rates in first generation migrants compared with the prevailing rates in Japan; these changes are even greater in later generations of migrants (*4*). Some of the data for these past studies have come from Hawaii. Previously, however, analyses by place of birth could only be done for cancer deaths; yet, incidence statistics can be more informative, since survival rates are high for certain important sites such as the prostate. Furthermore, in the past, there were no population-based data in Hawaii on exposure factors related to cancer (such as diet and smoking) which could also be examined by place of birth.

During the past few years, we have been collecting both of these types of data in Hawaii. For this report, we will compare incidence rates for major sites of cancer among Japanese in Japan, migrants from Japan to Hawaii (Issei), and their descendants born in Hawaii (Nisei, Sansei, *etc.*). For the latter two population groups, we will also examine the incidence rates in relation to their respective patterns of exposure to smoking, drinking, and dietary factors.

METHODS

The incidence statistics cover the period 1973–1977, and were developed from data in the Hawaii Tumor Registry (HTR) and the Hawaii Department of Health (DOH). The HTR was established in 1960, and in 1973 became a member of the SEER Program (Surveillance, Epidemiology and End Results) of the National Cancer Institute. Since 1973, the HTR has been routinely recording place of birth on the cases of cancer it identifies in the State of Hawaii. The registry also records the race or ethnicity of each case.

The distribution of the Japanese population in Hawaii by place of birth was developed from an ongoing survey of the state's resident population. This survey, conducted by the DOH, samples about 2% of the households in Hawaii each year, and interview data are collected from all residents in the households surveyed. This survey has been the basis for determining ethnic distributions of the population of Hawaii in recent years, since the racial provisions of the decennial U.S. censuses are no longer adequate for this purpose.

For the comparisons of incidence statistics between Hawaii and Japan, cancer rates for Miyagi Prefecture for the years 1968–1971 as reported in Ref. *6* were used. The Miyagi registry was selected since it is the oldest registry in Japan, and rates do not vary greatly among registries in Japan for most sites. None of the results or conclusions in the present study would have been altered if we had used data from the Osaka registry, for example. Because there is a particular interest in prostate cancer in Hawaii Japanese, it is possible that Japanese men in Hawaii are being "overdiagnosed" for this cancer relative to men in Japan; thus, differences in incidence between the two locations could be exaggerated. For this site only, we excluded cases diagnosed at autopsy from the tabulations in this report.

The data on smoking, drinking, eating, and other demographic factors were obtained from a special questionnaire which we added in 1975 to the DOH survey mentioned above. This questionnaire included information on height, weight, alcohol use, cigarette use, and food consumption (frequency of intake only), and is being administered to all persons age 18 or older.

Since 1977, we have also been administering a detailed quantitative food intake schedule to a subset of volunteers from this survey group; this subset includes all persons who are 45 years of age or older. This information is collected by trained interviewers on our own staff, who assess usual weekly food consumption patterns for each individual surveyed. Quantitation is facilitated by the use of a series of colored photographs that show food samples in three different commonly-eaten portion sizes. This method had been described elsewhere (*7*, *8*).

Because the quantitative food and nutrient intakes were available only on persons age 45 and older (the ages at which most cancers occur), the comparisons of exposure factors with cancer incidence data were based on truncated age-adjusted rates. These rates were calculated from age-specific data, using the standard world population age 45 and older (*6*).

RESULTS

Table 1 shows the age distribution of first and later generations of Japanese in Hawaii. (For convenience, Nisei is used to refer to all second and subsequent generations. Many of the younger Japanese in Hawaii today, of course, are Sansei, and some are even fourth generation.) The age distributions of these two groups are quite different, with more Nisei in the younger ages and more Issei in the older ages. This reflects the pattern of Japanese migration to Hawaii, most of which took place from around the turn of the century until the first World War.

Because of these differences in age-distribution between the two groups, particularly the high proportion of Issei over age 60, we were concerned about the

TABLE 1. Estimated Japanese Population in Hawaii (1975) by Nativity and Sex

Age group	Male		Female	
	No.	%	No.	%
U. S.-born (Nisei)				
0–19	33,463	30.8	29,169	28.1
20–39	30,665	28.2	29,427	28.4
40–59	34,083	31.4	34,486	33.3
≥60	10,420	9.6	10,588	10.2
Total	108,631	100.0	103,670	100.0
45 and older	37,425	34.5	38,373	37.0
Japan-born (Issei)				
0–19	2,387	23.2	2,621	14.8
20–39	2,371	23.1	3,848	21.8
40–59	1,149	11.2	4,791	27.1
≥60	4,371	42.5	6,425	36.3
Total	10,278	100.0	17,685	100.0
45 and older	5,124	49.9	9,135	51.7

TABLE 2. Comparison of Cancer Incidence Rates among Three Japanese Populations: Issei in Hawaii, Nisei in Hawaii, and Japanese in Japan[a,b]

Cancer site	Average annual incidence per 100,000					
	Females			Males		
	Japan	Hawaii Issei	Hawaii Nisei	Japan	Hawaii Issei	Hawaii Nisei
Esophagus	4.1 (168[c])	0.6 (4)	0.3 (2)	12.9 (406)	8.5 (13)	2.0 (13)
Stomach	40.1 (1,621)	22.9 (60)	13.2 (75)	84.6 (2,749)	46.9 (93)	28.5 (149)
Colon	5.4 (213)	19.4 (54)	18.8 (113)	5.6 (178)	29.2 (54)	28.0 (152)
Rectum	5.0 (200)	9.2 (33)	7.9 (52)	6.8 (222)	24.0 (49)	17.6 (105)
Liver	0.6 (25)	6.5 (23)	6.1 (30)	1.8 (58)	6.9 (24)	8.5 (47)
Pancreas	4.5 (179)	4.9 (19)	3.5 (19)	7.3 (233)	10.2 (19)	8.0 (40)
Bladder	1.3 (53)	5.9 (19)	2.3 (12)	3.7 (114)	10.0 (24)	9.7 (50)
Lung	7.0 (275)	17.6 (45)	9.1 (53)	20.0 (632)	47.2 (99)	28.0 (160)
Breast	13.0 (511)	35.9 (41)	57.2 (407)	—	—	—
Cervix (invasive)	13.8 (563)	12.6 (22)	5.2 (27)	—	—	—
Corpus uteri	1.3 (47)	15.4 (20)	20.3 (140)	—	—	—
Prostate[d]	—	—	—	2.7 (87)	35.2 (87)	21.1 (81)
Thyroid	2.1 (84)	5.4 (10)	6.3 (39)	0.6 (23)	4.8 (7)	4.4 (22)
Leukemias	3.8 (138)	2.6 (7)	6.0 (28)	4.6 (160)	6.7 (9)	6.2 (30)
Lymphomas (including Hodgkin's disease)	1.5 (58)	6.4 (12)	4.6 (28)	2.7 (89)	5.7 (12)	4.4 (28)

a) All rates are age-adjusted to the world population standard (*6*). b) Data from Japan are for the period 1968–1971; data from Hawaii are for the period 1973–1977. c) Number of cases for the period in b) which were used in computing the rate. d) Excludes cases first diagnosed at autopsy in Hawaii (see text).

use of age-adjusted rates in our tabulations of cancer statistics. If the incidence of a given cancer were higher for Issei at one age but higher for Nisei at another, the choice of a standard population could determine which group showed the higher age-adjusted rate. Accordingly, we compared our age-adjusted rates with those for a single age range (55–74 years) where a large proportion of cancer occurs, as well as with our overall impressions from examining all age groups separately, and found that the adjusted rates computed from the world population standard summarized the data satisfactorily. This is primarily a reflection of the general consistency in the relationship of the age-specific rates among groups.

Table 2 compares the cancer incidence rates for males and females of three different population groups: Japanese in Japan (Miyagi Prefecture), first generation Japanese migrants to Hawaii (Issei), and Japanese born in Hawaii (Nisei). Several different patterns are apparent in these data:

1. Rates in Hawaii higher than in Japan
   a) Nisei rates higher than Issei rates
      Sites: breast (female), corpus uteri
   b) Nisei rates lower than Issei rates
      Sites: lung, prostate
   c) Nisei rates equal to Issei rates

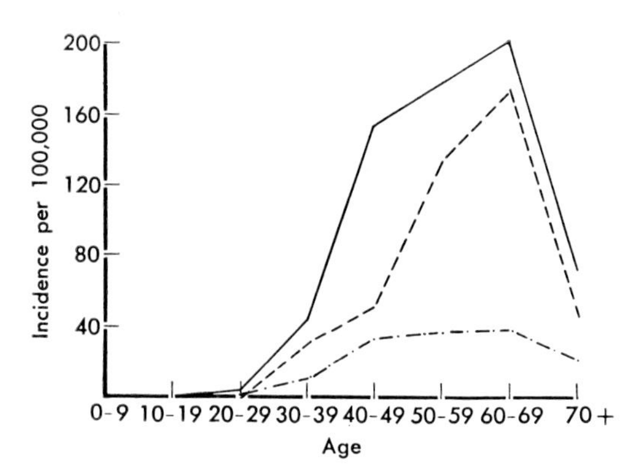

FIG. 1. Age-specific breast cancer incidence by nativity among Japanese women in Hawaii (1973–1977) and Japan (1968–1971). —— Hawaii Nisei; - - - Hawaii Issei; -·-·- Japan.

Sites: colon, rectum, liver, bladder, thyroid, lymphomas

2. Rates in Hawaii lower than in Japan
   a) Nisei rates lower than Issei rates
      Sites: esophagus, stomach, cervix
3. No apparent difference in rates between Hawaii and Japan
   Sites: pancreas, ?leukemias

Two other features of Table 2 are notable: (1) relationships among the three populations are consistent for males and females, and (2) with the exception of thyroid cancer, female rates are always lower than corresponding male rates.

Age-specific incidence rates for breast cancer among the three population groups are shown in Fig. 1. The gradient in rates from lowest in indigenous Japanese to intermediate in Issei and highest in Nisei, which was seen in the age-adjusted rates of Table 2, is consistent throughout the age range.

Table 3 presents data on food and nutrient intakes, on smoking and drinking habits, and on height and weight for the Japanese in Hawaii. The first three food items are reported as frequencies only, and the Issei show more frequent consumption of all items. That this does not necessarily imply greater quentitative intakes of these items is shown by comparison of the frequency and qantitative intakes for preserved meats. Whereas the Issei men report somewhat greater frequencies of intake of these meats, the Nisei men actually consume greater quantities. The nutrient data do not show dramatic differences between Issei and Nisei within sex groups, except for carbohydrate intake among women (probably a reflection of the greater rice consumption by the Issei women).

The smoking data show higher proportions of cigarette smokers among Issei than among Nisei. The daily use by smokers, however, shows more consumption by Nisei than Issei women, although there is little difference between the men. The alcohol data show higher rates of regular use by Issei of both sexes. However, among consumers, more beer is drunk by Nisei than Issei men. The consumption of beer by Japanese women is not great for either group. There is little difference in reported height and weight between the Issei and Nisei 45 years and older, although the Nisei women are somewhat heavier (about 1 kg on average).

TABLE 3. Selected Characteristics of Japanese in Hawaii by Nativity[a)]

	Females		Males	
	Issei	Nisei	Issei	Nisei
Diet	N=77	N=344	N=44	N=323
Foods:				
Pickled vegetables (times/week)	4.1	2.8	3.1	2.8
Fresh fruit (times/week)	6.7	6.2	6.8	5.8
Preserved meats (times/week)	2.1	2.0	2.8	2.4
Preserved meats (g/week)	115.5	112.8	169.3	191.6
Dried/salted fish (g/week)	4.4	1.9	4.5	3.3
Nutrients:				
Total fat (g/day)	55.3	52.5	60.9	66.4
Saturated fat (g/day)	19.7	17.9	22.0	22.8
Protein (g/day)	54.6	50.0	64.0	64.4
Carbohydrate (g/day)	175.2	146.9	196.5	194.8
Vitamin C (g/week)	1.7	1.8	1.1	2.1
Alcohol				
% regular drinkers[b)]	11.2	7.5	46.5	37.3
Beer consumption by drinkers (cans/week)	2.8	1.6	6.8	8.4
Cigarette smoking				
% regular smokers[c)]	27.6	19.7	56.2	50.9
Daily use by smokers (No. of cigarettes)	12.4	18.6	24.9	24.4
Anthropometry				
Height (cm)	154.7	154.4	166.4	165.1
Weight (kg)	52.5	53.3	65.3	65.0

[a)] All data are mean values which have been age-adjusted to the world population standard 45 years and older (*6*). [b)] Defined as drinking on average once a week or more. [c)] Defined as at least one cigarette per day (includes both current and ex-smokers).

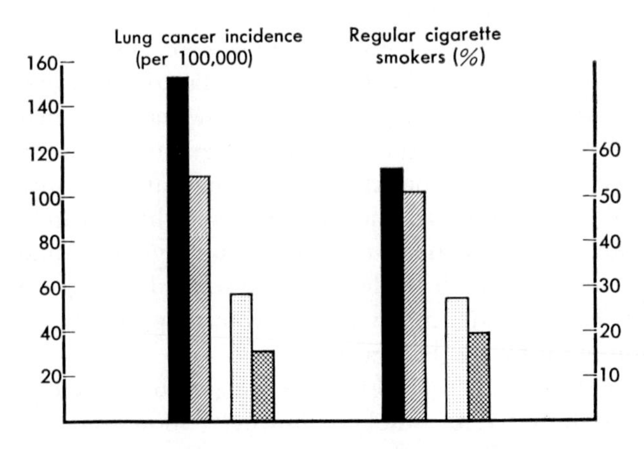

FIG. 2. Cigarette smoking and lung cancer incidence by nativity in Hawaii Japanese (45 years and older, age-adjusted). ■ Issei males; ▨ Nisei males; □ Issei females; ▩ Nisei females.

Figure 2 compares the lung cancer incidence rates among the four nativity-sex groups with their respective cigarette smoking rates. (Note that the cancer rates here are truncated and therefore different from those in Table 2.) The correspondence of the lung cancer pattern with that for cigarette smoking is seen in both the higher male than female rates and the higher Issei than Nisei rates for each sex.

In Fig. 3, esophageal and rectal cancer incidence rates are compared with the rates for alcohol use. Both esophageal and rectal cancer rates correspond with the pattern for alcohol use.

Figure 4 compares dietary fat intake with colon, breast, and prostate cancer incidence rates. The fat intake pattern corresponds with that for colon cancer incidence in both sexes, although differences in the incidence rates are slight. Neither

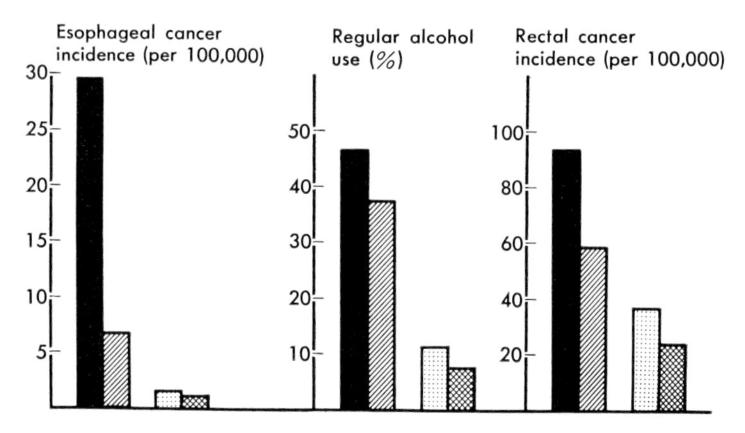

FIG. 3. Alcohol use and cancer incidence by nativity in Hawaii Japanese (45 years and older, age-adjusted). Symbols are the same as in Fig. 2.

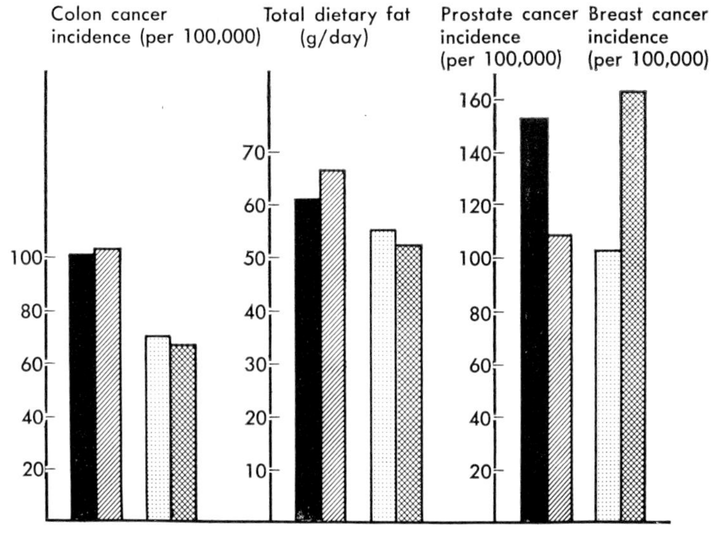

FIG. 4. Dietary fat intakes and cancer incidence by nativity in Hawaii Japanese (45 years and older, age-adjusted). Symbols are the same as in Fig. 2.

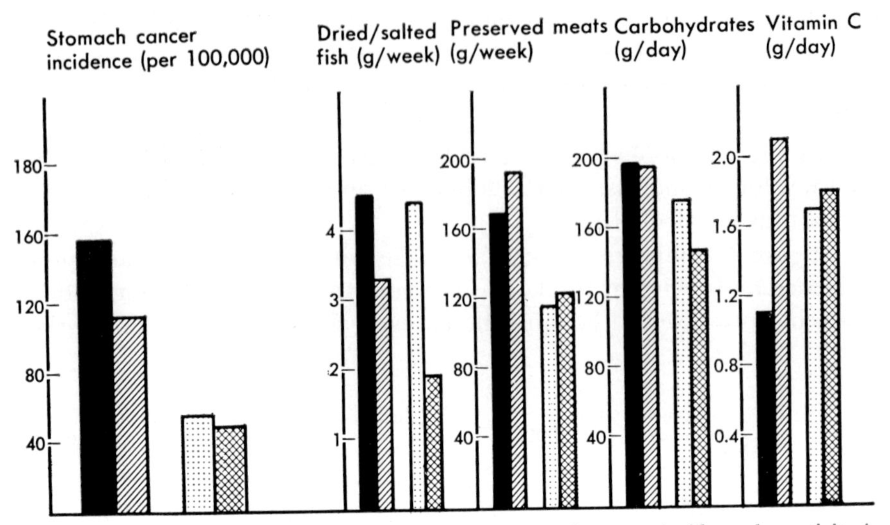

FIG. 5. Intake of selected dietary components and stomach cancer incidence by nativity in Hawaii Japanese (45 years and older, age-adjusted). Symbols are the same as in Fig. 2.

female breast cancer nor prostate cancer incidence shows a similar pattern to that for dietary fat intake.

Several dietary items of interest relative to the occurrence of stomach cancer are shown in Fig. 5. Issei of both sexes consume greater quantitites of salted fish and carbohydrate than corresponding Nisei groups (although the carbohydrate differences among men are minimal), and this correlates with the greater risk for gastric cancer among the Issei. Conversely, the Issei consume less vitamin C (intakes include supplemental vitamins) and less preserved meats than the Nisei in each sex group.

DISCUSSION

The reliability of the results reported here depends heavily on the data of Table 1 which shows the distribution of the Japanese population in Hawaii by nativity. The statewide health survey from which these estimates were developed has been conducted for about 10 years, and uses appropriate statistical methods to develop its annual household sample. We have no reason to question the validity of peoples' responses to the question on nativity, and certainly the data look reasonable in terms of what we know about the history of Japanese migration to Hawaii. The results also depend on the reliability of the tumor registry's reporting of place of birth. The registry obtains this information from the hospital charts, which record the patients' self-declared nativity. The registry had birthplace recorded on 97.0% of the Japanese patients in this report. The unknown birthplace rate exceeded 4% for only three of the sites: lymphomas (7.5%), esophagus (9.4%), and cervix uteri (14.3%). Thus, for the latter sites, the Issei-Nisei relationships could be less reliable, if those with unknown birthplace are not distributed between the two groups similarly to those with known birthplace. Since the cases with unknown birthplace result in an underestimation of the Hawaii rates, they also affect the Japan-Hawaii comparisons. However, most of the inter-country differences in Table 2 are considerable, and would not likely be accounted for by this small error.

For 10 of the sites in Table 2 (67%), Hawaii rates are higher than those in Japan. These variations presumably reflect differences in exposure to environmental factors that contribute to cancer risk, as well as to possible selection factors affecting migrants. For 6 of these 10 sites (colon, rectum, liver, bladder, thyroid, and lymphomas), the Issei incidence rates are comparable to those of the Nisei. Since most migrants were adults, this suggests that for these sites, changes in exposure to certain environmental factors in adulthood can have a major influence on the development of cancer. Such factors may be promoters rather than initiators. On the other hand, Issei rates are intermediate to those of indigenous Japanese and Nisei for breast and corpus uteri cancers, indicating a greater influence of early-life exposures (probably initiating factors). Lung and prostate are interesting sites in that the Issei rates are actually higher than the Nisei rates. Perhaps in this case, selection for those who migrate or the stress of migration itself and its aftermath in some way leads to heavier exposure to causal factors. This is consistent with the data on cigarette smoking, which show that a higher proportion of Issei than Nisei have ever been cigarette smokers. Similar findings of higher rates for lung cancer among first generation migrants have been reported by others (*3*, *9*).

Only 3 sites show lower risks in Hawaii Japanese than in Japan. For esophageal and stomach cancers, there is a clear gradient in rates from Japan to the second generation migrants in Hawaii, suggesting a more dominant role for influences in early life. In the case of cervical cancer, exposure at younger ages appears to be totally determinative, since no risk reduction is seen in Hawaii Japanese until the second generation. The lower risk of invasive cervical cancer in Nisei than Issei women could also reflect diagnostic rather than exposure factors. If more Nisei than Issei women are screened regularly for cervical cancer, more of their cancers would be detected at the *in situ* stage.

Findings generally similar to those of Table 2 have been seen by others in mortality studies of migrants. Haenszel (*4*) reported higher cancer mortality rates for Japanese in the U.S. than in Japan, with Issei rates higher than Nisei rates. Buell (*9*) examined data on the Japanese in California and found the same pattern of higher lung cancer rates among Issei than Nisei that we observed in the incidence data of this report. Staszewski (*3*) studied cancer mortality among Polish migrants to the U.S. The high stomach cancer rates seen in Poland persisted among migrants, whereas the low mortality rates for breast, colon, and rectal cancers in Poland were reflected in much higher rates among the migrants. Lung, larynx, and esophageal cancer mortality rates among migrants actually exceeded U.S. white rates.

In an incidence study, Tulchinsky (*2*) reported higher gastric cancer rates in migrants to Israel from Europe than from Asia or Africa. The rates were also higher in recent immigrants than in veteran immigrants, a pattern somewhat analogous to that for the Issei and Nisei in Hawaii. For the decade 1960–1970, Buell (*10*) reported that the incidence of breast cancer among Japanese-American women in California had approached the U.S. white rate much more closely than it had in Hawaii. Although Buell found no significant increase in California Nisei rates over Issei rates for the age group 35–64 years, the data in Table 2 continue to show a notable Nisei-Issei differential in Hawaii. This is also reflected in the age-specific incidence data of Fig. 1. In addition, Fig. 1 shows for both groups of Hawaii women the decline in rates in the older ages that is characteristic of Japan but not of the high risk western countries (*11*). The extent of this decline in the Issei and Nisei, however, suggests at least in part a cohort effect. Also notable in Fig. 1 is the absence of the so-called "menopausal hook," again a characteristic of high-risk populations, such as U.S. whites. This persistence among the Nisei of certain characteristics of the disease as it occurs in Japan is interesting, since Nisei rates are approaching those of Caucasian women in Hawaii.

The data of Table 3 show that for most exposure items, male-female differences are greater than Issei-Nisei differences within each sex. This is quite consistent with the incidence patterns of Table 2 and Figs. 2–5. Since environmental factors, including smoking and diet, have been implicated in many cancers, one would expect differences among these migrant groups in such exposures to be in the same direction as their altered cancer risks. Figure 2 shows that for the known risk factor of cigarette smoking, there is such a consistency among these groups: following the lung cancer pattern, men smoke more than women in all cases, and Issei smoke more than Nisei for both sexes. (The incidence pattern for squamous cell carcinoma of the lung shows identical relationships.) This correspondence of a known risk factor with observed incidence patterns gives us some confidence that other associations in these data may also be valid.

Both cigarette smoking and alcohol use have been associated with esophageal cancer (*12*, *13*), and the patterns of use of both factors among the Japanese in Hawaii agree with the esophageal cancer incidence rates in Fig. 3. The smoking and alcohol patterns also correspond with the rectal cancer incidence rates in Fig. 3; tobacco and cigarette use have been associated with rectal cancer in other studies (*14*, *15*). Beer consumption in particular has been reported to correlate with rectal cancer mortality (*16*, *17*). Recent data on nitrosamines in beer (*18*) suggest a possible etiologic mechanism for this association. However, the beer consumption data of Table 3 show that while men drink more beer than women and have more rectal cancer (Fig. 3), the Issei-Nisei patterns do not correspond in men.

Hirayama (*19*) has reported on dietary fat intake in Japan, where the average *per capita* consumption (both sexes combined) is 52 g per day. This is lower than the male intake values in Table 3 and about the same as the female intakes. Correlations of dietary fat and colon cancer have been reported (*20*, *21*), and the data of Fig. 4 support the hypothesis that dietary fat is etiologic in this disease. Not only are the male-female and within-sex fat intake values in the same directions as the colon cancer incidence rates, but also the small Issei-Nisei differences in fat intake within each sex are matched by small differences in colon cancer incidence. The saturated fat data of Table 3 show similar relationships.

Haenszel (*22*) reported an association of colon cancer with the frequency of consumption of beef, which could, of course, reflect fat intake. Although we have collected data on frequency of beef consumption, we have not yet examined the quantitative beef intakes for Issei and Nisei separately. Haenszel also found a weaker association between colon cancer and starches. The carbohydrate data of Table 3 are not as consistent with the colon cancer incidence rates in Fig. 4 as are the dietary fat data.

Dietary fat consumption has also been correlated with breast and prostate cancers (*23–25*). Although endocrine factors are of great importance in these cancers (*11*, *26*), diet could also contribute to the cancer risks, particularly as a promoting factor. The dietary fat data of Fig. 4 do not support this association, however. If a dietary component such as fat does contribute to the risk for these cancers, it may exert its major influence so early in life that the detection of dietary differences among adults would only be relevant to the extent it reflected childhood differences. The anthropometric data of Table 3 (based on height and weight as stated, not measured) do show that the Nisei women are a little heavier though not taller, than the Issei women, whose breast cancer incidence is lower. This is consistent with the reported association of body mass indicators with breast cancer risk (*27*).

The prostate cancer incidence rates for both Issei and Nisei men in Hawaii are much higher than in Japan. However, subclinical prostate cancer occurs as frequently in Japan as in Hawaii, although the proliferative type of latent carcinoma is more common among Hawaii Japanese (*28*). Based on our data, differences in dietary fat intake would not seem to provide an explanation for the variation in clinical rates that occurs between Japan and Hawaii despite apparently equal predispositions for the disease in both locations.

Several different dietary factors have been associated with stomach cancer *29–34*). Positive associations have been reported for starches (*30*, *31*), and for dried/salted fish and pickled vegetables in Hawaii but not Japan (*32*, *33*). Protective effects have been reported for milk, lettuce, and fresh fruits (*29*, *30*, *32*, *34*). Since endogenous formation of nitrosamines from precursor nitrites and secondary amines has been suggested as a causal mechanism for this cancer (*35*), the positive association with dried/salted fish could be explained by its high concentration of nitrite and secondary amines (*36*, *37*). Preserved or cured meats contain nitrite additive, as well as nitrate which can be reduced to nitrite (*38*), although most nitrite in the stomach probably results from production in saliva (*39*). Ascorbate blocks the nitrosamine reaction, which suggests that the negative association of gastric cancer with fresh fruits could be due to their vitamin C content (*40*).

Although the patterns in Fig. 5 for dried/salted fish and for vitamin C are consistent within each sex with these mechanisms, they are not consistent between men and women. For example, although Nisei men report a higher vitamin C intake than both Nisei and Issei women, their stomach cancer incidence rate is higher than those of both groups of women. Of course, these simple correlations cannot adequately describe a complicated mechanism that entails endogenous as well as exogenous factors that are both predisposing and protective for the disease.

In general, the fingings for Issei and Nisei in Hawaii show considerable correspondence between the exposure data (smoking, drinking, diet) and the cancer incidence rates for selected sites. These results strengthen our belief that aspects of personal lifestyle encompass the critical factors related to the occurrence of many cancers. The comparisons between Japan and Hawaii suggest that changes in exposure patterns even in adult life can alter cancer risks. Additional studies of Japanese migrants in Hawaii and elsewhere should further our understanding of the etiology of cancer.

ACKNOWLEDGMENT

This work was supported in part by Grant #1 NO1 CA 15655 and #1 RO1 CA 20897, and by Contract #NO1-CP-53511 from the U.S. National Cancer Institute.

REFERENCES

1. Haenszel, W. Cancer mortality among the foreign-born in the United States. J. Natl. Cancer Inst., *26*: 37–132, 1961.
2. Tulchinsky, D. and Modan, B. Epidemiological aspects of cancer of the stomach in Israel. Cancer, *20*: 1311–1317, 1967.
3. Staszewski, J. and Haenszel, W. M. Cancer mortality among the Polish-born in the U.S. J. Natl. Cancer Inst., *35*: 291–297, 1965.
4. Haenszel, W. and Kurihara, M. Studies of Japanese migrants. I. Mortality from cancer and other diseases among Japanese in the United States. J. Natl. Cancer Inst., *40*: 43–68, 1968.
5. Kreiebel, D. and Jowett, D. Stomach cancer mortality in the North Central States: High risk is not limited to the foregin-born. Nutr. Cancer, *1* (2): 8–12, 1979.
6. Waterhouse, J., Muir, C., Correa, P., and Powell, J. (eds.). Cancer Incidence in Five Continents, vol. III. IARC Scientific Publication No. 15, International Agency for Research on Cancer, Lyons, 1976.
7. Hankin, J. H., Rhoads, G. G., and Glober, G. A. A dietary method for an epidemiologic study of gastrointestinal cancer Am. J. Clin. Nutr., *28*: 1055–1061, 1975.
8. Nomura, A., Hankin, J., and Rhoads, G. G. The reproducibility of dietary intake data in a prospective study of gastrointestinal cancer. Am. J. Clin. Nutr., *29*: 1432–1436, 1976.
9. Buell, P. and Dunn, J. E. Cancer mortality among Japanese Issei and Nisei of California. Cancer, *18*: 656–664, 1965.
10. Buell, P. Changing incidence of breast cancer in Japanese-American women. J. Natl. Cancer Inst., *51*: 1479–1483, 1973.
11. MacMahon, B., Cole, P., and Brown, J. Etiology of human breast cancer: A review. J. Natl. Cancer Inst., *50*: 21–42, 1973.
12. Rothman, K. and Keller, A. The effect of joint exposure to alcohol and tobacco on risk of cancer of the mouth and pharynx. J. Chron. Dis., *25*: 711–716, 1972.
13. Wynder, E. L. and Mabuchi, K. Cancer of the gastrointestinal tract. I. Esophagus: Etiological and environmental factors. J. Am. Med. Assoc., *226*: 1546–1548, 1973.
14. Doll, R. and Peto, R. Mortality in relation to smoking: 20 years' observation on male British doctors. Br. Med. J., *2*: 1525–1536, 1976.
15. Wynder, E. L. and Bross, I. J. A study of etiological factors in cancer of the esophagus. Cancer, *14*: 389–413, 1961.
16. Enstrom, J. E. Colorectal cancer and beer drinking. Br. J. Cancer, *35*: 674–683. 1977.

17. Breslow, N. E. and Enstrom, J. E. Geographic correlation between cancer mortality rates and alcohol-tobacco consumption in the United States. J. Natl. Cancer Inst., *53*: 631–639, 1974.
18. Speigelhalder, B., Eisenbrand, G., and Preussmann, R. Contamination of beer with trace quantities of N-nitroso-dimethylamine. Food Cosmet. Toxicol., *17*: 29–31, 1979.
19. Hirayama, T. Epidemiology of cancer of the stomach with special reference to its recent decrease in Japan. Cancer Res., *35*: 3460–3463, 1975.
20. Hill, M. J., Crowther, J. S., Drasar, B. S., Hawksworth, G., Aries, V., and Williams, R.E.O. Bacteria and aetiology of cancer of large bowel. Lancet, *i*: 95–100, 1971.
21. Wynder, E. L. and Shigematsu, T. Environmental factors of cancer of the colon and rectum. Cancer, *20*: 1520–1561, 1967.
22. Haenszel, W., Berg, J. W., Segi, M., Kurihara, M., and Locke, F. B. Large bowel cancer in Hawaii Japanese. J. Natl. Cancer Inst., *51*: 1765–1779, 1973.
23. Armstrong, B. and Doll, R. Environmental factors and cancer incidence and mortality in different countries, with special reference to dietary practices. Int. J. Cancer, *15*: 617–631, 1975.
24. Drasar, B. S. and Irving, D. Environmental factors and cancer of the colon and breast. Br. J. Cancer, *27*: 167–172, 1973.
25. Howell, M. A. Factor analysis of international cancer mortality data and per capita food consumption. Br. J. Cancer, *29*: 328–336, 1974.
26. Weiner, J. M., Marmorston, J., Stern, E., and Hopkins, C. E. Urinary hormone metabolites in cancer and benign hyperplasia of the prostate: A multivariate statistical analysis. Ann. N.Y. Acad. Sci., *125*: 974–983, 1966.
27. deWaard, F. and Baanders-van Halewijn, E. A. A prospective study in general practice on breast cancer risk in postmenopausal women. Int. J. Cancer, *14*: 153–160, 1974.
28. Akazaki, K. and Stemmermann, G. N. Comparative study of latent carcinoma of the prostate among Japanese in Japan and Hawaii. J. Natl. Cancer Inst., *50*: 1137–1144, 1973.
29. Hirayama, T. Epidemiology of stomach cancer. Gann, *11*: 3–19, 1971.
30. Graham, S., Lilienfeld, A. M., and Tidings, J. E. Dietary and purgation factors in the epidemiology of gastric cancer. Cancer, *20*: 2224–2234, 1967.
31. Modan, B., Lubin, F., Barell, V., Greenberg, R. A., Modan, M., and Graham, S. The role of starches in the etiology of gastric cancer. Cancer, *34*: 2087–2092, 1974.
32. Haenszel, W., Kurihara M., Locke, F. B., Shimizu, K., and Segi, M. Stomach cancer in Japan. J. Natl. Cancer Inst., *56*: 265–278, 1976.
33. Haenszel, W., Kurihara, M., Segi, M., and Lee, R.K.C. Stomach cancer among Japanese in Hawaii. J. Natl. Cancer Inst., *49*: 969–988, 1972.
34. Bjelke, E. Epidemiologic studies of cancer of the stomach, colon, and rectum, with special emphasis on the role of diet. Scand. J. Gastroenterol., *9* (Suppl 31): 1–253, 1974.
35. Lijinsky, W. Nitrosamines and nitrosamides in the etiology of gastrointestinal cancer. Cancer, *40*: 2446–2449, 1977.
36. Ishidate, M., Tanimura, A., Ito, Y., Sakai, A., Sakuta, H., Kawamura, T., Sakai, K., Miyazawa, F., and Wada, H. Secondary amines, nitrites and nitrosamines in Japanese foods. *In*; W. Nakahara *et al.* (eds.), Topics in Chemical Carcinogenesis, pp. 313–322, Japan Scientific Societies Press, Tokyo / University Park Press, Baltimore, 1972.

37. Lijinsky, W. and Epistein, S. S. Nitrosamines as environmental carcinogens. Nature, *225*: 21–23, 1970.
38. White, J. W. Relative significance of dietary sources of nitrate and nitrite. J. Agric. Food Chem., *24*: 886–891, 1975.
39. Tannenbaum, S. R., Sinskey, A. J., Weisman, M., and Bishop, W. Nitrite in human saliva: its possible relationship to nitrosamine formation. J. Natl. Cancer Inst., *53*: 79–84, 1974.
40. Mirvish, S. S., Wallcave, L., Eagen, M., and Shubik, P. Ascorbate-nitrite reaction: Possible means of blocking the formation of carcinogenic *N*-nitroso compounds. Science, *177*: 65–68, 1972.

GENETIC AND ENVIRONMENTAL FACTORS IN EXPERIMENTAL AND HUMAN CANCER,
H. V. GELBOIN ET AL. (EDS.), JAPAN SCI. SOC. PRESS, TOKYO, PP. 341–350, 1980

Shifts in Cancer Mortality from 1920 to 1970 among Various Ethnic Groups in Hawaii

Tomio HIROHATA

Department of Public Health, School of Medicine, Kurume University, Fukuoka, Japan

Abstract: The present study investigated shifts in mortality of seven selected sites of cancer for the 50-year period between 1920 and 1970 among major ethnic groups in Hawaii (Caucasian, Japanese, Hawaiian, male Chinese). Age-group-specific and age-adjusted mortality rates were computed for each decennial census year from 1920 to 1970.

Cancer mortality varied greatly among ethnic groups. The time trend differed markedly for various types of cancer. When all malignancies were combined, the mortality increased with time for males, but slightly decreased for females, due to a large decline of gastric and uterine cancers. The decrease in gastric cancer mortality was striking for both sexes and all races. Although gastric cancer mortality among Japanese in Hawaii was only 40–50% that of Japan (1970), birth cohort analysis indicated a continued decline in the future.

Uterine cancer (primarily cervix uteri) was another site of cancer which showed a marked decrease during the 50 years. Cancers of the colon (male), pancreas, prostate, and breast (female) showed substantial increases. Cancers of the lung (particularly for males) and ovary showed a striking increase. These findings are indicative of great environmental impacts on the causation of human cancer.

The present study deals with shifts in cancer mortality for seven selected sites from 1920 to 1970 among major ethnic groups in Hawaii. Hawaii is known as a "human laboratory" for epidemiologists, because of its multi-ethnic, mixed population. Various ethnic groups in Hawaii show different levels of risk for various diseases, which provides an unusual opportunity for the investigation of the etiology of diseases. Residents of Hawaii are essentially migrants and their descendants, hence, shifts in mortality of an ethnic group in Hawaii as compared to the country of origin provide an additional advantage for investigating the effects of environmental factors in the etiology of disease.

The author has reported on mortality patterns from various causes among major ethnic groups in Hawaii in 1960 as compared to their countries of origin

(*1*). Further, the author has been analyzing, in cooperation with the Department of Health in Hawaii and the Epidemiology Program, Cancer Center of Hwaii, the mortality pattern for 80 selected causes among major ethnic groups. The present study has focused on one aspect of this mortality analysis, namely, shifts in cancer mortality for the 50-year period between 1920 and 1970.

MATERIALS AND METHODS

Major ethnic groups selected for the present analysis were Japanese, Caucasians, Hawaiians, and male Chinese. There are five major ethnic groups in Hawaii. Among them, Filipino and female Chinese were not included in this study because too small a number of deaths was observed in earlier years (less than 50 for all cancer deaths for 1930 and/or 1920). Seven major sites of cancer with sufficient numbers of deaths were chosen from among 35 sites of cancer which were initially examined for further analysis. The seven sites selected were: stomach, large intestine (excluding rectum), pancreas, lung, female breast, uterus, and prostate.

The annual number of deaths was averaged for each 5 years centering on the decennial census between 1920 and 1970. Five- and 10-year age-group-specific rates, as well as age-adjusted rates with 95% confidence intervals, were computed. The age adjustment was done by the direct method using the world population as the standard (*2*).

In order to adjust for periodic changes in the International Classification of Diseases (ICD), all deaths were re-coded to the 7th ICD revision by the Department of Health of Hawaii.

RESULTS

Time trends in cancer mortality of all sites are shown in Fig. 1. Due to chance variation, the rates fluctuated substantially. On the whole, however, females of all ethnic groups (Caucasian, Japanese, Hawaiian) showed a slight reduction in rates over the 50 years from 1920 to 1970, whereas males of all ethnic groups (Caucasian, Japanese, Hawaiian, Chinese) showed a marked increase with time. The reduction in rates for females was brought about by great decreases in gastric cancer and uterine cancer.

A great reduction in gastric cancer was also seen for males. This was offset, however, by large increases in other sites of cancer, such as cancer of the lung. Nevertheless, it was interesting to note that both Japanese and Chinese males showed a slight decrease from 1950 to 1970 due to the decrease in gastric cancer.

Figure 2 shows the time trend for gastric cancer. The reduction in rates was very striking for both sexes and for all races. For example, the rates for Caucasian males declined from 53.48 in 1920 to 12.51 in 1970, *i.e.*, a 77% reduction. The corresponding figures for Caucasian females were 35.51 and 5.77, *i.e.*, an 84% reduction. The rates for Japanese males decreased from the highest value of 98.65 in 1930 to 29.95 in 1970, *i.e.*, a 70% reduction. For Japanese females, the cor-

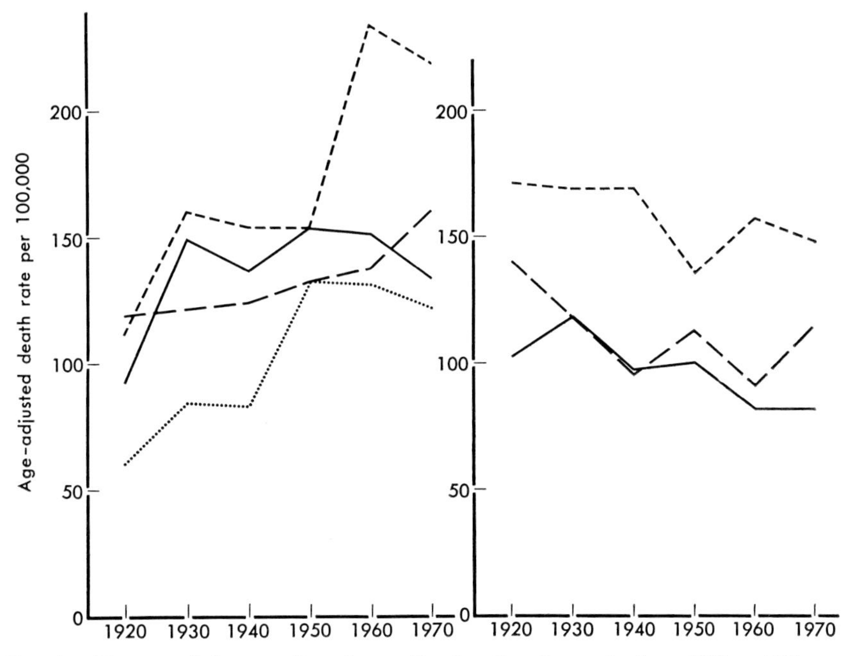

FIG. 1. Time trends in age-adjusted mortality for all malignancies from 1920 to 1970 among major ethnic groups in Hawaii. Left, male; right, female. — — Caucasian; —— Japanese; ······ Chinese; - - - Hawaiian.

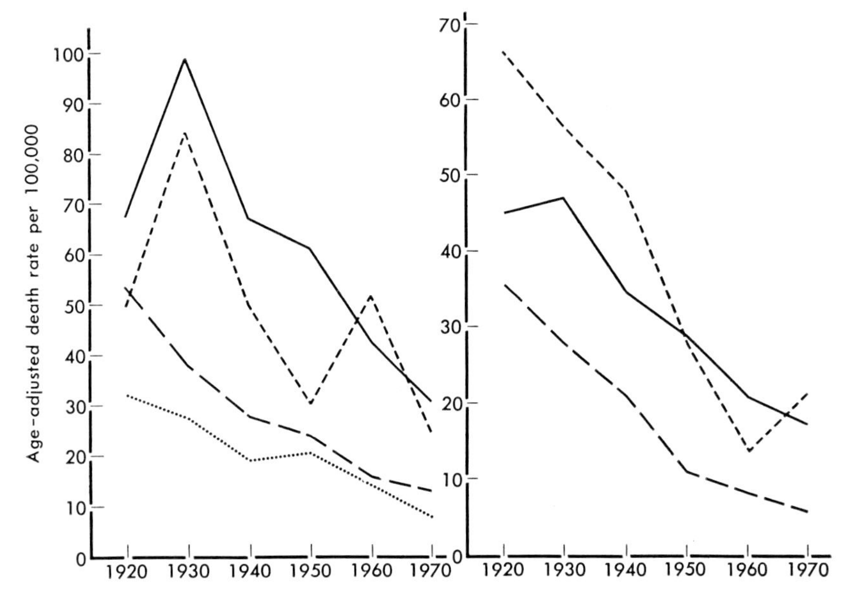

FIG. 2. Time trends in age-adjusted mortality for gastric cancer from 1920 to 1970 among major ethnic groups in Hawaii. Left, male; right, female. Symbols are the same as in Fig. 1.

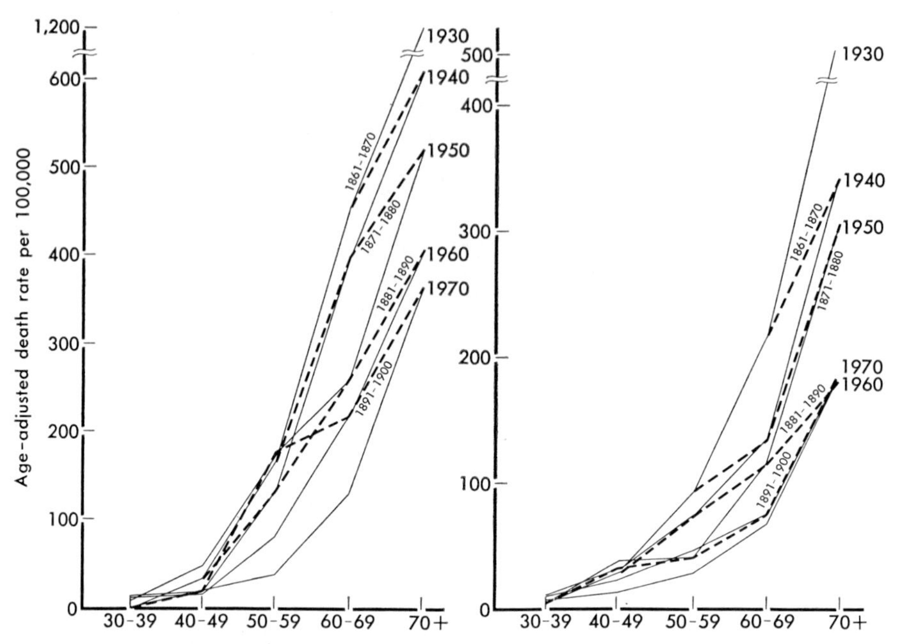

Fig. 3. Birth cohort analysis for gastric cancer among Japanese in Hawaii, 1930–1970. Solid lines are age distribution curves and bold broken lines are birth-cohort curves. Left, male; right, female.

responding figures were 46.95 (1930) and 17.06 (1970), *i.e.*, a 64% reduction within 40 years.

In contrast to Japanese immigrants in Hawaii, indigenous Japanese in Japan did not show any signs of decrease in gastric cancer mortality until about 1965–1970. Thus, the Japanese in Hawaii who were exposed to new environments (western types of diet, *etc.*) began to show a decline in gastric cancer mortality far earlier than the Japanese in Japan. This is interesting with regard to the study of the etiology of gastric cancer. The notable decreasing trend among Japanese in Hawaii can be expected to continue, as the birth cohort analysis in Fig. 3 indicates. For successive birth cohorts, based on 10-year intervals of birth, younger cohorts showed greater reductions in rates of gastric cancer than older cohorts. Thus, the already lower rates of gastric cancer among Japanese in Hawaii compared with those in Japan (45% for males and 49% for females) can be expected to show a further decline in the future.

Chinese (males), another oriental population, showed rates of gastric cancer that were as low as those of Caucasians. Hawaiians showed very high rates, which is in concert with the high rates among Maoris in New Zealand who are also Polynesians and closely related to the Hawaiians.

Figure 4 shows time trends for colon cancer. For males, increasing trends were evident for all ethnic groups. Mortality from colon cancer among Japanese in Japan was only about 25% of the U.S. White rate in 1966–67 (*2*). In Hawaii, however, the Japanese showed a marked increase over the years and their colon

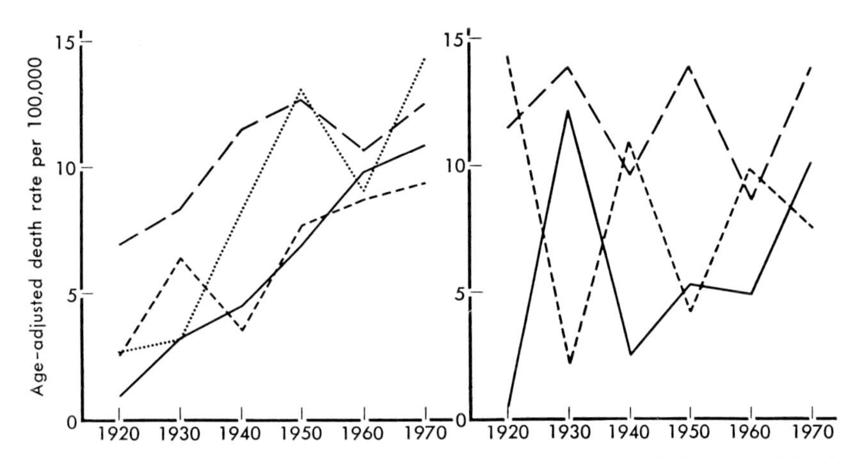

FIG. 4. Time trends in age-adjusted mortality from colon cancer. Left, male; right, female. — — Caucasian; —— Japanese; ······ Chinese; - - - Hawaiian.

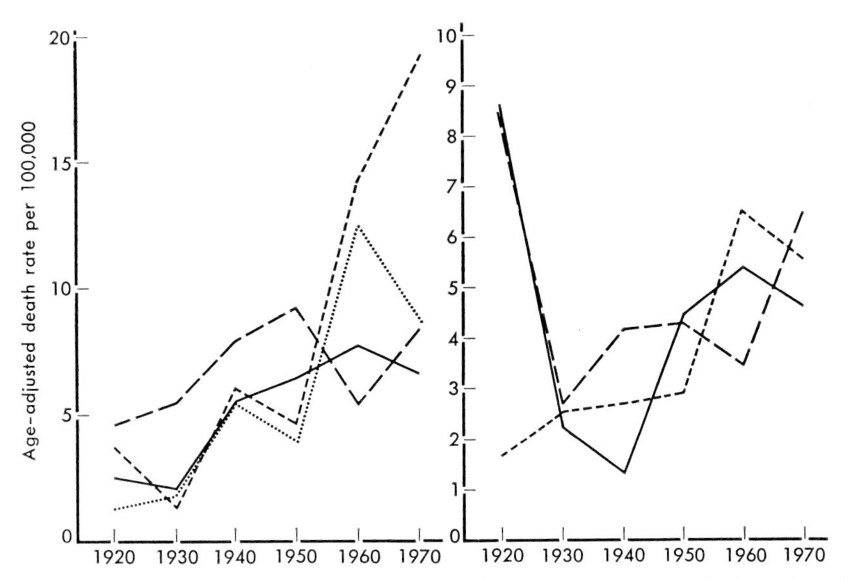

FIG. 5. Time trends in age-adjusted mortality from pancreas cancer. Left, male; right, female. Symbols are the same as in Fig. 4.

cancer mortality became nearly comparable with the Caucasian rate in 1970. For females, the trends were erratic due to small numbers. In contrast to males, no increasing trends were evident.

Figure 5 shows time trends for pancreas cancer. Among males, except for a drastic increase for Hawaiians from 1950 to 1970, moderate increases were observed for the other three ethnic groups over the study period (1920 to 1970). The rates for females were erratic due to small numbers, especially in the earlier years. The general tendency, however, was a substantial increase with time.

Figure 6 shows rates for lung cancer. The increasing trends with time were

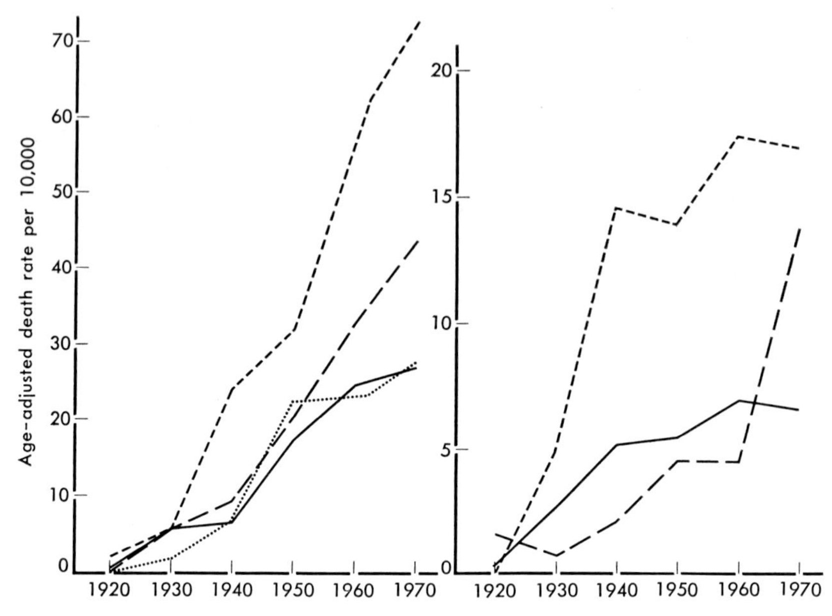

FIG. 6. Time trends in age-adjusted mortality from lung cancer. Left, male; right, female. Symbols are the same as in Fig. 4.

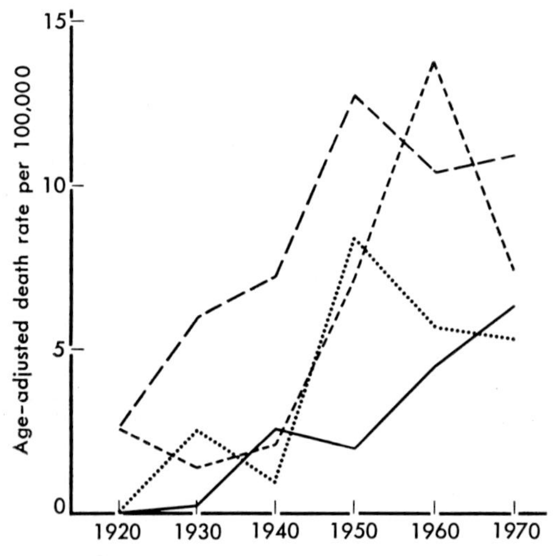

FIG. 7. Time trends in age-adjusted mortality from prostate cancer. Symbols are the same as in Fig. 4.

very marked and this was particularly so for males. For both sexes, Hawaiians showed much higher rates than other ethnic groups. The epidemic of lung cancer seen in many countries of the world was also evident in Hawaii.

Figure 7 shows the time trend for prostate cancer mortality. On the whole, an increasing tendency was observed for all ethnic groups. Caucasians had much higher rates than other ethnic groups. The two oriental populations, Chinese and

Japanese, showed rapid increases over the years and the rate for Japanese in 1970 was more than 3 times the rate of indigenous Japanese in Japan.

Figure 8 shows the mortality rates for female breast cancer. Japanese women showed far lower mortality rates from breast cancer than Caucasian or Hawaiian women. For all these ethnic groups, breast cancer showed some increases during the 50-year study period.

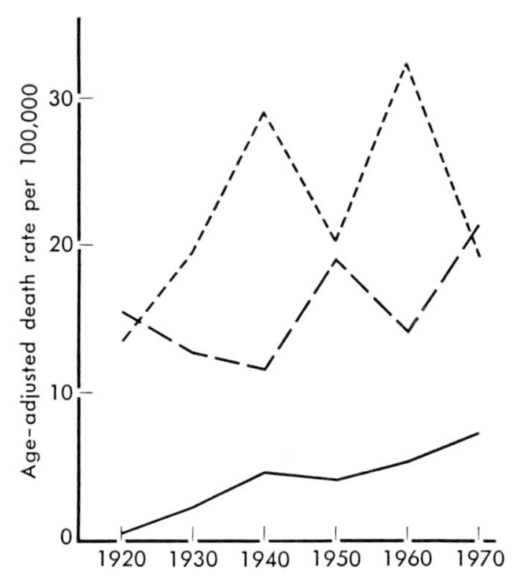

FIG. 8. Time trends in age-adjusted mortality from cancer of the female breast. — — Caucasian; —— Japanese; - - - Hawaiian.

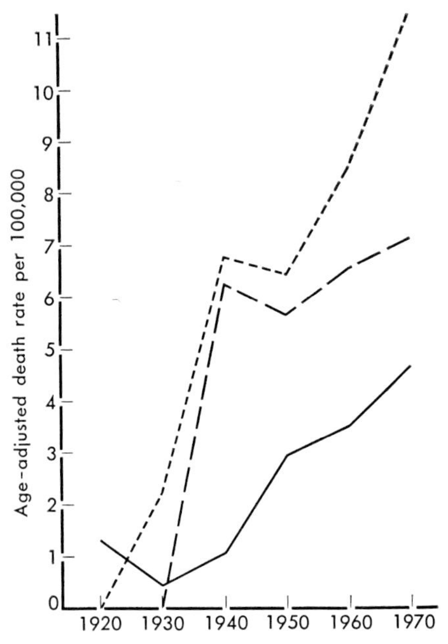

FIG. 9. Time trends in age-adjusted mortality from ovarian cancer. Symbols are the same as in Fig. 8.

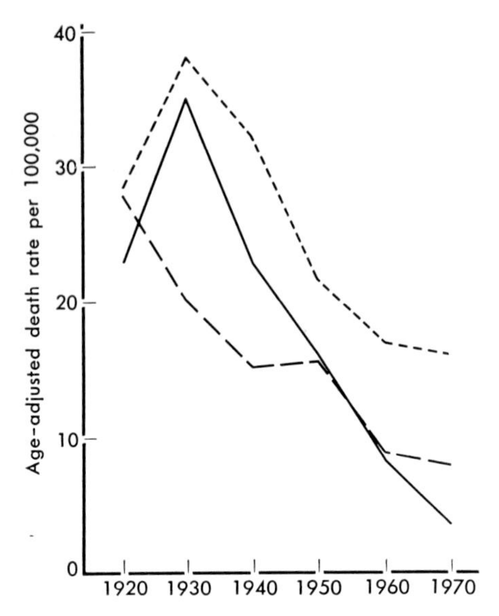

FIG. 10. Time trends in age-adjusted mortality from uterine cancer. Symbols are the same as in Fig. 8.

Figure 9 shows the data for ovarian cancer; all three ethnic groups showed striking increases with time. The rate for the Japanese was much lower than those for the Caucasians or Hawaiians, but was about 2.5 times the rate for Japan in 1970.

Figure 10 shows time trends for uterine cancer, which includes cervix uteri, corpus uteri, and uterus unspecified. It is probable that cancer of the cervix uteri comprised the vast majority of uterine cancer, particulary in the earlier years. As shown in the figure, the decreases in uterine cancer rates for all ethnic groups for the 50-year period were quite remarkable. For example, the rates for Caucasian women dropped from 28.03 in 1920 to 7.84 in 1970, *i.e.*, a 72% reduction. As stated earlier, the decrease in uterine cancer mortality together with that of gastric cancer, was largely responsible for the overall decrease of cancer of all sites observed in women of the three ethnic groups from 1920 to 1970.

DISCUSSION

Before these cancer mortality statistics were accepted at face value, it seemed necessary to examine whether they were reflecting some artificial factors rather than representing true risks for cancer. As stated earlier, changes in the ICD over the years were adjusted by re-coding all cancer deaths to the 7th ICD revision. Accuracy of diagnosis on death certificates was an additional problem. As all races lived in the restricted area of the Hawaiian Islands, the problem of possible differences in medical practice between ethnic groups was far less acute than it is for racial comparisons between countries. Among various causes of deaths, cancer diagnosis was found to be relatively accurate (*3*). Further, cancer sites on the surface of the body were diagnosed very accurately. However, the accuracy of diagnosis for cancers of deep organs was rather poor. On the whole, when autopsy diagnosis was compared with clinical diagnosis, it was clear that cancer was underdiagnosed (*3, 4*). Thus, more cancer deaths would be expected with the passage of time due to improvements in medical diagnosis. In view of this, the true decline for gastric cancer and uterine cancer from 1920 to 1970 could in fact be greater than is presented in Figs. 2 and 10.

Among the seven major sites of cancer examined, the reductions in mortality from gastric and uterine cancers over the years were truly remarkable. There must have been a great reduction in exposure to the causative factors for gastric cancer and uterine cancer. The Japanese population in Japan has an unusually high mortality from gastric cancer, even today. Therefore, identification of environmental factors that brought about a large decline in gastric cancer among the Japanese in Hawaii is very important. In this connection, risk factors suspected epidemiologically—a lower intake of fruit, vegetables, and milk, and a higher intake of starchy foods, dried and salty fish, *etc.* (*5*)—are interesting factors for consideration. Further, the fact that some traditional Japanese food items such as bracken fern, broiled fish, *etc.* were found to be carcinogenic for laboratory animals, is noteworthy.

Those cancer sites which showed increases with time are more difficult to interpret because of possible medical improvements in finding larger numbers of cancer patients with the passage of time. However, the constant, marked increases noted for lung cancer, prostate cancer, ovarian cancer, *etc.*, probably represent true increases in risk and indicate accompanying increases in exposure to environmental, causative factors.

ACKNOWLEDGMENTS

The present study was made possible through the cooperation and assistance of many individuals. Special appreciation is expressed to Dr. Burch and the late Mr. Bennet at the Hawaii State Department of Health; Drs. Kolonel and Inaba, Mr. Johnson, and Ms. Siu at the Epidemiology Program, Cancer Center of Hawaii; the late Dr. Dickinson at the School of Public Health, University of Hawaii; and Mr. Shibata at the Department of Public Health, Kurume University, School of Medicine.

REFERENCES

1. Hirohata, T. Mortality experience of various ethnic groups in Hawaii with particular reference to the comparison with their countries of origin. Jpn. J. Public Health, *21*: 17–26, 1974 (in Japanese).
2. Segi, M. and Kurihara, M. Cancer Mortality for Selected Sites in 24 Countries, No. 6, Japan Cancer Society, Tokyo, 1972.
3. Jablon, S., Angevine, D. M., Matsumoto, Y. S., and Ishida, M. On the significance of cause of death as recorded on death certificates in Hiroshima and Nagasaki, Japan. *In*; Epidemiological Study of Cancer and Other Chronic Diseases. Natl. Cancer Inst. Monogr. No. 19, pp. 445–465, 1966.
4. Hiyoshi, Y., Omae, T., Takeshita, M., Ueda, K., Hirota, Y., Katsuki, S., Tanaka, K., and Enjoji, M. Malignant neoplasms found by autopsy in Hisayama, Japan,

during the first ten years of a community study. J. Natl. Cancer Inst. *59*: 13–19, 1977.

5. Hirohata, T., Tomita, Y., and Shibata, A. Epidemiology of diet, nutrition and cancer. J. Jpn. Soc. Food Nutr., *33*: 1–7, 1980 (in Japanese).

GENETIC AND ENVIRONMENTAL FACTORS IN EXPERIMENTAL AND HUMAN CANCER,
H. V. GELBOIN ET AL. (EDS.), JAPAN SCI. SOC. PRESS, TOKYO, PP. 351–357, 1980

Clinical Clues to Interactions in Carcinogenesis

Robert W. MILLER
National Cancer Institute, Bethesda, Md. 20205, U.S.A.

Abstract: Clinical observations have led to recognition that carcinogenesis can be enhanced by interactions of the host, chemicals, physical agents, and viruses in various combinations. When such occurrences are seen at the bedside, specimens should be obtained for laboratory studies which can provide new understanding of the carcinogenic process.

Observations at the bedside of peculiarities in the occurrence of cancer may serve as human models of carcinogenesis for which animal models are not yet known. Cells or body fluids from affected persons are available for study in laboratories to provide new understanding of the biology of cancer.

Host-Chemical Interactions

From bedside observations in Boston, diethylstilbestrol (DES) was discovered in 1971 to cause a cancer in young women that ordinarily develops only among the elderly (*1*). The frequency of the neoplasm, clear-cell adenocarcinoma of the vagina or cervix, affects about one in 1,000 females exposed to DES in the first 17 weeks of gestation (*2*). (The gestational age of susceptibility differs from that in animal experimentation with other transplacental carcinogens, which produce their effects when given only late in pregnancy (*3*).) Why does human exposure to DES cause cancer in only 1 of 1,000 females and exhibit no dose-response effect? The explanation may be differences in host susceptibility, in the intrauterine environment or a combination of the two.

Any chemical that crosses the placenta and is known to induce human cancer may be carcinogenic to the offspring. Thus, exposure of the pregnant woman to benzene might be leukemogenic to her child, but no such cases have been identified. The same is true for *l*-phenylalanine mustard (melphalan), an alkylating agent used for chemotherapy, immunosuppressive drugs used in patients with renal transplantation and vinyl chloride which was formerly used as a propellant in hair-sprays.

Diphenylhydantoin (DPH) crosses the placenta and sometimes produces the fetal hydantoin syndrome (*4*), characterized by mid-facial skeletal hypoplasia that causes the face to grow disproportionately, and absence or hypoplasia of the fingernails and toenails. Among people who receive DPH to control epilepsy, lymphoma apparently occurs excessively, and if the drug is withdrawn in time, the tumor may regress spontaneously (pseudolymphoma) (*5*). We expected, therefore, that lymphoma would soon be observed in children with the fetal hydantoin syndrome. Much to our surprise, an alumnus of our Branch reported in a letter to Lancet in 1976 that a child with the syndrome had developed neuroblastoma (*6*). Within a few weeks a second such case was reported from the United States (*7*). Our rule of thumb is that if three pairs of rare occurrences are observed in the United States in a short period of time, the association is not due to chance. We advertised for a third case, and one was reported from Chicago (*8*). The child's mother was not only epileptic but also alcoholic, so the child had what appeared to be overlapping syndromes due to alcohol *and* DPH. He developed ganglioneuroblastoma at 35 months of age. We have since heard of another child in Utah with fetal hydantoin syndrome and congenital neuroblastoma (*9*). Thus DPH represents the second known transplacental chemical to induce cancer in human beings.

The mother of the last-mentioned child had been on DPH since 5 years of age, and had had three previous children who were affected with neither the syndrome nor the neoplasm. Thus an interaction must be involved to explain why the last child was affected but the earlier ones were not. Among the possibilities are genetic susceptibility of the child or changes due to maternal age, diet, smoking habits, alcohol use and/or infection, other diseases (*e.g.*, hypertension), prescribed drugs, *etc.*

Host-Viral Interactions

In 1969 a family was seen at the National Cancer Institute because three children had developed acute myelogenous leukemia (AML), as had three maternal relatives. Two others had malignant reticuloendotheliosis (*10*). Studies of skin fibroblasts in culture, infected with simian virus 40 (SV40) revealed that the cells transformed at 50 times the normal rate for the only surviving child with AML, her mother and her older sister. The transformation rates were normal for her twin brothers and father. We suspected that the older sister, who was healthy at that time, was at increased risk of AML. In 1976 she returned to NCI because she too had developed AML, as if the test predicted it (*11*). The skin fibroblasts were recently studied for DNA-repair defects by Paterson *et al.* (*12*), and a small but reproducible deficiency has been identified in the three members of the family who showed increased transformation of fibroblasts after infection with SV40 in culture. Urine was collected from the mother, 5,000 ml, lyophilized and evaluated for mutagenicity in the Ames test by M. Nagao of Dr. Sugimura's laboratory. Some mutagenic activity was detected, the nature of which is unknown. The mother had earlier been treated for carcinoma of the cervix and was apparently cured. She has recently developed carcinoma of the rectum. The laboratory tests indicate a host susceptibility to a virus that is known to be oncogenic (but not leukemogenic) in newborn hamsters.

A causal relation between host susceptibility and a specific virus has been indicated by bedside observations made by Purtilo *et al.* (*13*). They have found that in certain families more than one boy is affected with fatal infectious mononucleosis, agammaglobulinemia or a lymphoproliferative disorder, and concluded that an X-linked gene predisposes to the effects of Epstein-Barr virus on B lymphocytes.

Host-Physical Agent Interactions

In 1965 Good and his associates (*14*) from bedside observations pieced together evidence for an association between inborn immunological deficiency disorders and lymphoma. They also predicted that these disorders would occur with acute lymphocytic leukemia (ALL), although no such cases were known. We reasoned that if ataxia-telangiectasia (AT) occurred excessively with ALL we should be able to find such cases in a series of about 500 children in the National Childhood Leukemia Survey conducted by NCI in 1958–59 (*15*). In those years AT was not yet well known, so we looked for the occurrence of leukemia with other congenital neurological disorders. Immediately we found a pair of sibs with ataxia (Friedreich's?) and ALL (*16*). The Survey records showed that a third sibling also had ataxia and was still alive. The family lived in Oregon, and a telephone call to F. Hecht eventually led to the diagnosis of AT in the surviving member, and photographs that showed one of the younger sibilings clearly had telangiectasia of the bulbar conjunctivae (*17*). Cytogenetic studies of the surviving sibling revealed chromosomal instability on culture (*17*) and later clonal evolution of a D/D translocation (*18*). Within 4 years the clone increased from almost 0 to 78% of the peripheral lymphocytes. The study was terminated with the patient's death from pulmonary infection.

Meanwhile, other bedside observations led to individual case-reports of two patients with AT and lymphoma, who developed severe acute reactions to conventional doses of radiotherapy (*19*, *20*). Ten years later another such case occurred in Birmingham, England, where clinic and laboratory scientists were in close touch. Skin fibroblasts studied in culture showed diminished cell survival in proportion to radiation dose (*21*), due to a DNA-repair defect (*22*). Clinical observations thus led to recognition of clonal evolution of cytogenetically abnormal lymphocytes in patients with a heritable disorder, immunodeficiency, radiation sensitivity, and predisposition to cancer. In this mixture is indisputable evidence of an interaction between host and a physical agent.

Chemical and Viral or Immunological Interactions

Perhaps the best example to date of a chemical-viral interaction in carcinogenesis, apart from laboratory experimentation, is not from human studies but from an observation in a domestic animal. Scottish cattle develop squamous cell carcinoma of the gastrointestinal tract after grazing in fields containing bracken fern, in which a naturally occurring mutagen/carcinogen is known to exist. The animals also develop papillomas of the esophagus due to a virus morphologically indistinguishable from bovine cutaneous papilloma virus. In neighboring areas free of bracken fern, the cattle develop papillomas of the esophagus, but do not develop squamous cell carcinoma (*23*).

In patients who have received renal transplants and are heavily immunosuppressed with drugs a marked excess of lymphoma develops, often primary in the brain, beginning as soon as 3 months after transplantation (*24*). Cancers of the hepatobiliary system and skin occur in less marked excess (*24*). A scattering of patients given immunosuppressive drugs for medical disorders have developed lymphomas primary in the brain (*25*), but the excess is well below that observed after renal transplantation. The difference suggests that foreign tissue plus immunosuppression enhances the development of lymphomas. No instance of transplacental cancer induction has been reported after immunosuppression during pregnancies in women who have received renal transplants (*26*).

An interacting circumstance in which a virus might be causally related to lymphoma could be sought in children with congenital immunodeficiency disorders. When such children are seen, inquiry should be made concerning their exposure to cats. If such history is obtained, appropriate virological studies of the child and the cats should be made to determine if feline leukemia virus might have played a role.

Chemical-Physical Agent Interactions

The markedly enhanced risk of bronchogenic carcinoma from the interaction of cigarette-smoking and asbestos exposure is too well known (*27*) to be recapitualted here. In the United States an estimated 10,000 schools built between 1946 and 1973 have asbestos-sprayed ceilings, which have deteriorated with time (*28*). Some children exposed recently to asbestos fibers as the ceilings in schoolrooms deteriorate are now beginning to smoke cigarettes. Clinicians should be on the lookout for cases of bronchogenic carcinoma in young adults, so inquiry can be made about smoking in relation to asbestos exposures at school. In non-smokers mesothelioma from such exposure is a possibility, as indicated by a family in which asbestos fibers brought home on the father's workclothes apparently was responsible for the development of mesothelioma in his wife and 34-year-old daughter decades later (*29*).

Host-? Interactions

Absences of certain cancers in special groups indicate host resistance for reasons as yet unknown. Blacks in the United States and in Africa have a near absence of Ewing's tumor (*30*), and no peak at 4 years of age for childhood leukemia of the lymphocytic type (*31*). Chinese and Japanese have a near absence of chronic lymphocytic leukemia (*32*). These ethnic groups are apparently resistant to the causes of these neoplasms, for they do not develop them even after migration to countries in which the neoplasms occur with much greater frequency (*33*).

Identical twins with the 11p- chromosomal deletion now consistently found in the Wilms' tumor-aniridia syndrome have been seen in Chicago. Only one developed the tumor although the twins are now well beyond the age of occurrence of nephroblastoma (*34*). Other children have been reported with the deletion, but have not developed the tumor (*35*). Clearly something more than the marked genetic predisposition is required for tumorigenesis.

Interactions: Clues from Geographic Variations

Geographic variations in the frequencies of tumors may indicate the presence of a co-carcinogenic factor or an anticarcinogen. The association of Burkitt's lymphoma in Africa with holoendemic malaria (*36*) is well known. Not so well known is the near absence of neuroblastoma in the same and neighboring regions, even when sought by pathologists who are experienced in the diagnosis. Neuroblastoma regresses spontaneously in infants with Type IVS disease (*37*), and among fetuses and infants under 3 months of age with *in-situ* lesions (*38*, *39*). In Africa possibly a hyperimmunity of the mother passively protects the fetus or infant by causing regression of *in-situ* neuroblastoma.

CONCLUSION

A wide array of clinical observations has led to new understanding of interactions in the development of human cancers. It is important for physicians and others to be alert continuously for peculiarities in the occurrence of cancer in the patient, family or community, which present opportunities for laboratory research in the carcinogenic process.

REFERENCES

1. Herbst, A. L., Ulfelder, H., and Poskanzer, D. C. Adenocarcinoma of the vagina. N. Engl. J. Med., *284*: 878–881, 1971.
2. Herbst, A. L., Poskanzer, D. C., Robboy, S. J., Friedlander, L., and Scully, R. E. Prenatal exposure to stilbestrol: A prospective comparison of exposed female offspring with unexposed controls. N. Engl. J. Med., *292*: 334–339, 1975.
3. Rice, J. M. An overview of transplacental chemical carcinogenesis. Teratology, *8*: 113–126, 1973.
4. Hanson, J. W. and Smith, D. W. The fetal hydantoin syndrome. J. Pediatr., *87*: 285–290, 1975.
5. Hoover, R. and Fraumeni, J. F., Jr. Drugs. *In*; J. F. Fraumeni, Jr. (ed.), Persons at High Risk of Cancer: An Approach to Cancer Etiology and Control, pp. 185–198, Academic Press, New York, 1975.
6. Pendergrass, T. W. and Hanson, J. W. Fetal hydantoin syndrome and neuroblastoma. Lancet, *ii*: 150, 1976.

7. Sherman, S. and Roizen, N. Fetal hydantoin syndrome and neuroblastoma. Lancet, *ii*: 517, 1976.
8. Seeler, R. A., Israel, J. N., Royal, J. E., Kaye, C. I., Rao, S., and Abulaban, M. Ganglioneuroblastoma and fetal hydantoin-alcohol syndromes. Pediatrics, *63*: 524–527, 1979.
9. Allen, R. W., Jr., Ogden, B., Bentley, F. L., and Jung, A. L. Fetal hydantoin syndrome, neuroblastoma and hemorrhagic disease in a neonate. JAMA, in press.
10. Snyder, A. L., Li, F. P., Henderson, E. S., and Todaro, G. J. Possible inherited leukaemogenic factors in familial acute myelogenous leukaemia. Lancet, *i*: 586–589, 1970.
11. McKeen, E. A., Miller, R. W., Mulvihill, J. J., Blattner, W. A., and Levine, A. S. Familial leukaemia and SV40 transformation. Lancet, *ii*: 310, 1977.
12. Paterson, M. C., Smith, P. J., Bech-Hansen, N. T., Smith, B. P., and Sell, B. M. Gamma-ray hypersensitivity and faulty DNA repair in cultured cells in humans exhibiting familial cancer proneness. *In*; Radiation Research. Proceedings of the Sixth International Congress of Radiation Research, Tokyo, Japan, 1979, pp. 484–495, Japanese Association for Radiation Researches, Tokyo, 1980.
13. Purtilo, D. T., DeFlorio, D., Jr., Hutt, L. M., Bhawan, J., Yang, J.P.S., Otto, R., and Edwards, W. Variable phenotypic expression of an X-linked recessive lymphoproliferative syndrome. N. Engl. J. Med., *297*: 1077–1081, 1977.
14. Peterson, R.D.A., Cooper, M. D., and Good, R. A. Disorders of the thymus and other lymphoid tissues. *In*; A. G. Steinberg and A. G. Bearn (eds.), Progress in Medical Genetics, vol. 4, pp. 1–31, Grune & Stratton, New York, 1965.
15. Miller, R. W. Down's syndrome (mongolism), other congenital malformations and cancers among the sibs of leukemic children. N. Engl. J. Med., *268*: 393–401, 1963.
16. Miller, R. W. Syndrome delineation and other uses of epidemiology. Pediatr. Clin. North Am., *15*: 387–394, 1968.
17. Hecht, F., Koler, R. D., Rigas, D. A., Dahnke, G. S., Case, M. P., Tisdale, V., and Miller, R. W. Leukaemia and lymphocytes in ataxia-telangiectasia. Lancet, *ii*: 1193, 1966.
18. Hecht, F., McCaw, B. K., and Koler, R. D. Ataxia-telangiectasia—Clonal growth of translocation lymphocytes. N. Engl. J. Med., *289*: 286–291, 1973.
19. Gotoff, S. P., Amirmokri, E., and Liebner, E. J. Ataxia telangiectasia. Neoplasia, untoward response to X-irradiation, and tuberous sclerosis. Am. J. Dis. Child., *114*: 617–625, 1967.
20. Morgan, J. L., Holcomb, T. M., and Morrissey, R. W. Radiation reaction in ataxia telangiectasia. Am. J. Dis. Chid., *116*: 557–558, 1968.
21. Taylor, A.M.R., Metcalfe, J. A., Oxford, J. M., and Harnden, D. G. Is chromatid-type damage in ataxia telangiectasia after irradiation at G_0 a consequence of defective repair? Nature, *260*: 441–443, 1976.
22. Paterson, M. C., Smith, B. P., Lohman, P.H.M., Anderson, A. K., and Fishman, L. Defective excision repair of γ-ray-damaged DNA in human (ataxia telangiectasia) fibroblasts. Nature, *260*: 444–447, 1976.
23. Jarrett, W.F.H., McNeil, P. E., Grimshaw, W.T.R., Selman, I. E., and McIntyre, W.I.M. High incidence area of cattle cancer with a possible interaction between an environmental carcinogen and a papilloma virus. Nature, *274*: 215–217, 1978.
24. Hoover, R. Effects of drugs—Immunosuppression. *In;* H. H. Hiatt, J. D. Watson, and J. A. Winsten (eds.), Origins of Human Cancer: Book A, Incidence of Cancer

in Humans, Cold Spring Harbor Conferences on Cell Proliferation, vol. 4, pp. 369–379, Cold Spring Harbor Laboratory, New York, 1977.

25. Varadachari, C., Palutke, M., Climie, A.R.W., Weise, R. W., and Chason, J. L. Immunoblastic sarcoma (histiocytic lymphoma) of the brain with B cell markers: Case report. J. Neurosurg., *49*: 887–892, 1978.
26. Cederqvist, L. L., Merkatz, I. R., and Litwin, S. D. Fetal immunoglobulin synthesis following maternal immunosuppression. Am. J. Obstet. Gynecol., *129*: 687–690, 1977.
27. Selikoff, I. J. and Hammond, E. C. Multiple risk factors in environmental cancer. *In*; J. F. Fraumeni, Jr. (ed.), Persons at High Risk of Cancer: An Approach to Cancer Etiology and Control, pp. 467–483, Academic Press, New York, 1975.
28. Spooner, C. M. Asbestos in schools—A public health problem. N. Engl. J. Med., *301*: 782–784, 1979.
29. Li, F. P., Lokich, J., Lapey, J., Neptune, W. B., and Wilkins, E. W., Jr. Familial mesothelioma after intense asbestos exposure at home. J. Am. Med. Assoc., *240*: 467, 1978.
30. Miller, R. W. Contrasting epidemiology of childhood osteosarcoma, Ewing's tumor and rhabdomyosarcoma. Natl. Cancer Inst., Monogr., *57*, in press.
31. Miller, R. W. Environmental agents in cancer. Yale J. Biol. Med., *37*: 487–501, 1965.
32. Miller, R. W. Ethnic differences in cancer occurrence: Genetic and environmental influences with particular reference to neuroblastoma. *In*; J. J. Mulvihill, R. W. Miller, and J. F. Fraumeni, Jr. (eds.), Genetics of Human Cancer, pp. 1–14, Raven Press, New York, 1977.
33. Elizaga, F. V. and Oishi, N. Chronic lymphocytic leukemia in Japanese in Hawaii. Hawaii Med. J., *36*: 169–171, 1977.
34. Francke, U., Riccardi, V. M., Hittner, H. M., and Borges, W. Interstitial del(11p) as a cause of the aniridia-Wilms tumor association: Band localization and a heritable basis. Am. J. Hum. Genet., *30*: 81A, 1978.
35. Riccardi, V. M., Sujansky, E., Smith, A. C., and Francke, U. Chromosomal imbalance in the aniridia-Wilms' tumor association: 11p interstitial deletion. Pediatrics, *61*: 604–610, 1978.
36. O'Conor, G. T. Persistent immunologic stimulation as a factor in oncogenesis, with special reference to Burkitt's tumor. Am. J. Med., *48*: 279–285, 1970.
37. Schwartz, A. D., Dadash-Zadeh, M., Lee, H., and Swaney, J. J. Spontaneous regression of disseminated neuroblastoma. J. Pediatr., *85*: 760–763, 1974.
38. Beckwith, J. B. and Perrin, E. V. *In situ* neuroblastoma: A contribution to the natural history of neural crest tumors. Am. J. Pathol., *43*: 1089–1104, 1963.
39. Turkel, S. B. and Itabashi, H. H. The natural history of neuroblastic cells in the fetal adrenal gland. Am. J. Pathol., *76*: 225–244, 1974.

Closing Remarks

Dr. Harry V. Gelboin

In being asked to make some comments at this time, I am most fortunate that the task required was not one to summarize this Symposium but rather to simply draw some concluding remarks. To summarize a Symposium of the breadth and scope of subjects that we have covered during these three days would indeed have been a most formidable and perhaps unproductive task.

In listening to the unfolding of subjects that we have all heard, I was reminded and encouraged by a discussion of the psychology of creativity by Arthur Koestler in his book titled "The Act of Creation." Koestler points out that the creative process whether it be in humor and wit, in art, or in science, are all likely to operate by a similar mechanism. The mechanism involves a sub-conscious connection or jump between circuits of information in the brain that are not ordinarily or commonly connected. Thus this creative jump is neither conscious or a product of logic or reasoning. In other words, one does not decide on what day or at what time one is going to give birth to a creative thought or a creative idea. It just happens, often at odd places and at odd times. However, a key requirement for the creative jump, is the presence of appropriate information circuits in the mind. These information packets we generally call "knowledge" or "information."

In order to illustrate the point, Koestler retells the classic story of the ancient scientist, Archimedes. At that time, the emperor received a gift, a beautiful large intricately carved crown. The emperor was very impressed with the gift, but being wise, was also sceptical. He had a question which needed an answer. "Was the crown really made of gold?" The king naturally sent for Archimedes, the most prominent scientist of his time, and posed the question to him. "Is the crown made of gold?" Archimedes mediated and puzzled a while. He knew the density of gold, and if he could melt down the crown into a cube and hence measure both the weight and the volume, he would have the answer. Archimedes, however, was also wise enough to know that this scientific procedure was not one to be readily acceptable to the emperor, so he really did not have an answer to the problem. Later that night, Archimedes went home, put the question out of his mind and had a fine dinner with his wife and family. After dinner, he decided to take a bath, perhaps a very hot bath, Japanese style. As he stepped into and settled in the tub of hot water, he saw the water rise. There was an instantaneous flash in his mind and he shouted "Eureka" and he immediately knew he had the answer for the emperor. Simply measure the volume of water displaced by the crown and calculate its density. The point of the story is clear. The creative moment was the result of a connection between two very different packets or circuits of information. I do hope and believe that this story has relevance to this Symposium.

In developing the program for this Symposium, the organizing committee and its advisors tried to choose areas of research which we believed to have good scientific content as well as to be key areas of relevance to understanding carcinogenesis and to the more specific subject of this Symposium, "The Genetic and Environmental Factors in Carcinogenesis."

After hearing the content and high quality of the papers presented during these three days, I am convinced that our choices of both subjects and participants were indeed optimal for our purpose. In the course of the meeting we have heard about specific environmental factors which may be related to the etiology of cancer and their possible modes of action. We have heard discussions on the mixed-function oxidases which are the metabolic interface between environmental chemicals and the human organism and the various factors both environmental and genetic which can modify both drug and carcinogen activation and disposition. We have had thorough discussions of the process of DNA repair both from the standpoint of the basic molecular biology and its function in the repair of DNA damaged by ionizing or ultraviolet radiation or by environmental chemicals. Related to this but also quite independent there have been penetrating discussions of research concerned with the genetics and cell and molecular biology of diseases in which there is a high risk or proneness to cancer. These studies are indeed not only important for understanding basic mechanisms of cell biology but may be models useful for extrapolation to the more common forms of human cancers. There have been most stimulating discussions of the role of the immune system in ultraviolet carcinogenesis, leukemogenesis and in more specialized tumors. Further we heard the results of a variety of epidemiologically based studies using different approaches; twin populations, isolated populations with particular life styles, family studies, cancer patterns in migrant populations with particular environmental experiences such as radiation, or smoking or particular physiological experiences such as age at first childbirth, or special exposure to certain drugs or chemicals.

Thus the packages or circuits of information we have obtained have indeed been diverse. It is our hope that this diveristy will set the necessary stage in each of us for the creative insights needed to understand the seemingly very complex process of carcinogenesis.

In closing, I represent many different populations and sub-groups. First and foremost, on behalf of all of the participants, I express thanks to the Princess Takamatsu for her kind and noble patronage of this Symposium and support of cancer research. We thank all of her associates, scientific advisors and contributors to the Princess Takamatsu Cancer Research Fund for their many efforts. We wish to thank all of the support personnel for their smooth efficiency and graciousness which so well characterizes the Japanese people, the timekeepers, who were so efficient and I am sure were trained at the Olympic games, the projectionists who thought so well of one of Dr. Martin's slides that they refused to move it on until all understood it completely, all of those at the outside desk who were so helpful with our many problems, both trivial and important, and all the other staff, visible and those behind the scenes, who helped make this Symposium so successful and gratifying.

Personally, I would like to thank my colleagues on the Organizing Committee for the privilege of working with them, Dr. MacMahon, Dr. Matsushima, Dr. Sugimura, Dr. Takayama and Dr. Takebe.

On behalf of the Organizing Committee, I thank all of the participants, both speakers and attendees for your presence and your participation. I wish you all good health and success in your future endeavors, and finally I would like to close with a thought which I believe is applicable to our work and task. It is a thought of Rabbi Tarfon, a Talmudic sage of 2,000 years ago, who said,

"The day is short, and the work is great, and the labourers are sluggish, and the reward is much, and the Master of the House is urgent."

He went on to say,

"It is not your duty to complete the task, but neither are you free to desist from it."

Thank you all.

Subject Index